Springer Series in Applied Biology

W0042373

The 4-Quinolones: Anti Bacterial Agents in Vitro

Springer Series in Applied Biology

Series Editor: Prof. Anthony W. Robards PhD, DSc, FIBiol

Proposed future titles:

Food Freezing in the 1990's and Beyond
Ed. W. B. Bald

Biodegradation of Natural and Synthetic Materials
Ed. W. B. Betts

Critical Loads of Environmental Pollutants
Ed. M. J. Chadwick

Separation and Immobilisation of Biomolecules
Ed. U. B. Sleytr

Gene Transfer in Eukaryotic Cells
Ed. J. R. Warr

The 4-Quinolones:
Anti Bacterial Agents in Vitro

Edited by G.C.Crumplin

Springer-Verlag
London Berlin Heidelberg New York
Paris Tokyo Hong Kong

Geoff Crumplin
Department of Biology, University of York, York YO1 5DD

Series Editor
Professor Anthony William Robards, BSc, PhD, DSc, DipRMS, FIBiol
Director, Institute for Applied Biology, University of York, York YO1 5DD, UK

Cover Illustration: Electron micrograph of uropathogenic *Escherichia coli* showing Type 1 pili

ISBN-13: 978-1-4471-3451-0   e-ISBN-13: 978-1-4471-3449-7
DOI: 10.1007/978-1-4471-3449-7

British Library Cataloguing in Publication Data
The 4-Quinolones: antibacterial agents in vitro,
1. Medicine. Drug therapy. Quinolones
I. Crumplin, Geoffrey 1946-   615'.3

Library of Congress Cataloging-in-Publication Data
The 4-quinolone: antibacterial agents in vitro/edited by G. C. Crumplin
        p.cm.-(Springer series in applied biology)
    Includes bibliographical references.

    1. Quinolone antibacterial agents - Physiological effect. 2. Microbial sensitivity tests.
    3. DNA topoisomerase II.
    I. Crumplin, G. C. (Geoff C.), 1946- . II. Series.
RM666.Q55A14 1990   89-26200
615'.7 - dc20   CIP

© Springer-Verlag London Limited 1990
Softcover reprint of the hardcover 1st edition 1990

Set by Institute for Applied Biology, University of York

Foreword from Series Editor

Biology, and its applications for the benefit of mankind, represents one of the most crucial and rapidly moving areas of scientific study as we enter the last decade of the 20th century. The Institute for Applied Biology represents one of Britain's largest and best integrated teams of research biologists to be found within a single academic department, while Springer-Verlag is one of the world's outstanding publishing houses in the area of science and medicine. The combination of these two forces leads to the publication of the "Springer Series in Applied Biology" which will become a major reference work in its own subject area.

As is to be expected from a large and multidisciplinary Department, the range of seminars will extend from the study of foams to ecotoxicology and from biodegradation to genetic engineering. The aim will be to keep abreast of topics that have a special, applied, and contemporary interest.

Up to four volumes will be published each year through the editorial office in York. Modern methods of manuscript assembly will streamline the publication process without losing quality and, crucially, will allow the books to be available in the shops within four to five months of the actual seminar. In this way authors will be able to publish their most recent work without fear that it will, as so often happens, become outdated during an overlong period between submission and publication.

Our aim is to maintain the highest standards of both science and publication quality while minimising the delay between submission and publication. While the choice of subjects for seminars is made by our own editorial board, we are also always pleased to receive suggestions from external sources or, indeed, to consider the organisation and publication of seminars on topics other than those of direct interest to the Institute for Applied Biology.

The applications of biology are fundamental to the continuing welfare of all people, whether by protecting their environment or by ensuring the health of their bodies. This series aims to become an important means of disseminating the most up-to-date information in this field.

York, November 1989   A. W. Robards

Editor's Preface

The 4-quinolone antibacterial agents arose by Serendipity, and to a large extent have evolved in the same manner to the highly potent antibacterial agents which are now being advocated as potential 'drugs of first choice' for the treatment of a wide range of bacterial infections. It is now clearly established that they kill susceptible bacterial cells as a consequence of a unique mechanism of action upon what is a unique target molecule within the cell. However, fundamental biological studies of the effects of the 4-quinolones upon susceptible cells (even those carried out before the target molecule had been identified) have shown that the sheer diversity of effects upon cellular metabolism, control, and behaviour is limited only by the number of biological processes in the cell which involve DNA. Such a diversity of effects is not associated with any other group of antibacterial agents.

The *in vitro* evaluation of 4-quinolones represents an important component of the development programme leading up to possible clinical use, as well as a means of monitoring and developing clinical usefulness. The scientific meeting on which this book is based was convened in order to provide a forum for the bringing together of many of the *in vitro* study procedures so that an overall philosophical question might be addressed. This philosophical question arises out of the accepted belief that the 4-quinolones are an unusual (if not unique) group of antibacterial agents and may be phrased as - "Are the 4-quinolones just like other antibacterials?"

The evaluation and clinical use of antibacterial agents in the 1980s is carried out in the context of the principles of chemotherapy which were defined as long ago as 1912 by Paul Ehrlich. Up until the advent of the 4-quinolones, antibacterial agents have apparently conformed to these principles by functioning as fairly straightforward selective toxins, and techniques of experimentation and interpretation have become standardized. An antibacterial agent (such as a b-lactam, a sulphonamide, macrolide, aminoglycoside or tetracycline) is described as an agent which either inhibits the growth of susceptible bacterial cells or which renders the bacteria non-viable. In contrast 4-quinolone agents have been variously described as;-bactericidal agents, mutagens, anti-mutagens, modifiers of gene expression control, promoters or inhibitors of genetic transposition, DNA-damaging agents, plasmid-curing agents, recombinogens, anti-recombinogens, inhibitors of DNA-synthesis, inhibitors of RNA synthesis etc... Some of these descriptions conform to the basic definition of an antibacterial agent, but some clearly do not.

It is of course convenient to maintain continuity of perception and methodology, but once in a while we have to ensure that what we are doing is justified. It is just possible that the 4-quinolone antibacterial agents represent not only a good opportunity to try and justify uniformity, but also that they may represent an instance where uniformity of treatment might not be fully justified. This Advanced Study Seminar was convened as a forum to relate what we do and do not know about the 4-quinolones *in vitro*, and to discuss how this knowledge might allow us to ensure that data derived from *in vitro* studies might relate to the clinical application of the 4-quinolones.

Geoff Crumplin

Contents

Contributors

Mr. J. Baranowski
Department 47N, Building AP9A Abbott Laboratories, Abbott Park,
Illinois 60064, USA

Dr. J. F. Barrett
The R. W. Johnson Pharmaceutical Research Institute, Raritan,
NJ 08869 0602, USA

Dr. J. M. Besterman
Department of Molecular Biology, Burroughs Wellcome Company,
3030 Cornwallis Road, Research Triangle Park, NC 27709, USA

Ms. B. R. Brescia
Miles Inc, Pharmaceutical Division, 400 Morgan Lane, West Haven,
CT 06516, USA

Dr. R. A. Celesk
Miles Inc, Pharmaceutical Division, 400 Morgan Lane, West Haven,
CT 06516, USA

Dr. D. T. W. Chu
Department 47N, Building AP9A Abbott Laboratories, Abbott Park,
Illinois 60064, USA

Dr. P. Courvalin
Institut Pasteur, 28 Rue du Dr Roux, 75724 Paris, Cedex 15, France

Dr. G. Crumplin
Department of Biology, University of York, York Y01 5DD

Dr. E. Derlot
Institut Pasteur, 28 Rue du Dr Roux, 75724 Paris, Cedex 15, France

Mr. P. V. Devasthale
Department of Medicinal Chemistry, Kansas University, Lawrence,
KS 66045, USA

Dr. L. P. Elwell
Department of Molecular Genetics and Microbiology, Burroughs Wellcome
Company, 3030 Cornwallis Road, Research Triangle Park, NC 27709, USA

Dr. T. D. Gootz
Departments of Immunology and Infectious Diseases and Genetic
Toxicology, Pfizer Central Research, Pfizer Inc, Eastern Point Road, Groton,
CT 06340, USA

Mr. G. T. Grimes
Department of Drug Safety Evaluation, The Wellcome Foundation Ltd,
Temple Hill, Dartford, Kent DA1 5AH

Mr. P. Hallett
Department of Biochemistry, University of Leicester, Leicester LE1 7RH

Dr. H. E. Holden
Departments of Immunology and Infectious Diseases and Genetic
Toxicology, Pfizer Central Research, Pfizer Inc, Eastern Point Road, Groton,
CT 06340, USA

Mr. L. S. Holmes
Department of Molecular Sciences, The Wellcome Research Laboratories,
Beckenham, Kent BR3 3BS

Dr. D. C. Hooper
Infectious Disease Unit, Massachusetts General Hospital, Boston,
Massachusetts 02114, USA

Dr. A. T. Hudson
Research Division, The Wellcome Research Laboratories, Beckenham,
Kent BR3 3BS

Dr. D. J. C. Knowles
Beecham Pharmaceuticals Research Division, Brockham Park, Betchworth,
Surrey RH3 7AJ

Dr. C. Lee
Department 47N, Building AP9A Abbott Laboratories, Abbott Park,
Illinois 60064, USA

Dr. C. S. Lewin
Bacteriology Department, Medical School, Edinburgh University,
Teviot Place, Edinburgh EH8 9AG

Ms. B. L. M. Masecar
Miles Inc, Pharmaceutical Division, 400 Morgan Lane, West Haven,
CT 06516, USA

Dr. A. Maxwell
Department of Biochemistry, University of Leicester, Leicester LE1 7RH

Dr. P. R. McGuirk
Departments of Immunology and Infectious Diseases and Genetic
Toxicology, Pfizer Central Research, Pfizer Inc, Eastern Point Road, Groton,
CT 06340, USA

Dr. A. Mehlert
Department of Biochemistry, University of Dundee, Dundee DD1 4HN

Prof. L. A Mitscher
Department of Medicinal Chemistry, Kansas University, Lawrence,
KS 66045, USA

Mr. I. Morrissey
Microbiology Section, Department of Pharmaceutics, School of Pharmacy,
University of London, 29-39 Brunswick Square, London WC1N 1AX

Prof. H. C. Neu
Columbia University, 630 West 168, New York, N.Y. 10032, USA

Dr. M. Nuss
Department 47N, Building AP9A Abbott Laboratories, Abbott Park,
Illinois 60064, USA

Mr. M. Odell
Department of Biology, University of York, York YO1 5DD

Dr. J. J. Plattner
Department 47N, Building AP9A Abbott Laboratories, Abbott Park,
Illinois 60064, USA

Dr. C. Poyart-Salmeron
Institut Pasteur, 28 Rue du Dr Roux, 75724 Paris, Cedex 15, France

Dr. V. A. Ray
Departments of Immunology and Infectious Diseases and Genetic
Toxicology, Pfizer Central Research, Pfizer Inc, Eastern Point Road, Groton,
CT 06340, USA

Dr. N. J. R. Robillard
Miles Inc, Pharmaceutical Division, 400 Morgan Lane, West Haven,
CT 06516, USA

Dr. L. L. Shen
Department 47N, Building AP9A Abbott Laboratories, Abbott Park,
Illinois 60064, USA

Prof. J. T. Smith
Microbiology Section, The School of Pharmacy, Brunswick Square, London
WC1N 1AX

Dr. J. Tadanier
Department 47N, Building AP9A Abbott Laboratories, Abbott Park,
Illinois 60064, USA

Dr. L. M. Walton
Department of Molecular Genetics and Microbiology, Burroughs Wellcome
Company, 3030 Cornwallis Road, Research Triangle Park, NC 27709, USA

Dr. J. S. Wolfson
Infectious Disease Unit, Massachusetts General Hospital, Boston,
MA 02114, USA

Ms. R. M. Zavod
Department of Medicinal Chemistry, Kansas University, Lawrence,
KS 66045, USA

Chapter 1

Quinolones as Broad-spectrum Agents

H. C. Neu

In a few short years, the new quinolone antimicrobial agents have been established as the therapy of many infections. It is reasonable to question whether this class of antimicrobial agent can truly be considered broad-spectrum in the way that the penicillins, cephalosporins, tetracyclines, chloramphenicol, and trimethoprim/sulfamethoxazole are. This report will review the characteristics of the new and future quinolones which show that these agents are broad-spectrum compounds.

What does the term broad-spectrum mean in the 1980's and in the next decade? In the 1940's penicillin could well have been considered a broad-spectrum agent since it inhibited the major pathogens of that era. In 1946 after World War II the major pathogens in the community and hospital were the hemolytic streptococci such as *Streptococcus pyogenes, Streptococcus pneumoniae, Staphylococcus aureus, Neisseria meningitidis, Neisseria gonorrhoeae*, and *Treponema pallidum*. Sepsis due to the *Enterobacteriaceae* was rare, and penicillin G actually achieved urinary concentrations that inhibited many *Escherichia coli* the major cause of urinary infections, then as now. Infection due to *Pseudomonas aeruginosa* was distinctly rare, and the important nosocomial pathogens of today such as *Enterobacter cloacae* and *Serratia marcescens* were unheard of. Anaerobic species such as the Bacteriodes were not adequately understood, and penicillin G killed the dreaded anaerobe *Clostridium perfringens*.

The definition of broad-spectrum is elusive. Should an agent be designated as broad-spectrum only if it inhibits virtually all aerobic and anaerobic bacteria? If an agent inhibits staphylococci, *Enterobacteriaceae*, and Pseudomonads is it broad-spectrum? Agents such as the first generation cephalosporin cefazolin inhibit hemolytic streptococci, *Strep. pneumoniae, S. aureus, Klebsiella* species, *E. coli*, and *Proteus mirabilis*. Today as many as 40% of *E. coli* and 20% of *K. pneumoniae* are resistant to first generation cephalosporins. Many "broad-spectrum" second generation cephalosporins such as cefuroxime and cefamandole fail to inhibit *Enterococcus faecalis, Ent. cloacae, S. marcescens, Ps.aeruginosa*, and *Bact. fragilis*.

Aminoglycosides such as gentamicin, netilmicin, and amikacin do not inhibit *Strep. pneumoniae*, most enterococci, hemolytic streptococci, and anaerobic species. Nonetheless, aminoglycosides are usually considered broad-spectrum agents.

It was the aminoglycosides which refashioned our concepts about the meaning of broad-spectrum since these agents kill at concentrations close to, or identical to, their inhibitory concentration. The third generation cephalosporin agents, aztreonam, and the carbapenem, imipenem, also made it evident that mere inhibitory activity was not adequate in evaluating an antimicrobial agent. Today it is essential to correlate minimal inhibitory concentrations (MICs) with the pharmacological properties of the agent (Neu 1985). However, our knowledge of this correlation remains to some degree, in the dark. Should the serum level of a compound be four, eight or 16-fold above the MIC or minimal bactericidal concentration (MBC) of an organism? Should the fold level above the MIC/MBC be of free drug, or is it adequate to calculate activity based on peak levels (Neu 1986)? Should the tissue levels of drug be four or eight-fold above the MIC and the urine concentrations eight, 16, or 32-fold above the MIC? Does the effect of inoculum size, pH, ionic content found *in vitro* affect the breadth of spectrum of an agent? Can an agent be considered broad-spectrum for urinary infections, but of restricted spectrum for organisms when they cause infection in the lung?

Microbiological Aspects of Quinolones which Relate to Broad-spectrum

There are commercially available in the United States only two quinolones, norfloxacin and ciprofloxacin. In Europe other agents such as enoxacin, pefloxacin, and ofloxacin are also available. Soon to be available are fleroxacin, lomefloxacin and temafloxacin (Fig. 1.1). The activity of the agents varies (Table 1.1) with only ciprofloxacin, ofloxacin and temafloxacin having activity against *Strep. pneumoniae*, *Strep pyogenes*, and some degree of activity against enterococci (Eliopoulos and Eliopoulos 1989). Antistaphylococcal activity is excellent for most urinary isolates and when infection occurs in soft tissue or, in the case of ciprofloxacin and ofloxacin, bone.

All of the agents have inhibitory activity against the *Enterobacteriaceae*, particularly urinary pathogens including *Ps. aeruginosa*, but the activity of a number of the agents is inadequate to treat infections due to *Ps. aeruginosa* when the infection is in the lung, bone or other tissue site (Neu 1989). All of the agents also inhibit *Neisseria* species, *N. meningitidis*, and *N. gonorrhoeae*, *Haemophilus* species, and *Moraxella* (Branhamella) *catarrhalis*.

The activity of the newer quinolones against species such as Brucella, Mycoplasma, Ureaplasma, Chlamydia, Legionella species, and mycobacteria is considerably less than that against the aforementioned species (Phillips *et al.* 1988). Nonetheless the quinolones achieve high concentrations in phagocytic cells which are well above the MBCs (Gerding and Hitt 1989).

There are under study a number of quinolone agents which have much greater activity against streptococci, *Strep. pneumoniae*, staphylococci, and anaerobic species. Agents such as tosufloxacin,spafloxacin, AM 1091, WIN 57273 (Figs. 1.1 and 1.2), inhibit *S. aureus* at concentrations of 0.008 to 0.5µg/ml, and *Strep. pyogenes* and *Strep. pneumoniae* at 0.008 to 0.25µg/ml which is many-fold below that of ciprofloxacin and ofloxacin (Table 1.2).These agents also inhibit *Bact. fragilis* at <2µg/ml. In the case of WIN 57273 and spafloxacin, there is some loss of activity against some of the members of the *Enterobacteriaceae* and definitely less activity against *Ps. aeruginosa*.

Table 1.1. Comparative *In Vitro* Activity of Quinolone Antibiotics

Comparative in Vitro Activity of Quinolone Antibiotics

ORGANISM	MINIMUM INHIBITORY CONCENTRATIONS (MIC90) (µg/mL)							
	Ciprofloxacin	Norfloxacin	Ofloxacin	Enoxacin	Lomefloxacin	Fleroxacin	Temafloxacin	Tosufloxacin
Escherichia coli	0.12	0.25	0.25	0.25	0.25	0.25	0.25	0.12
Klebsiella pneumoniae	0.12	0.5	0.25	0.25	0.5	0.5	0.5	0.06
Enterobacter spp	0.12	1	0.25	0.5	0.5	0.5	2	0.25
Citrobacter spp	0.12	1	0.25	0.25	1	0.5	2	0.12
Serratia marcescens	0.25	2	2	2	2	2	4	0.25
Shigella spp	0.12	0.25	0.25	0.25	0.25	0.25	0.5	0.12
Proteus mirabilis	0.12	0.25	0.25	0.25	0.25	0.25	0.5	0.12
Proteus, other	0.25	0.5	0.5	0.5	1	0.5	1	0.12
Morganella morganii	0.12	0.5	0.12	2	0.5	0.5	1	0.25
Providencia spp	0.25	2	0.5	2	0.25	0.25	1	0.12
Yersinia enterocolitica	0.12	0.25	0.12	0.25	4	4	4	0.25
Pseudomonas aeruginosa	0.5	8	4	2	4	4	0.25	0.12
Acinetobacter spp	1	8	1	8	4	2	4	0.5
Staphylococcus aureus methicillin-susceptible	0.5	1	0.5	2	2	2	1	0.12
Staphylococcus aureus methicillin-resistant	0.5	4	0.5	4	2	2	0.5	0.12
Staphylococcus epidermidis	0.5	2	2	8	2	2	0.5	0.25
Enterococci	2	8	4	6	1	2	0.5	0.5
Streptococcus (group A)	1	4	2	8	16	8	2	0.12
Streptococcus (group B)	2	8	2	16	8	8	0.5	0.25
Streptococcus pneumoniae	<0.06	8	2	16	16	16	1	0.25
Haemophilus influenzae	<0.06	<0.12	<0.12	<0.12	8	8	0.25	<0.12
Neisseria gonorrhoeae	<0.06	<0.12	<0.12	<0.12	0.12	0.12	0.12	<0.12
Neisseria meningitidis	<0.06	<0.12	<0.12	<0.12	0.12	0.12	0.12	<0.12
Campylobacter spp	0.05	0.05	1	1	0.12	0.12	0.12	<0.12
Bacteroides fragilis	16	>32	8	>32	64	—	4	—
Bacteroides melaninogenicus	8	>32	8	>32	16		4	2
Peptostreptococcus	4	>32	2	>32	16		4	2
Mycobacterium tuberculosis	1	—	1	—	—		4	—
Branhamella catarrhalis	<0.12	<0.12	<0.12	<0.12	0.12		<0.12	<0.12
Brucella spp	2	8	2	8	8		0.25	0.25
Listeria monocytogenes	1	4	2	8	—		0.25	0.25
Corynebacterium JK	<0.12	4	1	8	8		0.25	0.25
Legionella spp	2	0.5	<0.25	0.5	—		0.25	0.25
Chlamydia trachomatis	0.5	8	1	8	0.25		0.12	0.25
Mycoplasma hominis	2	32	1	64	1		—	0.5
Ureaplasma	1	32	2	64	—		—	—
Mycoplasma pneumoniae		8	1	8	—		—	—

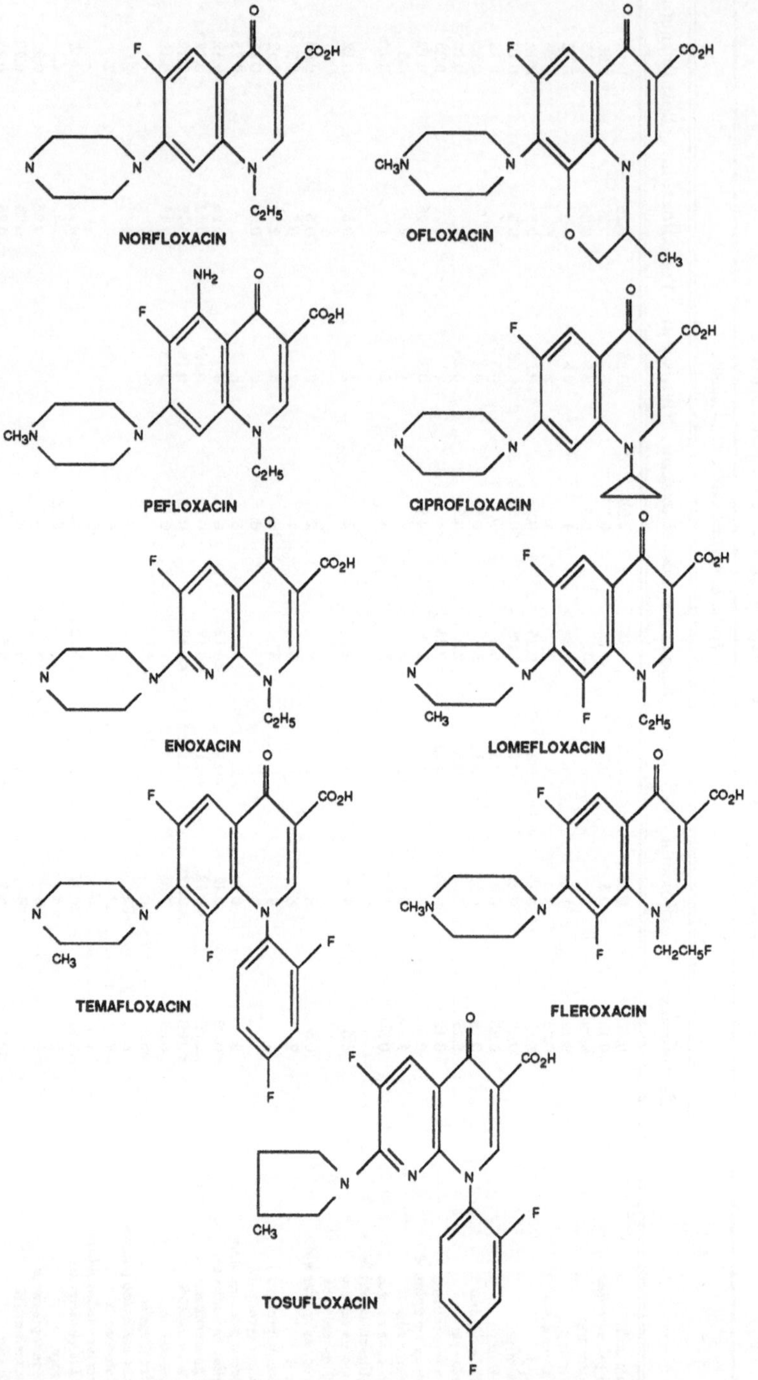

Fig. 1.1. Structures of the new generation of 4-quinolones which are already either in clinical use or likely to enter use in the near future.

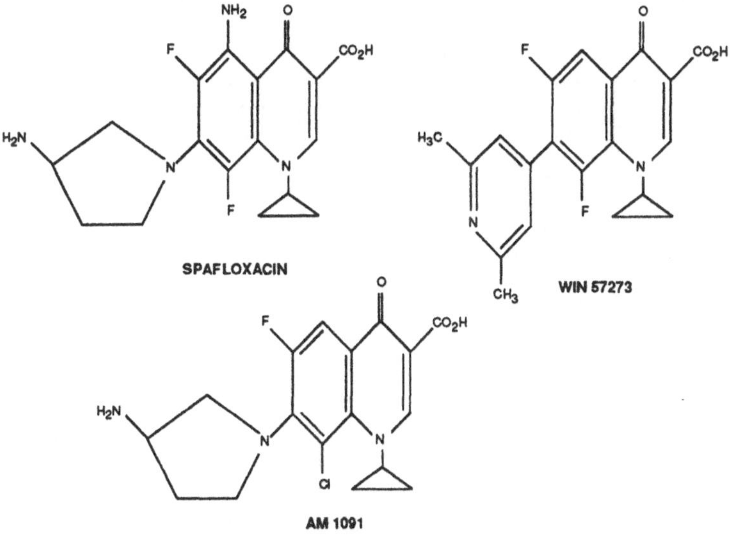

Fig. 1.2. Structures of three more recent 4-quinolones which are being investigated for potential clinical usefulness.

Do these changes in activity make the agents more "broad-spectrum" or less broad-spectrum than the older, new quinolones? No, they do not. The new agents may inhibit some Gram-positive species at lower concentrations, but thus far the new agents do not inhibit the new quinolone-resistant, methicillin-resistant *S. aureus* or the methicillin-resistant coagulase-negative staphylococci. These agents also do not inhibit ciprofloxacin-resistant *Ps. aeruginosa* or *S. marcescens*. Indeed we have shown that *Enterobacteriaceae*, Pseudomonas, staphylococci selected for resistance to ciprofloxacin or ofloxacin by repeated passage in the drug have MICs beyond concentrations which could be achieved with the new agents (Neu *et al.* 1989). Thus at present we have agents which appear to be as broad-spectrum as possible against *Enterobacteriaceae* and *Ps. aeruginosa*.

Other Microbiological Factors which Relate to their Broad-spectrum Activity

In general quinolones kill bacteria at the same or at most only two-fold above their inhibitory concentration. There is minimal inoculum effect with the quinolones until one gets above 10^7 CFU. These two factors are quite important when one considers the blood levels achieved with quinolones in comparison to the penicillins and cephalosporins (Bergen 1988). Quinolones also produce a significant post-antimicrobial effect at concentrations only several-fold above the MIC for both Gram-positive and Gram-negative species (Chin and Neu 1987). This fact has a major impact on the frequency with which the drugs need to be administered. Quinolones also damage many bacteria in their resting stage which does not occur with ß-lactams or most other classes of antimicrobial agents (Zeiler and Voigt 1987). Finally, the new quinolones can be combined with other classes of antimicrobial agents without deleterious effect on the antimicrobial activity and without any increased toxicity to the patient (Neu 1989b).

Table 1.2. New Quinolones which attempt to make the agents broad-spectrum

Organism	Agent	Range ug/ml	MIC 90%
Staphylococcus aureus	Ciprofloxacin	0.25-2	1
	Ofloxacin	0.25-2	1
	WIN 57273	0.004-1	0.015
	Spafloxacin	0.03-0.25	0.25
	AM 1091	0.015-0.06	0.06
	Temafloxacin	0.25-1	0.5
	Tosufloxacin	0.015-0.5	0.12
Streptococcus pyogenes	Ciprofloxacin	0.25-2	2
	Ofloxacin	0.25-2	2
	WIN 57273	0.008-0.06	0.06
	Spafloxacin	0.12-0.5	0.5
	AM 1091	0.03-1	0.25
	Temafloxacin	0.25-0.5	0.5
	Tosufloxacin	0.06-1	0.12
Streptococcus pneumoniae	Ciprofloxacin	0.5-2	2
	Ofloxacin	0.5-2	2
	WIN 57273	≤0.015-0.03	0.03
	Spafloxacin	0.12-0.5	0.5
	AM 1091	-	-
	Temafloxacin	0.5-1	1
	Tosufloxacin	0.03-0.25	0.25
Enterobacter faecalis	Ciprofloxacin	0.5-2	2
	Ofloxacin	1-4	2
	WIN 57273	0.015-0.06	0.06
	Spafloxacin	0.25-0.5	0.5
	AM 1091	0.06-2	2
	Temafloxacin	0.5-2	2
	Tosufloxacin	0.12-0.5	0.5
Bacteroides fragilis	Ciprofloxacin	2-16	16
	Ofloxacin	2-16	8
	WIN 57273	0.12-2	1
	Spafloxacin	1-2	2
	AM 1091	0.12-1	0.5
	Temafloxacin	1-4	4
	Tosufloxacin	0.5-8	2
Escherichia coli	Ciprofloxacin	<0.008-0.06	0.015
	Ofloxacin	0.03-0.25	0.12
	WIN 57273	0.06-0.25	0.25
	Spafloxacin	0.008-0.12	0.03
	AM 1091	0.015-0.25	0.12
	Temafloxacin	0.015-0.12	0.12
	Tosufloxacin	0.015-0.06	0.06
Pseudomonas aeruginosa	Ciprofloxacin	0.03-0.5	0.5
	Ofloxacin	0.5-4	4
	WIN 57273	0.5-8	8
	Spafloxacin	0.25-2	2
	AM 1091	0.25-1	1
	Temafloxacin	0.5-4	4
	Tosufloxacin	0.03-1	0.5

Although the current quinolones do not inhibit anaerobic streptococci and other anaerobic species, this microbiological defect may actually be a benefit since the colonization resistance of the intestine is maintained by preserving the anaerobic streptococci, Bacteroides, lactobacilli, and other intestinal flora while enteric pathogens are eliminated. The true microbiological defects of quinolones are their decreased activity in the presence of high magnesium concentrations or in the presence of an acidic environment. It is unclear whether resistant mutants occur more easily at high Mg^{2+} concentrations or at an acid pH. But recent work by Crumplin and colleagues (this conference) have shown that the frequency of resistant mutants goes from 10^{-10} to 10^{-7} or 10^{-6} if the selection for mutants occurs at 32°C. This may have real clinical significance since it appears that resistant mutants are more easily isolated from decubitus ulcers and other wounds which would be at such temperatures rather than at 37°C as is the rest of the body. Resistant bacteria occur in situations such as in the presence of indwelling urethral catheters. Here an organism would at the penile urethra be at room temperature in the presence of an acid pH and a high Mg^{2+} concentration.

Pharmacology of Quinolones

The currently available fluoroquinolones vary in their oral bioavailability from 50% to 100% (Bergen 1988; Drusano 1987). Ciprofloxacin is available as a parenteral agent, and ofloxacin should also be available as a parenteral within a short time. Lomefloxacin, fleroxacin, and temafloxacin are being evaluated for parenteral administration. Absorption of all of the fluoroquinolones is markedly reduced by ingestion with aluminium and magnesium compounds. Reduction in absorption when ingested with calcium salts occurs, but to a lesser degree. The fluoroquinolones are well distributed to body sites of infections with excellent concentrations in bronchial mucosa (Gerding and Hitt 1989). The high concentrations achieved in lung and bronchial tissue probably contributes to the effectiveness of the agents against respiratory pathogens such as *Strep. pneumoniae* which have MICs near the peak blood levels. The intracellular concentrations contribute to the activity of the agent against Mycoplasma, Legionella, and other pathogens.

What is unclear about the intracellular concentrations of fluoroquinolones is what level of intracellular concentration in an acid vacuole is needed to destroy a microorganism. Ciprofloxacin, norfloxacin, and ofloxacin have easily cured infection due to Salmonella which have MICs $\leq 0.12\mu g/ml$. But an organism such as *Brucella mellitensis* with an MIC of $1\mu g/ml$ often relapses even though the intracellular concentrations of an agent such as ciprofloxacin are $5\mu g$ - $7\mu g$. Fluoroquinolones hold promise as treatment of mycobacterial infections, but it is unlikely that they will be successful as monotherapy, and combination with other agents which penetrate macrophages will be required. How effective quinolones would be for treatment of osseous tuberculosis is unknown, but it is possible that they would be useful alternatives in patients unable to take standard agents or patients infected with isoniazid-rifampin resistant isolates.

The long serum half-life of the fluoroquinolones is of major importance in the activity of the quinolones since it seems that maintenance of concentrations above the MIC and MBC of an organism is important in clearing organisms and the longer the exposure the longer the post-antibiotic effect which is critical in preventing re-growth of bacteria. The secretion of fluoroquinolones in salivary, lacrimal, and bowel fluids provides bactericidal concentrations against upper airway colonizers such as *N. meningitidis* and *S. aureus,* vaginal *E. coli,* bowel mucosal pathogens such as Shigella, Salmonella, Campylobacter, Aeromonas, Vibrios, and *E. coli.*

The high bowel concentrations combined with the low MICs and low frequency of spontaneous mutation of these species to resistance has so far prevented the appearance of resistance. However, *Campylobacter pylori* protected in the mucosa of the stomach become resistant. If too low concentrations of fluoroquinolones are ingested, resistance may appear in species of bacteria causing diarrhœa.

Currently the intravenous doses of ciprofloxacin are 200mg or 300mg every 12 hours. The oral bioavailability of ciprofloxacin is 65% to 70%. Thus after a 500 mg oral dose, more drug is bioavailable than occurs with 300 mg intravenous doses. Use of the lower intravenous doses in treatment of serious respiratory illness due to Pseudomonas and even Klebsiella potentially will cause selection of resistant isolates. Fear of the potential toxicity of fluoroquinolones has caused inappropriately low intravenous dosing of ciprofloxacin. In the case of ofloxacin which is 100% bioavailable, it is reasonable to use the same intravenous dose as oral dose. This will also be true for temafloxacin and lomefloxacin which are completely absorbed orally.

Clinical Aspects of Fluoroquinolones with Regard to Broad-spectrum Activity

If fluoroquinolones are broad-spectrum agents, should they be appropriate therapy in virtually any infection? No. There should be appropriate use of fluoroquinolones. Fluoroquinolones are excellent treatment for urinary tract infections either uncomplicated problems such as cystitis or pyelonephritis in which oral therapy with a nonrenal toxic agent that has high bactericidal concentrations in the medullary area of the kidney is appropriate (Naber 1989). The quinolones will kill those bacterial cells which are slowly dividing and which can relapse with ß-lactam therapy. In complicated urinary infections fluoroquinolones are also the appropriate agents since these patients are often only temporarily cured due to the intrinsic defect, stones, bladder diverticulum, cystocoel, or need to use a urethral catheter. The fluoroquinolones sterilize the intestinal flora, periurethral areas so that there is not a rapid superinfection which frequently occurs with other agents (Hooten *et al.* 1989).

Sexually transmitted diseases such as gonorrhea and chancroid respond to low doses of the new fluoroquinolones (Dallabetta and Hook 1989). It is interesting that longer half-life may influence activity against Chlamydia as illustrated by the higher cures with ofloxacin (Oriel 1989). Others of the newer fluoroquinolones that show promise for treatment of Chlamydia would be agents such as temafloxacin, fleroxacin, and lomefloxacin. But the appearance of new macrolides may be a better answer to the Chlamydia problem than the quinolones.

Fluoroquinolones are excellent therapy for gastrointestinal infections since they eliminate both luminal organisms and those organisms that have penetrated below the mucosa (Dupont 1989). Brief therapy is possible, allowing rapid restoration of the normal bowel flora. It is probable that fluoroquinolones should become the main therapy of diarrhœal disease. Should fluoroquinolones be used in pediatrics where diarrhœa is actually life-threatening (Fontaine 1989)? The answer is complex since short therapy of two to three days would not be toxic in view of the extensive use of nalidixic acid in this population. Rather than toxiciy, a more serious consideration in the use of oral fluoroquinolones in pædiatric diarrhœa would be that physicians and patients would lose sight of the major therapy of infantile diarrhœa, namely oral fluid replacement as advocated by the WHO.

Oral quinolones are appropriate agents to use as therapy of Gram-negative osteomyelitis in which the osteomyelitis is rarely of hematogenous source except in narcotic addicts and is due primarily to contiguous infection. Whether the quinolones will be as effective as ß-lactams in therapy of osteomyelitis due to *S. aureus* is problematic (Waldvogel 1989). Current ongoing clinical studies suggest that cure of staphylococcal osteomyelitis with ciprofloxacin or ofloxacin is comparable to what has been achieved with oral agents such as cloxacillin, flucloxacillin, or cephalexin (Lee *et al.* 1990).

Of major interest will be whether new compounds such as temafloxacin, spafloxacin, or tosufloxacin which are more active than ciprofloxacin and ofloxacin against streptococci, and which inhibit anaerobic species such as *Bact. fragilis*, will be more effective in treating diabetic foot, skin-structure infections, and osteomyelitis in this clinical population.

Whether an oral quinolone could be taken for the life of a patient who has an infected prosthesis which cannot be removed is unknown. The two potential problems are toxicity and development of bacterial resistance. We have treated patients with infected knee prostheses for over two years without adverse effects.

Skin-structure and post-operative wound infections represent clinical situations in which parenteral quinolones may have a role as initial therapy to be followed by the oral agents. Studies with ciprofloxacin intravenously and then orally have shown that it is equivalent to parenteral cefotaxime or ceftazidime (Neu 1989c). It is conceivable that the quinolone agents in development with superior Gram-positive and anti-anaerobic activity will make it possible to use oral agents to treat wound infections that heretofore required combination therapy or the broad-spectrum activity of an agent such as imipenem.

The improved Gram-positive activity of the new oral fluoroquinolones raises serious questions about the use of the agents to treat staphylococcal endocarditis (Bayer 1989). This disease is not common, but the long parenteral therapy is costly. Animal experiments show that the currently available fluoroquinolones which are not as active as spafloxacin or tosulfloxacin result in a greater reduction in colonies in cardiac valvular vegetations. It is unfortunate that the legal climate in the United States will make it difficult to determine the efficacy of the fluoroquinolones in this infection, but studies in the United Kingdom, France, Germany, and Switzerland may be able to establish this use.

The most controversial clinical use of the quinolones has been in the treatment of respiratory infections (Scully 1989; Thys *et al.* 1989). Should quinolones be used as initial therapy of bacterial exacerbations of bronchitis or of pneumonia? The studies of the agents so far examined demonstrate that ciprofloxacin, ofloxacin, and temafloxacin are equivalent to oral agents such as ampicillin, amoxicillin, cephalexin, cefaclor, and macrolides (Scully 1989). Nonetheless the aforementioned agents, with the exception of cefaclor, are much less expensive and have much better *in vitro* activity against *Strep. pneumoniae*.

Studies by our group have shown that intravenously administered ofloxacin or ciprofloxacin will cure bacteremic pneumococcal pneumonia even in patients with serious underlying illness such as HIV infection (Kuschner *et al.* 1989).

The major concern about use of quinolones in treatment of respiratory infections is not for those mentioned, but for the use of agents such as enoxacin, fleroxacin, and lomefloxacin which are highly effective against *Haemophilus influenzae* and *Moraxella* (Branhamella) *catarrhalis*. These agents are well absorbed, inhibit these two species at ≤0.25 µg/ml, but the clinical studies are done in institutions or situations in which carefully performed Gram stains done by trained personnel have excluded *Strep. pneumoniae* as the etiologic agent of the pneumonia or bacterial bronchitis. Most practicing physicians do not do cultures or perform Gram stains. Thus caution is needed in the use of oral fluoroquinolones to treat lower respiratory infections.

There are studies of ciprofloxacin and ofloxacin that show the efficacy of these two agents in serious hospital pneumonia due to multiply-resistant *Enterobacteriaceae* or *Ps. aeruginosa* (Neu 1989c). Ciprofloxacin intravenously alone or intravenous/oral has proved equal to intravenous ceftazidime in treatment of serious respiratory infections. Should the fluoroquinolones be used as single agents, or should they be used in combination with other agents? My opinion would be that the fluoroquinolone should be used with another agent until the pathogens are identified. However, I doubt that combination of a fluoroquinolone with a penicillin, cephalosporin, or with an aminoglycoside will delay the emergence of resistance. The microbiological concept is that reduction of the number of colony forming units would make it less likely for a resistant clone to emerge. There is no clinical evidence to support this concept, and in our hands combination of drugs in *in vitro* studies did not prevent emergence of resistance to either compound. The major reason to use two agents would be to rapidly reduce the CFU and decrease the inflammatory response which is not only beneficial but is also destructive. Also should there be a resistant organism present in the infection, two agents prevent early failure or progress of the lung destruction.

It will be interesting to see whether new fluoroquinolones with anti-anaerobic activity will be useful in treatment of aspiration pneumonitis.The ability to use an oral agent when dealing with the mixed aerobic-anaerobic aspiration pneumonitis would be a major cost-saving to the health care system. Comparative trials of agents such as temafloxacin, tosufloxacin, spafloxacin, and WIN 57273 may reveal that these agents are equivalent to drugs such as clindamycin and superior to metronidazole which has proved less successful due to its poorer activity against anaerobic streptococci.

In the neutropenic patient there are three uses of the quinolones. The first is as prophylaxis to prevent sepsis and other infection from an intestinal source by aerobic Gram-negative bacilli (Rosenberg-Arska *et al.* 1989; Winston 1989). This use has been established for norfloxacin, ciprofloxacin, ofloxacin, enoxacin, and pefloxacin. The best results to date have been with ciprofloxacin since it has had the best activity against Gram-positive bacteria. Newer agents which have better Gram-positive activity with modest reduction in anti-Pseudomonas activity may be even better.

A second use of the parenteral forms of the fluoroquinolones is treatment of the febrile neutropenic patient. Studies with intravenously administered ciprofloxacin by Rolston *et al.* (1989) and by Brown and colleagues (1989) have demonstrated that ciprofloxacin is equivalent to other programs for treatment of suspected infection and for Gram-negative bacteremia in the neutropenic cancer patient. The incidence of Gram-positive infections and the failure of quinolones in such infections indicate that classes of antibiotics with superior activity against streptococcus and other Gram-positive organisms should be used. The third way the fluoroquinolones could be used would be as oral therapy in the febrile neutropenic patient who has responded to initial therapy by becoming afebrile but has remained neutropenic (Flaherty *et al.* 1989). Previous studies have shown that neutropenic patients should receive antimicrobial therapy for a minimum of seven to ten days or until the white count has returned to more than 1000 per mm^3 since a number susbsequently become septic if antibiotics are discontinued when the patient becomes afebrile and no initial cultures were positive.

Other considerations for the use of fluoroquinolones are as prophylaxis in surgical patients (Nord 1989). There is minimal data on the use of the fluoroquinolones as prophylaxis. There is a rationale for their use in patients undergoing urological surgery, and it may be rational to use them for patients undergoing biliary tract procedures.

Whether the newer agents with anti-anaerobic activity could be used as prophylaxis before colonic surgery will have to be established. Likewise the newer agents might be appropriate for oral treatment of gynæcological infections, but only controlled clinical trials will establish this use.

Future Areas of Quinolone Use

The synthesis of the quinolones mentioned in this article which have increased Gram-positive and/or anti-anaerobic activity, raises major questions about the use of this class of compound where these species are predominant. Future comparative studies of one quinolone against another are necessary. Unlike the situtation which existed for ß-lactams and aminoglycyosides in which the later compounds were markedly superior to the early agents, the first of the fluoroquinolones have extremely good activity against Enterobacteriaceae, *Ps. aeruginosa, Haemophilus, Neisseria, Moraxella* species and a number of intracellular organisms such as Mycoplasma and Legionella. It is doubtful that one agent can be shown to be superior to another in treatment of infections due to most of the aforementioned organisms. In the case of Chlamydia, *Strep. pneumoniae*, hemolytic streptococci, staphylococci, and anaerobes, there is room for improvement and new generations of quinolones, or those under study, may well prove to be superior compounds.

Clinical studies of the fluoroquinolones as treatment of *Mycoplasma pneumoniae*, Legionella, and *Chlamydia pneumoniae* pulmonary infections are needed since the extant data is inadequate. More use of the agents as prophylaxis of vascular surgery procedures with the agents in development is justified since such use will not result in more resistance. Finally, the new agents need to be studied as follow-up therapy of surgical and gynæcological infections.

Conclusions

The current quinolones are truly broad-spectrum agents that can effectively eradicate many of the most common infections which afflict man. However, better use of the current and future agents will be necessary to avoid the widescale resistance which will develop if the compounds are improperly used in clinical settings in which resistance easily develops.

References

Bayer AS (1989) Treatment of experimental and human bacterial endocarditis and meningitis with quinolone antimicrobial agents. In: Wolfson JS, Hooper DC (eds) Quinolone Antimicrobial Agents, Washington DC, American Society for Microbiology pp 213-232

Bergan T (1988) Pharmacokinetics of fluorinated quinolones. In: Andriole VT (ed) The Quinolones, London Academic Press, pp 119-154

Brown AE, Smith G (1989) Intravenous ciprofloxacin as treatment of sepsis in selected patients with neoplastic diseases. Am J Med 87: (Suppl 5a) 266F-268F

Chin NX, Neu HC (1987) Post-antibiotic suppressive effect of ciprofloxacin against Gram-positive and Gram-negative bacteria. Am J Med 82 (4A):58-62

Dallabetta GA, Hook EW III (1989) Treatment of sexually transmitted diseases with quinolone antimicrobial agents. In:Wolfson JS, Hooper DC (eds) Quinolone Antimicrobial Agents, Washington DC, American Society for Microbiology, pp 125-142

Drusano GL (1989) Pharmacokinetics of the quinolone antimicrobial agents. In: Wolfson JS, Hooper DC (eds) Quinolone Antimicrobial Agents, Washington DC, American Society for Microbiology, pp 71-106

Dupont HL (1989) Quinolone antimicrobial agents in the management of bacterial enteric infections. In: Wolfson JS, Hooper DC (eds) Quinolone Antimicrobial Agents, Washington DC, American Society for Microbiology pp 167-176

Eliopoulos GM, Eliopoulos CT (1989) Quinolone antimicrobial agents activity *in vitro*. In: Wolfson JS, Hooper DC (eds) Quinolone Antimicrobial Agents, Washington DC, American Society for Microbiology, pp 35-90

Flaherty J, Edin B, Waitley D, George D, Sexton B, Anow P, O'Keefe P, Weinstein RA (1989) Multicenter, randomized trial of ciprofloxacin (IV/PO) plus azlocillin versus ceftazidime plus amikacin for empiric treatment of febrile neutropenic patients. Am J Med 87: (Suppl 5a) 278F-282F

Fontaine O (1989) Antibiotics in the management of Shigellosis in children: What role for the quinolones? Rev Infect Dis 11 (5):S145-S150

Gerding DN, Hitt JA (1989) Tissue penetration of the new quinolones in humans. Rev Infect Dis 11(5):S1046-S1057

Hooton TM, Latham RH, Wong ES, Johnson C, Roberts PL, Stamm WE (1989) Ofloxacin versus trimethoprim/sulfamethoxazole for treatment of acute cystitis. Antimicrob Agents Chemother 33:1308-1312

Kuschner R, Briones F, Scully BE, Neu HC (1989) Evaluation of IV ofloxacin as therapy of acute lower respiratory tract infections. Abstract #547, 29th Intersci Conv Antimicrob Agents Chemother, Houston, Texas, September

Lee T, Rosenwasser BE, Neu HC (1990) An oral antibiotic regimen for the treatment of chronic osteomyelitis. 57th Am Acad Ortho Surg, Abstract #487, February 13 1990

Naber KG (1989) Use of quinolones in urinary tract infections and prostatitis. Rev Infect Dis 11(5):1321-1337

Neu HC (1985) Antimicrobial activity, bacterial resistance, and antimicrobial pharmacology. Am J Med 78(6B):17-22

Neu HC (1986) Antimicrobial activity and pharmacokinetics, the importance of integration of these factors to achieve optimal chemotherapy. In: Genazzani E (ed) Pharmacokinetics and Antibiotic Efficacy. Masson, Milano, pp 77-89

Neu HC (1989a) Synergy of fluoroquinolones with other antimicrobial agents. Rev Infect Dis 11(5):S1025-S1035

Neu HC (1989b) The quinolones. Infect Dis Clin N Am 3:625-639

Neu HC (1989c) New oral and parenteral quinolones. Am J Med 87: (Suppl 5a) 283F-287F

Neu HC, Novelli A, Chin NX (1989) Comparative *in vitro* activity of a new quinolone AM-1091. Antimicrob Agents Chemother 33:1009-1018

Nord CE (1989) Surgical prophylaxis and treatment of surgical infections with quinolones. Rev Infect Dis 11(5): S1287-S1291

Oriel JD (1989) Use of quinolones in chlamydial infections. Rev Infect Dis 11(5):S1273-S1276

Phillips I, King A, Shannon K (1988) *In vitro* properties of the quinolones. In: Andriole VT (ed) The Quinolones. Academic Press, pp 83-118

Rolston KVI, Haron E, Cunningham C, Bodey GP (1989) Intravenous ciprofloxacin for infections in cancer patients. Am J Med 87: (Suppl 5a) 261F-265F

Rosenberg-Arska M, Dekker AVW, Verhoff J (1989) Prevention of infections in granulocytopenic patients by fluorinated quinolones. Rev Infect Dis 11(5):S1231-S1236

Scully BE (1989) Treatment of respiratory tract infections with quinolone antimicrobial agents. In: Wolfson JS, Hooper DC (eds) Quinolone Antimicrobial Agents. Washington DC, American Society for Microbiology, pp 143-166

Thys JP, Jacobs F, Monte S (1989) Quinolone in the treatment of lower respiratory tract infections. Rev Infect Dis 11 (Suppl):S1212-S1219

Winston DJ (1989) Use of quinolone antimicrobial agents in immunocompromised patients. In: Wolfson JS, Hooper DC (eds) Quinolone Antimicrobial Agents. Washington DC, American Society for Microbiology, pp 187-212

Waldvogel FA (1989) Use of quinolones for the treatment of osteomyelitis and septic arthritis. Rev Infect Dis 11(5):S1259-S1263

Zeiler HJ, Voigt WH (1987) Efficacy of ciprofloxacin in stationary-phase bacteria *in vivo*. Am J Med 82(4A):87-90

Discussion Summary

Professor Neu's presentation of data indicating that the new generation 4-quinolones were so potent that they can be regarded as being effective broad-spectrum agents, clearly indicated that the 4-quinolones could no longer be considered as limited spectrum agents. However, the clinical use of these agents in the treatment of serious infections, particularly in the hospital environment, provoked serious comment about the potential risks associated with their use in the treatment of *S. aureus* infections. Professor Neu pointed out that with this organism in hospitals, resistance to ciprofloxacin, ofloxacin and norfloxacin was appearing not only in simple *S. aureus* infections, but more worryingly in methicillin-resistant *S. aureus* (MRSA). In New York, virtually all hospitals have a *S. aureus* population of which between 1% and 25% are MRSA, and that this specific problem was being further exacerbated by the fact that many of the MRSAs were being found as being also resistant to aminoglycosides, macrolides and rifampicin, thus seriously reducing the therapeutic options available. It was also apparent that there are increasing numbers of reports of both MRSA and *S. aureus* as 4-quinolone-resistant, and that this increase appears to correspond with the increasing use of the agents in the chronic care of patients.

An important point raised with regard to the appearance of 4-quinolone-resistant *S. aureus* was the fact that in many cases the appearance of the resistant organisms was not apparently associated with the prior exposure of the patient to the 4-quinolones. This of course raised the semantic question of whether the 4-quinolone-resistant *S. aureus* were part of the original infecting population of bacteria, or whether these organisms arose, and were selected, during the course of the 4-quinolone therapy. The fact that such organisms were not detectable in samples taken before the 4-quinolone therapy seemed indicative of the fact that the resistance appeared during the treatment. This feature was apparently problematical in that 4-quinolone-resistant *S. aureus* are far from easy to isolate *in vitro* and so the ease with which they seem to appear in the clinical situation could not have been forseen. However, the fact that recent studies in Japan had indicated a possible correlation between the use of 4-quinolones in general practice and the incidence of MRSA in hospitals seemed not only contradictory to some American observations, but also clearly represented a worrying possibility.

It was thus concluded that there is a clear need for the further study of the potential use of the 4-quinolones in the treatment of problem organisms, like *S. aureus*, with specific reference to the question of the development of resistance to the 4-quinolones in such species. It was also suggested that it is perhaps time that we give serious consideration to our present strategy of always saving the most potent and effective antibacterial agents for use as last-resort treatment of problematical infections.

A corollary to this conclusion was the questioning of the practice of using reduced doses of 4-quinolones in an attempt to minimize the risks of adverse-reactions. Whilst it was generally agreed that this was commendable in principle, it represented a high-risk strategy in terms of the development and selection of organisms resistant to the 4-quinolones. Hence, the general conclusion of all participants was that the principal concern over the use of 4-quinolones as broad-spectrum antibacterial agents was not related to either their activity, or their pharmacology, but primarily to the risks of the selection of resistant organisms.

Chapter 2

Effects of Physiological Cation Concentration on 4-Quinolone Absorption and Potency

J. T. Smith

Introduction

Hoffken *et al.* (1985) found that when antacids containing magnesium and aluminium were orally co-administered with ciprofloxacin to volunteers, seriously reduced levels of the antibacterial appeared in the serum and urine. They concluded that since Ratcliffe and Smith (1983) had shown that magnesium antagonised the actions of 4-quinolones on bacteria, this ion, and possibly aluminium as well, could be complexing with ciprofloxacin in the gut, so reducing its enteral absorption. Subsequently the antagonistic role of oral aluminium in reducing the enteral absorption of ciprofloxacin in patients was identified by Golper *et al.* (1987) and reduced enoxacin absorption due to an aluminium-magnesium antacid was also reported by Shentag *et al.* (1987), who studied volunteers. Since urinary magnesium levels are known to antagonize nine out of ten different 4-quinolone antibacterials acting on *Escherichia coli* (Smith and Ratcliffe 1986), norfloxacin, ofloxacin and ciprofloxacin were tested against a total of eight different, bacterial species in the presence and absence of not only magnesium, but also aluminium, calcium, and iron, since these ions are also administered orally.

Materials and Methods

To test the effects of ions, minimum inhibitory concentrations (MIC's) in µg/ml of ofloxacin, norfloxacin and ciprofloxacin were determined on nutrient agar containing various salts. The inoculum was 20-100 colony forming units of *E. coli* KL16, *Enterobacter aerogenes* 418, *Serratia marcescens* 120, *Providencia stuartii* EN215, *Salmonella typhimurium* LT2, *Shigella sonnei* ZBX4, *Staphylococcus aureus* E3T or *Staphylococcus warneri* C14LN19.

The plates were incubated at 37°C overnight in the case of the Gram negative rods and for two days with the Staphylococci. The MIC was recorded as the lowest concentration of 4-quinolone which prevented bacterial growth. When the effects of multivalent metal ions were tested, each salt shown in the table was added to a final concentration of 5mM, and where necessary the pH of the nutrient agar was adjusted to 7.4 by the addition of M NaoH before the addition of the 4-quinolones.

Results

It was found that irrespective of the bacterial species tested, 15mM sodium chloride or 5mM sodium sulphate did not significantly alter the MIC of norfloxacin, ofloxacin

Table 2.1. Minimum Inhibitory Concentrations (µg/ml)

Drug	none	AlCl$_3$	FeCl$_3$	Addition FeSO$_4$	MgCl$_2$	CaCl$_2$
Escherichia coli						
Norfloxacin	0.04	0.75	0.5	1.5	0.075	0.04
Ofloxacin	0.03	0.3	0.3	0.5	0.1	0.03
Ciprofloxacin	0.004	1.5	0.2	0.3	0.015	0.01
Enterobacter aerogenes						
Norfloxacin	0.03	0.2	0.3	0.75	0.075	0.03
Ofloxacin	0.01	0.15	0.075	0.3	0.04	0.01
Ciprofloxacin	0.005	0.05	1	1.5	0.01	0.0075
Serratia marcescens						
Norfloxacin	0.075	2	3	7.5	0.2	0.075
Ofloxacin	0.1	2	1	3	0.3	0.1
Ciprofloxacin	0.03	0.75	1	5	0.075	0.05
Providencia stuartii						
Norfloxacin	0.75	10	20	40	3	0.75
Ofloxacin	0.4	3	1.5	5	0.75	0.4
Ciprofloxacin	0.2	4	4	10	0.4	0.3
Staphylococcus warneri						
Norfloxacin	0.75	10	7.5	10	3	0.75
Ofloxacin	0.2	2	0.75	1	0.5	0.2
Ciprofloxacin	0.2	3	3	5	0.5	0.3
Staphylococcus aureus						
Norfloxacin	0.75	10	5	7.5	3	0.75
Ofloxacin	0.1	0.75	0.5	0.75	0.3	0.1
Ciprofloxacin	0.1	0.75	0.75	0.75	0.3	0.15
Shigella sonnei						
Norfloxacin	0.03	0.3	0.75	0.5	0.075	0.03
Ofloxacin	0.03	0.15	0.5	0.3	0.1	0.03
Ciprofloxacin	0.0075	0.03	0.2	0.075	0.015	0.0075
Salmonella typhimurium						
Norfloxacin	0.075	0.5	1	0.75	0.15	0.075
Ofloxacin	0.04	0.2	0.5	0.3	0.1	0.04
Ciprofloxacin	0.0075	0.04	0.3	0.075	0.02	0.0075

Minimum inhibitory concentrations were determined on nutrient agar at pH 7.4 with and without each salt at a final concentration of 5mM or ciprofloxacin. However, it can be seen in Table 2.1 that aluminium, ferric or ferrous salts antagonised all 4-quinolones even more than magnesium chloride. The antagonisms observed can be attributed to the multivalent metal ions rather than to the sulphate or chloride anions due to the findings with sodium chloride and sodium sulphate. On average aluminium ions increased the MICs by 27 fold (range 4 to 375) – (see Table 2.2) whilst the mean MIC increases were 25 (range 5-200) for ferric ions – (see Table 2.3), 43 (range 8-300) for ferrous ions - (see Table 2.4 and 2.8) (range 1.9 to 4) for magnesium ions – (see Table 2.5). On the other hand, calcium ions (Table 2.6) did not affect the MIC's of norfloxacin or ofloxacin against any species. Whilst ciprofloxacin was antagonised by calcium ions with six out of the eight species tested, the effect was very weak (Table 2.6) because the average MIC increase was only 1.2 fold (range 1.0 - 2.5), which is even less than the antagonism of ciprofloxacin caused by magnesium ions where the average MIC increase was 2.8 fold. As regards any differences between the 4-quinolones tested, Table 2.7 shows that whilst calcium and magnesium caused similar antagonisms with all three drugs, aluminium, ferric and ferrous ions antagonised ciprofloxacin more than norfloxacin, whilst ofloxacin was antagonised least of all.

Table 2.2. MIC increases caused by 5mM Aluminium ions

	NOR	OFL	CIP	Mean
E. coli	19	10	375	135
E. aerogenes	7	15	10	11
S. marcescens	27	20	25	24
P. stuartii	13	8	20	14
Staph. warneri	13	10	15	13
Enteric pathogens				
Staph. aureus	13	8	8	9
Shig. sonnei	10	5	4	6
Salm. typhimurium	7	5	5	6
Mean	14	10	58	

NOR = norfloxacin, OFL = ofloxacin, CIP = ciprofloxacin
(also for Tables 2.3-2.6)

Table 2.3. MIC increases caused by 5mM Ferric ions

	NOR	OFL	CIP	Mean
E. coli	13	10	50	24
E. aerogenes	10	8	200	73
S. marcescens	40	10	33	28
P. Stuartii	27	4	20	17
Staph. warneri	10	4	15	10
Enteric pathogens				
Staph. aureus	7	5	8	6
Shig. sonnei	25	17	27	23
Salm. typhimurium	13	13	40	22
Mean	18	9	49	

Table 2.4. MIC increases caused by 5mM Ferrous ions

	NOR	OFL	CIP	Mean
E. coli	37	17	75	43
E. aerogenes	25	30	300	118
S. marcescens	100	30	167	99
P. stuartii	53	13	50	39
Staph. warneri	13	5	25	14
Enteric pathogens				
Staph. aureus	10	8	8	8
Shig. sonnei	17	10	10	12
Salm. typhimurium	10	8	10	9
Mean	33	15	82	

Table 2.5. MIC increases caused by 5mM Magnesium ions

	NOR	OFL	CIP	Mean
E. coli	1.9	3.3	3.8	3.0
E. aerogenes	2.5	4.0	2.0	2.8
S. marcescens	2.7	3.0	2.5	2.7
P. stuartii	4.0	1.9	2.0	2.6
Staph. warneri	4.0	2.5	2.5	3.0
Enteric pathogens				
Staph. aureus	4.0	3.0	3.0	3.3
Shig. sonnei	2.5	3.3	2.0	2.6
Salm. typhimurium	2.0	2.5	2.7	2.4
Mean	3.0	2.9	2.6	

Table 2.6. MIC increases caused by 5mM Calcium ions

	NOR	OFL	CIP
E. coli	1	1	2.5
E. aerogenes	1	1	1.5
S. marcescens	1	1	1.7
P. stuartii	1	1	1.5
Staph. warneri	1	1	1.5
Enteric pathogens			
Staph. aureus	1	1	1.5
Shig. sonnei	1	1	1
Salm. typhimurium	1	1	1
Mean	1	1	1.5

Discussion

These findings that aluminium, ferric, ferrous and magnesium ions antagonised norfloxacin, ofloxacin and ciprofloxacin acting on bacteria, accord well with later findings that these cations also prevent 4-quinolone uptake from the human gut (Hoffken *et al.* 1985). Shentag *et al.*(1987) also showed that when 4-quinolones were orally co-administered with antacids containing magnesium and aluminium, then the absorption of the 4-quinolones into the serum from the gut was seriously reduced in volunteers,whilst Fleming *et al.* (1986) found this also occurred in patients. Hoffken *et al.*(1985) suggested that as magnesium antagonises the quinolones acting on bacteria

(Ratcliffe and Smith 1983), a similar phenomenon may be hindering quinolone uptake from the human gut, or that the aluminium component could be responsible. This latter view seems to be correct because Golper *et al.* (1987) found that an antacid containing solely aluminium also seriously reduced the oral absorption of ciprofloxacin in patients.

Table 2.7. Mean MIC increases with 8 bacterial species caused by multivalent metal ions at 5mM

	NOR	OFL	CIP	Mean
Antacids				
Aluminium	14	10	58	27
Magnesium	3.0	2.9	2.6	2.8
Calcium	1	1	1.5	1.2
Oral iron				
Ferric	18	9	49	25
Ferrous	33	15	82	43
Mean	14	8	38	

Antacids not only contain magnesium or aluminium but can also contain calcium, whilst in anaemia and during pregnancy iron is another orally administered metal. Calcium had no effect on the potency of norfloxacin or ofloxacin, whilst ciprofloxacin was only weakly antagonised by calcium ions (Table 2.7). This agrees with the results of Fleming *et al.* (1986) who showed that a calcium antacid did not significantly reduce the enteral absorption of ciprofloxacin from patients. However, a contradictory result showing that calcium ions may impair ciprofloxacin absorption in humans has also been reported (Sahai *et al.* (1989). However, the latter study was in volunteers rather than in patients.

It is probable that the antagonism of 4-quinolones by multivalent metal ions occurs due to complexes forming which are less able to be taken up by bacteria than free drug. Similar complexes could also hinder the uptake of 4-quinolones from the human gastrointestinal tract. Good correlations have been observed with aluminium, which severely reduces both the antibacterial potency and the enteral absorption of 4-quinolones, and calcium which only has a weak effect with bacteria and no effect on the enteral absorption of ciprofloxacin in patients. This leaves magnesium to be assessed in humans because its antagonistic effect on bacteria is only marginally greater than that of calcium.

The results in Tables 2.3, 2.4 and 2.7 also show that iron in its ferrous state (as occurs in the gastric acid) or in its ferric state (as occurs at the higher pH values of the lower gastrointestinal tract) caused similar or greater antagonisms than those caused by aluminium, hence it would seem prudent to avoid oral iron preparations at times close to 4-quinolone oral therapy. In agreement subsequent pharmacokinetic studies have indicated that iron is able to reduce the enteral absorption of ciprofloxacin at least in volunteers (Polk *et al.* 1989).

It should be noted that oral iron is ingested particularly in those parts of the world where blood losses resulting from intestinal parasitic infestations are common. Unfortunately such patients are more liable to get severe enteric bacterial infections and hence are also likely candidates for oral 4-quinolone therapy.

It would seem that the antagonism of 4-quinolone antibacterial action *in vitro* could also occur within the gut lumen, hence oral antacids or oral iron preparations given during oral 4-quinolone therapy could seriously reduce the antibacterial potency of the drugs against enteric pathogens quite separately from any reduction in the uptake of the drugs into the serum which may also occur. The results with *Salmonella* and *Shigella* (Tables 2.1-2.6) show these genera too are as prone to interactions between multivalent metal ions and 4-quinolones as are other bacteria, so yet again these *in vitro* findings may forecast what may ensue clinically.

References

Fleming LN, Moreland TA, Stewart WK, Scott AC (1986) Ciprofloxacin and antacids. Lancet 2:294

Golper TA, Hartstein AI, Morthland VH, Christensen JM (1987) Effects of antacids and dialysate dwell times on multiple dose pharmacokinetics and oral ciprofloxacin in patients on continuous peritoneal dialysis. Antimicrob Agents Chemother 31:1787-1790

Hoffken G, Borner K, Glatzel PO, Koeppe P, Lode H (1985) Reduced enteral adsorption of ciprofloxacin in the presence of antacids. Eur J Clin Microb 4:345

Polk R, Healy D, Sahai J, Drwal L, Racht E (1989) Effect of ferrous sulphate and multivitamins with zinc on the absorption of ciprofloxacin in normal volunteers. Abstracts of the 29th Interscience Conference on antimicrobial agents and chemotherapy, abstract 212, p 136 (1989)

Ratcliffe NT, Smith JT (1983) Effects of magnesium on the activity of 4-quinolone antibacterial agents. J Pharm Pharmacol 35, p 61

Sahai J, Healy D, Stotka J, Polk R (1989) Influence of chronic administration of calcium on the bioavailability of oral ciprofloxacin. Abstracts of the 29th Interscience Conference on antimicrobial agents and chemotherapy, abstract 211, p 136

Schentag JJ, Sedman AJ, Wilton JH, Thomas DJ, Schultz RW, Kinkel AW, Grasela TH (1987) Interactions between enoxacin, ranitidine and antacids. Third European Congress of Clinical Microbiology, abstract 256, p 119

Smith JT, Ratcliffe NT (1986) Einfloss von pH-wert und magnesium auf die antibakterielle activitat von chinolonproparoten. Infection14, suppl 1:31-35

Discussion Summary

The clear evidence of the antagonism of the antibacterial activity of the 4-quinolones by metal ions clearly raised several possible explanations in the minds of the audience. Whilst it was generally accepted that the formation of essentially insoluble chelate complexes would inhibit the uptake of active drug, more concern was expressed over the possibility that the intracellular levels of such ions in the bacterial cell could themselves perturb the functioning of DNA gyrase. Such ionic changes affecting the functioning of DNA gyrase might further alter the possible interactions of the 4-quinolones with the gyrase:DNA system, and so our *in vitro* modelling of the interaction might not represent a valid model of the situation *in vivo*. Clearly some further study was required to try and understand the mechanisms which were operating in order to facilitate the development of appropriate *in vitro* methodology which will permit structure-activity studies useful in the further development of clinically useful molecules where such problems will be minimalized.

It was suggested that since the fact that the 4-quinolones chelate metal ions was known more than ten years ago we should have been able to anticipate the type of interaction showed by Professor Smith. However, the results obtained thus far, whilst clearly demonstrating antagonistic interactions, also showed that in the light of our present understanding of the chemistry of these compounds the degree of antagonism was still unpredictable.

The potential importance of these observations in the clinical management of infections in either patients on dialysis, and in patients (often of an older age-group) who routinely self-administer oral antacids was self-evident. It clearly represented an apparently unexpected consideration in the prescribing of the 4-quinolones since it might not only reduce the inhibitory activity of the 4-quinolones, but might also significantly alter the efficiency with which these agents kill susceptible bacteria (and maybe thus affect the efficiency with which the 4-quinolones select resistant organisms).

In this presentation the evidence presented did not seem to provide an answer to a long-established question, on the contrary it raised a new and important question requiring an answer in order to enable us to gain an understanding of the effects of these agents under physiological conditions.

Chapter 3

The Influence of Oxygen upon Bactericidal Potency

I. Morrissey, C. S. Lewin and J. T. Smith

Introduction

Varying the inoculum size by 10 000 fold has negligible effects on the minimum inhibitory concentrations (MICs) of 4-quinolones against *Escherichia coli* (Chin and Neu 1983; Smith 1984). MIC studies, however, only provide a measure of the inhibition of bacterial multiplication by 4-quinolones i.e. only provide information regarding the bacteriostatic activity of these drugs. The bacteriostatic and bactericidal activities of the 4-quinolones are quite distinct. For example, urinary magnesium concentrations have been shown to severely antagonise the bactericidal activity of 4-quinolones but have little effect on their bacteriostatic activity (Ratcliffe and Smith 1983). Conversely, low pH antagonises the bacteriostatic activity of ciprofloxacin more than its bactericidal activity (Ratcliffe and Smith 1985). Therefore it was decided to investigate the effect of initial inoculum size on the bactericidal activity of ofloxacin and ciprofloxacin on *E. coli* KL16 and *Staphylococcus aureus* E3T.

Preliminary reports of this work have been presented (Lewin and Smith 1986; Lewin and Smith 1987; Lewin *et al.* 1988).

Materials and Methods

Antibacterials: Ofloxacin (Hoechst) was firstly dissolved in 0.5M NaOH (0.02 ml mg⁻¹) and made up to volume in sterile distilled water. Ciprofloxacin (Bayer) was dissolved in sterile distilled water.

Media: Nutrient broth No 2 (Oxoid UK) and Liquid Thioglycollate Medium (Brewer's Medium) (Oxoid UK).

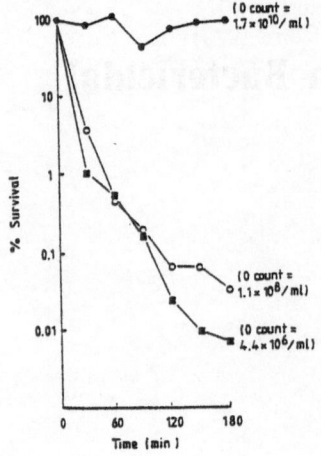

Fig. 3.1. Survival of *E. coli* KL16 treated with 0.9 µg ml⁻¹ ofloxacin. *E. coli* KL16 was grown overnight in nutrient broth at 37°C either statically or shaken at 150 cycles per minute. The static culture was diluted either 1 in 50 or 1 in 2 in nutrient broth to produce the two lower inoculum sizes used. The shaken culture was centrifuged and the bacteria resuspended to one twentieth of the original volume with fresh sterile nutrient broth. This concentrated suspension was then diluted 1 in 2 in nutrient broth to give the highest inoculum size. Each inoculum contained ofloxacin at a final concentration of 0.9 µg ml⁻¹. Incubation was continued statically in air at 37°C and survival estimated by viable counting on nutrient agar.

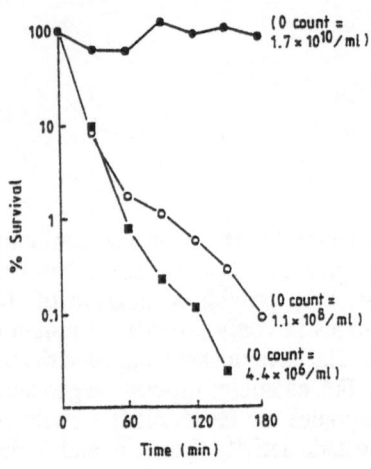

Fig. 3.2. Survival of *E. coli* KL16 treated with 0.15 µg ml⁻¹ ciprofloxacin. Similar conditions to those described in the legend to Fig. 3.1 were used except that *E. coli* KL16 was used with ciprofloxacin at a final concentration of 0.15 µg ml⁻¹.

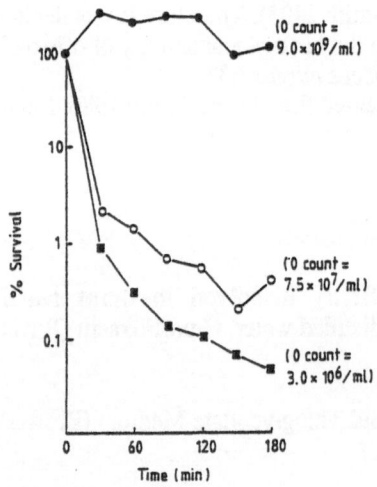

Fig. 3.3. Survival of *S. aureus* E3T treated with 5 µg ml⁻¹ ofloxacin. Similar conditions to those described in the legend to Fig. 3.1 were used except that *S. aureus* E3T was used with ofloxacin at a final concentration of 5 µg ml⁻¹.

Strains: E. coli KL16 (Hane and Wood 1969) and *S. aureus* E3T (Knox and Smith 1961).

Estimation of bactericidal activity: Reaction mixtures containing nutrient broth, 4-quinolone and bacteria at initial inoculum sizes of about 10^6, 10^8 and 10^{10} cfu ml^{-1} were prepared as follows: For experiments at 10^6 cfu ml^{-1}, 5 ml of sterile double-strength nutrient broth was added to sterile 1 oz bottles. Aqueous drug and sterile distilled water were added to 9.8 ml. 0.2 ml of a nutrient broth culture which had been incubated statically overnight at 37°C was then added to complete each reaction mixture. For experiments at 10^8 cfu ml^{-1}, 2.5 ml of sterile quadruple-strength nutrient broth was added to sterile 1 oz bottles. Aqueous drug and sterile distilled water were added to 5 ml. 5 ml of a nutrient broth culture which had been incubated statically overnight at 37°C was then added to complete each reaction mixture. For experiments at 10^{10} cfu ml^{-1}, 2.5 ml of sterile double strength nutrient broth was added to sterile 1 oz bottles. Aqueous drug and sterile distilled water were added to 5 ml. A 200 ml nutrient broth culture was shaken at 150 cycles per minute overnight at 37°C using a Gallenkamp orbital incubator then centrifuged at 6000 rpm for 15 min. The supernatant was discarded and the pellet resuspended in 10 ml of sterile nutrient broth. 5 ml of this concentrated suspension was then added to complete each reaction mixture.

The rate of kill by the 4-quinolones was estimated as follows. At 30 min intervals 0.1 ml samples were taken from the reaction mixtures and decimally diluted in sterile nutrient broth. 0.1 ml samples of these dilutions were spread onto nutrient agar plates which were incubated overnight at 37°C and the colonies counted.

When the rate of kill by the 4-quinolones was estimated under anaerobic conditions the overnight broth cultures used to inoculate the reaction mixtures were grown in Brewer's medium that had had a 2 cm layer of liquid paraffin added prior to being autoclaved. Brewer's medium overlaid in such a manner was steamed for 20 min to remove oxygen and used to make up the reaction mixtures. Bacteria and 4-quinolone were inoculated under the paraffin layer with a Hamilton syringe fitted with a 0.68 mm diameter needle.

Estimation of magnesium concentration: Atomic absorption spectroscopy (AAS) using a Perkin Elmer 280 spectrometer was used to determine the concentration of magnesium in nutrient broth. Standards were prepared using double-distilled water as a diluent and estimations of the unknown concentrations were made using dilutions that gave readings within the region of linear response (0.1 to 0.5 parts per million). The slit width and wavelength used were 0.7 nm and 285 nm, respectively, and the oxident flow and flame height were 50 and 3, respectively. The figures used corresponded to machine markings.

Results

When the initial inoculum size of *E. coli* KL16 was increased from 4.4 x 10^6 cfu ml^{-1} to 1.1 x 10^8 cfu ml^{-1} in nutrient broth a significant reduction in the bactericidal activity of both ofloxacin and ciprofloxacin, at their most bactericidal concentrations (0.9 and 0.15 µg ml^{-1}, respectively), occurred. However, this decrease in bactericidal activity was slight when compared to the effect of increasing the initial inoculum size to 1.7 x 10^{10} cfu ml^{-1} on the bactericidal activity of the 4-quinolones (Figs. 3.1 and 3.2). Both drugs were only bacteriostatic against this highest inoculum size of *E. coli* KL16.

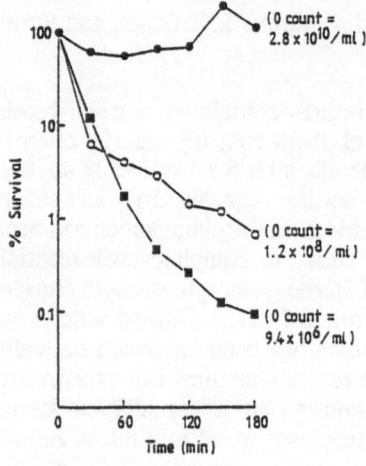

Fig. 3.4. Survival of *S. aureus* E3T treated with 5 μg ml⁻¹ ciprofloxacin. Similar conditions to those described in the legend to Fig. 3.1 were used except that *S. aureus* E3T was used with ciprofloxacin at a final concentration of 5 μg ml⁻¹.

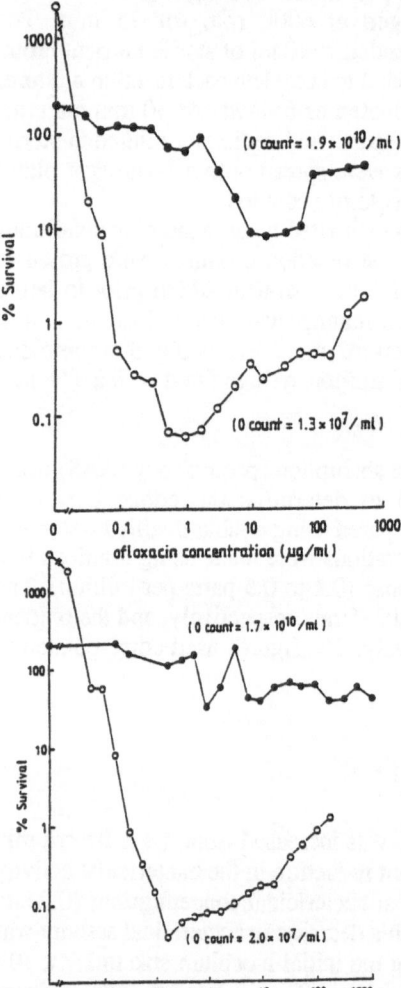

Fig. 3. 5. Survival of *E. coli* KL16 treated with ofloxacin for 3 h. *E. coli* KL16 was grown overnight in nutrient broth at 37°C either statically or shaken at 150 cycles per minute. The static culture was diluted 1 in 50 into nutrient broth containing different concentrations of ofloxacin to produce the lower inoculum size used. The shaken culture was centrifuged and the bacteria resuspended to one twentieth of the original volume with fresh sterile nutrient broth. This concentrated suspension was then diluted 1 in 2 into different concentrations of ofloxacin in nutrient broth to give the higher inoculum size. Incubation was continued statically in air at 37°C for 3 h when survival was estimated by viable counting on nutrient agar.

Fig. 3.6. Survival of *E. coli* KL16 treated with ciprofloxacin for 3 h. Similar conditions to those described in the legend to Fig. 3.5 were used except that ciprofloxacin was used.

Similar results were seen when ofloxacin and ciprofloxacin were tested against *S. aureus* E3T at their most bactericidal concentrations of 5 µg ml⁻¹ (Figs. 3.3 and 3.4). A significant reduction of kill was seen when the initial inoculum size was increased from 2×10^6 cfu ml⁻¹ to 1.2×10^8 cfu ml⁻¹. Again this reduction was not so dramatic as that seen at 2.8×10^{10} cfu ml⁻¹, where a bacteriostatic response was seen. Hence it would appear that the 4-quinolones are merely bacteriostatic against *E. coli* KL16 and *S. aureus* E3T when the initial inoculum size is high. It may be thought that this lack of bactericidal effect may be caused by the bacteria being unable to multiply at such high cell densities. However, the bacteria were suspended in fresh nutrient broth to allow multiplication to occur and indeed *E. coli* multiplied by a factor of 2.5 fold during 3 h in the absence of any 4-quinolone. *S. aureus*, on the other hand, did not multiply during the 3 h under similar conditions. Thus a high initial inoculum size prevents kill by the 4-quinolones, and this phenomenon occurs whether or not the bacteria can divide under similar conditions without 4-quinolone.

So far only the most bactericidal concentrations of each drug has been tested for an inoculum size effect. Therefore whether the effect persisted throughout the range of clinically achievable concentrations was investigated. When *E. coli* KL16 or *S. aureus* E3T were treated with ofloxacin or ciprofloxacin in nutrient broth, at an initial inoculum size of about 10^6 cfu ml⁻¹ and survival estimated after 3 h, a biphasic response was seen (Figs. 3.5-3.8). The bactericidal activity of the drugs increased with drug concentration until the most bactericidal concentration for the particular species was reached, beyond which their bactericidal activity then decreased.

On the other hand, at initial inoculum sizes of about 10^{10} cfu ml⁻¹, ciprofloxacin was essentially bacteriostatic over all the concentrations tested with both *E. coli* KL16 and *S.aureus* E3T (Figs. 3.6 and 3.8). The highest concentration tested, 900 µg ml⁻¹, exceeds its peak serum level by at least one hundred fold (Bergan *et al.* 1987) and is several times its maximal urine level (Gasser *et al.* 1987). Hence ciprofloxacin does not seem to be bactericidal in nutrient broth at any clinically achievable concentration when the initial inoculum size is high. Ofloxacin was also essentially bacteriostatic over all the concentrations tested with *E. coli* KL16 at an initial inoculum size of about 10^{10} cfu ml⁻¹. However, ofloxacin at concentrations above 15 µg ml⁻¹ began to show some bactericidal activity against *S. aureus* E3T at this inoculum size to produce a biphasic response similar to that seen at the lower initial inoculum size but with a most bactericidal effect occuring at 150 µg ml⁻¹. Thus ofloxacin seems to be more active than ciprofloxacin against *S. aureus* E3T at high initial inoculum size. However, this bactericidal activity occurs at concentrations higher than those clinically achievable in serum (Bergan *et al.* 1987) but could occur at maximal urine levels (Gasser *et al.* 1987).

Possible reasons as to why the bactericidal action of the 4-quinolones in nutrient broth was affected by initial inoculum size were then investigated.

Magnesium can antagonise the bactericidal activity of the 4-quinolones (Ratcliffe and Smith 1983) and therefore the concentrations of magnesium in the medium of nutrient broth cultures of *E. coli* KL16 after 3 h exposure to ofloxacin and ciprofloxacin were investigated. The results (Table 3.1) show that the magnesium concentrations from *E. coli* KL16 cultures with a high initial inoculum size were if anything slightly lower than those found in cultures with a low initial inoculum size. Hence increasing the initial inoculum size did not increase the magnesium concentration, which can therefore be ruled out as a causative factor for the inoculum size effect.

The hydrogen ion concentration can also affect the bactericidal activity of the 4-quinolones (Smith and Ratcliffe 1986). However, when the pH of nutrient broth

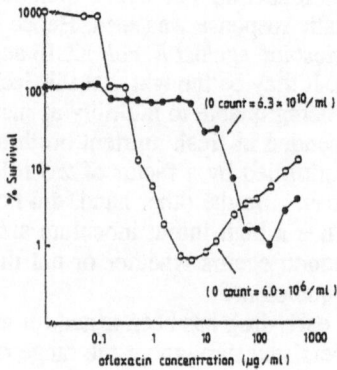

Fig. 3.7. Survival of *S. aureus* E3T treated with ofloxacin for 3 h. Similar conditions to those described in the legend to Fig. 3.5 were used except that *S. aureus* E3T was used with ofloxacin.

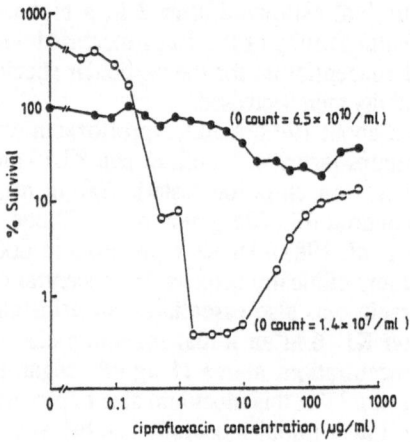

Fig. 3.8. Survival of *S. aureus* E3T treated with ciprofloxacin for 3 h. Similar conditions to those described in the legend to Fig. 3.5 were used except that *S. aureus* E3T was used with ciprofloxacin.

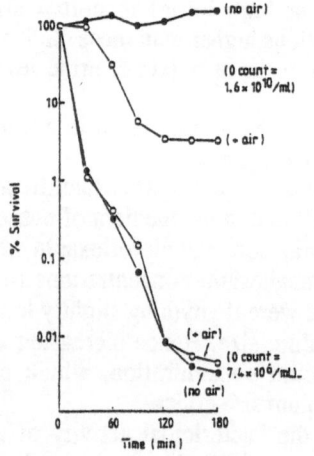

Fig. 3.9. Survival of *E. coli* KL16 treated with 0.9 μg ml^{-1} ofloxacin with and without air. *E. coli* KL16 was grown overnight in nutrient broth at 37°C either statically or shaken at 150 cycles per minute. The static culture was diluted 1 in 50 in nutrient broth. Half this culture was incubated statically in air at 37°C while air was bubbled through the remainder at a rate of 0.4 l min^{-1} at 37°C. The shaken culture was centrifuged and the bacteria resuspended to one twentieth of the original volume with fresh sterile nutrient broth. This concentrated suspension was then diluted 1 in 2 in nutrient broth. Half this culture was incubated statically in air at 37°C while air was bubbled through the remainder as above. Each inoculum contained .ofloxacin at a final concentration of 0.9 μg ml^{-1}. Survival was estimated by viable counting on nutrient agar.

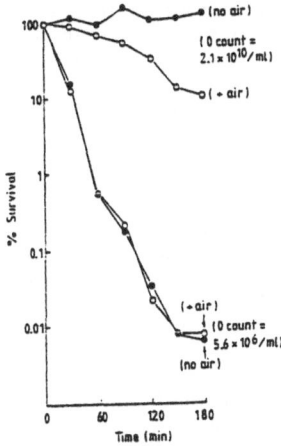

Fig. 3.10. Survival of *E. coli* KL16 treated with 0.15 μg ml^{-1} ciprofloxacin with and without air. Similar conditions to those described in the legend to Fig. 3.9 were used except that *E. coli* KL16 was used with ciprofloxacin at a final concentration of 0.15 μg ml^{-1}.

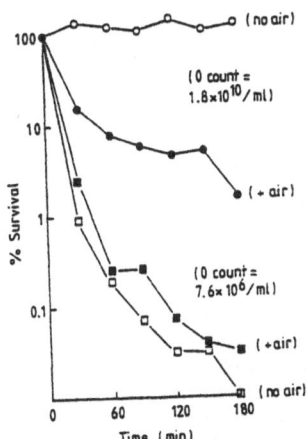

Fig. 3.11. Survival of *S. aureus* E3T treated with 5 μgml^{-1} ofloxacin with and without air. Similar conditions to those described in the legend to Fig. 3.9 were used except that *S. aureus* E3T was used with ofloxacin at a final concentration of 5 μg ml^{-1}.

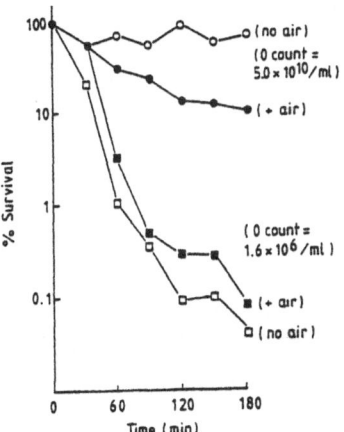

Fig. 3.12. Survival of *S. aureus* E3T treated with 5 μg ml^{-1} ciprofloxacin with and without air. Similar conditions to those described in the legend to Fig. 3.9 were used except that *S. aureus* E3T was used with ciprofloxacin at a final concentration of 5 μg ml^{-1}.

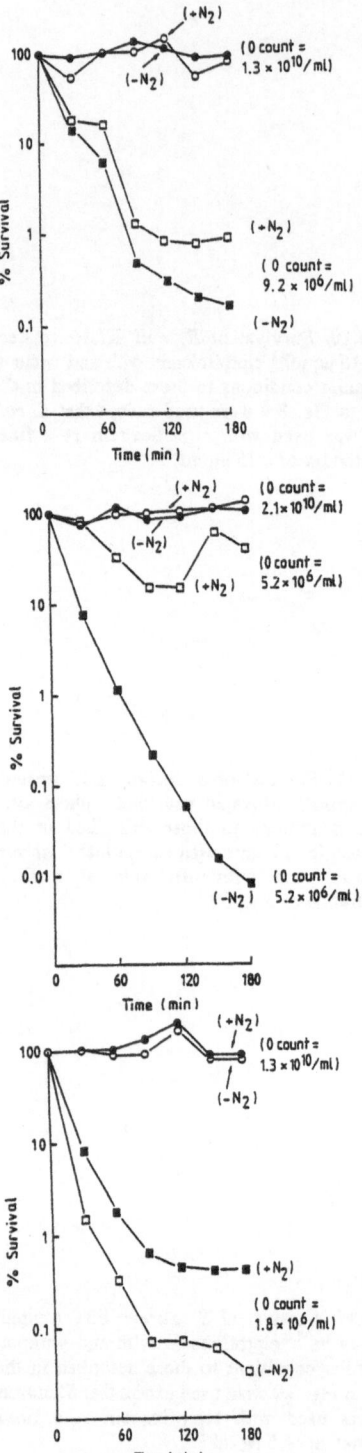

Fig. 3.13. Survival of *E. coli* KL16 treated with 0.9 µg ml^{-1} ofloxacin with and without nitrogen. *E. coli* KL16 was grown overnight in nutrient broth at 37°C either statically or shaken at 150 cycles per minute. The static culture was diluted 1 in 50 in nutrient broth. Half this culture was incubated statically in air at 37°C while nitrogen was bubbled through the remainder at a rate of 0.4 l min^{-1} at 37°C. The shaken culture was centrifuged and the bacteria resuspended to one twentieth of the original volume with fresh sterile nutrient broth. This concentrated suspension was then diluted 1 in 2 in nutrient broth. Half this culture was incubated statically in air at 37°C while nitrogen was bubbled through the remainder as above. Each inoculum contained ofloxacin at a final concentration of 0.9 µg ml^{-1}. Survival was estimated by viable counting on nutrient agar.

Fig. 3.14. Survival of *E. coli* KL16 treated with 0.15 µg ml^{-1} ciprofloxacin with and without nitrogen. Similar conditions to those described in the legend to Fig. 3.13 were used except that *E. coli* KL16 was used with ciprofloxacin at a final concentration of 0.15 µg ml^{-1}.

Fig. 3.15. Survival of *S. aureus* E3T treated with 5 µg ml^{-1} ofloxacin with and without nitrogen. Similar conditions to those described in the legend to Fig. 3.13 were used except that *S. aureus* E3T was used with ofloxacin at a final concentration of 5 µg ml^{-1}.

cultures of *E. coli* KL16 was measured after 3h exposure to the 4-quinolones no significant differences were observed between the pH values of cultures containing low or high initial inocula (Table 3.2). Therefore increasing the initial inoculum size did not affect the pH of the medium to an extent that could explain the phenomenon.

Drug destruction also did not account for the inoculum size effect observed with the 4-quinolones. No loss of drug was detected in a culture of *E. coli* KL16 containing an initial inoculum of 10^{10} cfu ml^{-1} after 3h exposure to 0.15 µg ml^{-1} ciprofloxacin.

At high initial inoculum sizes the conditions might be considered anaerobic due to the vast number of metabolising bacteria. As the effect of oxygen tension on the activity of the 4-quinolones was not known, aeration of cultures of a high initial inoculum size was tested. When the cultures were bubbled with air it was found that some of the bactericidal activity of ofloxacin and ciprofloxacin at their most bactericidal concentrations was restored at the high initial inoculum sizes of about 10^{10} cfu ml^{-1} for both *E. coli* KL16 and *S. aureus* E3T (Figs. 3.9-3.12 upper parts). On the other hand at the low initial inoculum sizes of about 10^6 cfu ml^{-1} with both species, aeration had no significant effect on the bactericidal activity of either 4-quinolone at their most bactericidal concentrations (Figs. 3.9-3.12 lower parts).

Table 3.1.

Drug Concentration		mM Mg at initial viable count of	
	(µg/ml)	10^6/ml	10^{10}/ml
ciprofloxacin	0.15	0.14	0.09
ofloxacin	0.90	0.13	0.11

Mg concentration of nutrient broth = 0.12 mM

Mg in medium after 3 h exposure of *E.coli* KL16 to 4-quinolones. *E. coli* KL16 was grown overnight in nutrient broth at 37°C either statically or shaken at 150 cycles per minute. The static culture was diluted 1 in 50 in nutrient broth to produce the lower inoculum size used. The shaken culture was centrifuged and the bacteria resuspended to one twentieth of the original volume with fresh sterile nutrient broth. This concentrated suspension was then diluted 1 in 2 in nutrient broth to give the higher inoculum size. Inocula contained either ofloxacin at a final concentration of 0.9 µg ml^{-1} or ciprofloxacin at 0.15 µg ml^{-1}. Incubation was continued statically in air at 37°C for 3 h after which magnesium concentration was estimated by atomic absorption spectroscopy.

Table 3.2.

Drug Concentration		pH of medium at initial viable count of	
	(µg/ml)	10^6/ml	10^{10}/ml
ciprofloxacin	0.15	6.9	6.9
ofloxacin	0.9	6.9	6.8

pH of medium after 3 h exposure of *E. coli* KL16 to 4-quinolones. Similar conditions to those described in the legend to Table 3.1 were used but the pH was measured after incubation.

This partial restoration of bactericidal activity at the high initial inoculum size by aeration could either be due to the agitation caused by the bubbles or due to an increase in the oxygen tension. To test these possibilities the aeration experiment was repeated bubbling with pure nitrogen instead of air. The upper parts of Figs. 3.13-3.16 show that ofloxacin and ciprofloxacin were devoid of bactericidal activity whether or not nitrogen was bubbled through cultures of *E. coli* KL16 and *S. aureus* E3T at high initial inoculum sizes. The lower parts of Figs 3.13-3.16, however, show with an initial inoculum size of about 10^6 cfu ml^{-1} with *E. coli* KL16 or *S. aureus* E3T that bubbling nitrogen severely reduced the bactericidal activity of either 4-quinolone at their most bactericidal concentrations. This is consistent with oxygen being required

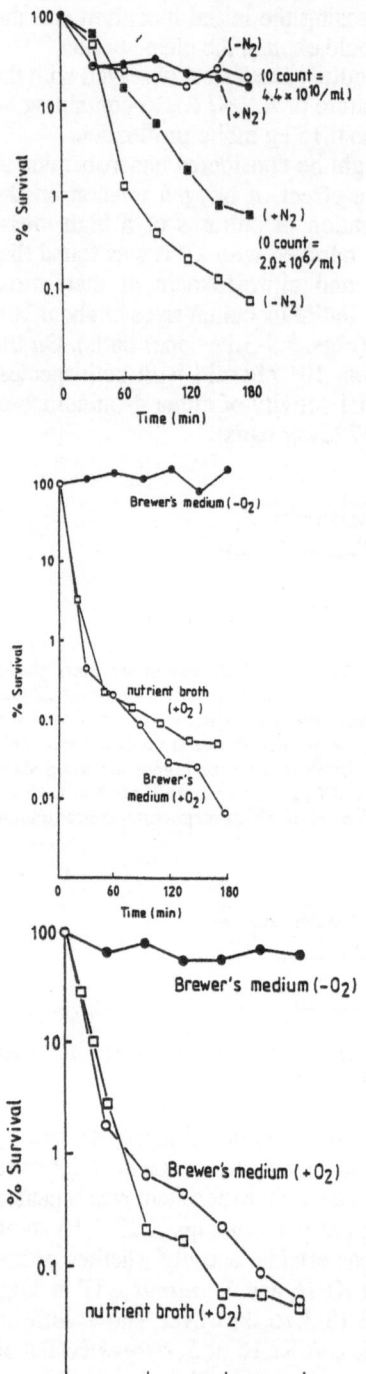

Fig. 3.16. Survival of *S. aureus* E3T treated with 5 µg ml⁻¹ ciprofloxacin with and without nitrogen. Similar conditions to those described in the legend to Fig. 3.13 were used except that *S. aureus* E3T was used with ciprofloxacin at a final concentration of 5 µg ml⁻¹.

Fig. 3.17. Survival of *E. coli* KL16 treated with 0.9 µg ml⁻¹ ofloxacin (0 count = 10⁶ cfu ml⁻¹). *E. coli* KL16 was grown overnight in nutrient broth at 37°C. Brewer's medium and nutrient broth were prepared in air and Brewer's medium covered with a 2 cm deep layer of liquid paraffin was steamed to remove oxygen and allowed to cool. Ofloxacin was added to give a final concentration of 0.9 µg ml⁻¹ and the bacterial culture was diluted 1 in 50 into these three media using Hamilton syringes with 0.68 mm diameter needles and incubation continued in air at 37°C. Survival was estimated by viable counting on nutrient agar.

Fig. 3.18. Survival of *E .coli* KL16 treated with 0.15 µg ml⁻¹ ciprofloxacin (0 count = 10⁶ cfu ml⁻¹). Similar conditions to those described in the legend to Fig. 3.17 were used except that *E. coli* KL16 was used with ciprofloxacin at a final concentration of 0.15 µg ml⁻¹.

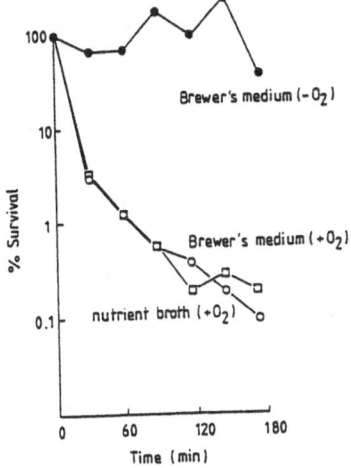

Fig. 3.19. Survival of S. aureus E3T treated with 5 μg ml⁻¹ ofloxacin (0 count = 10^6 cfu ml⁻¹). Similar conditions to those described in the legend to Fig. 3.17 were used except that S. aureus E3T was used with ofloxacin at a final concentration of 5 μg ml⁻¹.

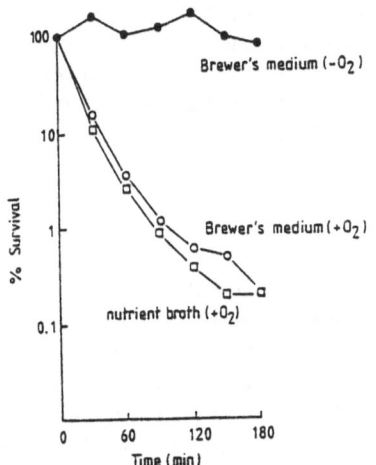

Fig. 3.20. Survival of S. aureus E3T treated with 5 μg ml⁻¹ ciprofloxacin (0 count = 10^6 cfu ml⁻¹). Similar conditions to those described in the legend to Fig. 3.17 were used except that S. aureus E3T was used with ciprofloxacin at a final concentration of 5 μg ml⁻¹.

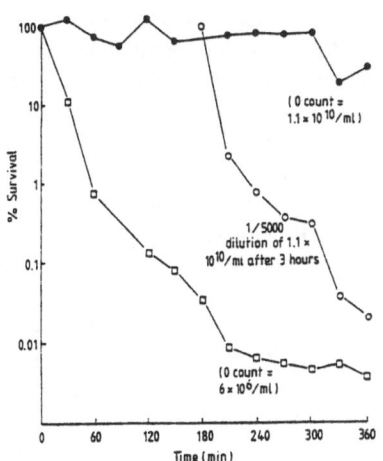

Fig. 3.21. Survival of E. coli KL16 treated with 0.9 μg ml⁻¹ ofloxacin. E. coli KL16 was grown overnight in nutrient broth at 37°C shaken at 150 cycles per minute. This was centrifuged and the bacteria resuspended to one twentieth of the original volume with fresh sterile nutrient broth. This concentrated suspension was then diluted 1 in 2 in nutrient broth containing a final ofloxacin concentration of 0.9 μg ml⁻¹. Incubation was continued statically in air at 37°C for 3 h. After 3 h an aliquot was taken and diluted 1 in 5000 in fresh nutrient broth containing 0.9 μg ml⁻¹ ofloxacin to give an initial inoculum size of 6 x 10^6 cfu ml⁻¹. Incubation was continued statically in air at 37°C for another 3 h. Survival was estimated by viable counting on nutrient agar.

for the bactericidal activity of 4-quinolones and that agitation caused by bubbling does not alleviate the inoculum size effect. As it seems that the 4-quinolones were not bactericidal under anaerobic conditions their efficacies were investigated under anaerobic conditions in Brewer's medium. Using *E. coli* KL16 and *S. aureus* E3T at low initial inoculum size of about 10^6 cfu ml^{-1} it can be seen in Figs. 3.17-3.20 that anaerobic conditions rendered both 4-quinolones bacteriostatic towards both species. Furthermore, in aerobic Brewer's medium both 4-quinolones were bactericidal with a similar rate of kill to that observed in nutrient broth exposed to air. The reasons why no kill was observed in anaerobic Brewer's medium but a rapid kill was observed in aerobic nutrient broth or Brewer's medium cannot be due to differences in the growth rates of the bacteria because similar growth rates were observed in all three media in the absence of a 4-quinolone; *E. coli* increasing by about 30-fold and *S. aureus* by about 4-fold.

An investigation was also carried out to see if the lack of bactericidal activity at a high initial inoculum size was permanent. From Fig.3.21 it can be seen that ofloxacin, at its most bactericidal concentration, was bacteriostatic against *E. coli* KL16 at an initial inoculum size of 1.1 x 10^{10} cfu ml^{-1} in nutrient broth exposed to air over 6h. After 3h an aliquot of this culture was diluted 1 in 5000 into fresh nutrient broth, containing ofloxacin at its most bactericidal concentration, to produce approximately 5 x 10^6 cfu ml^{-1}. Under these conditions the bactericidal activity was immediately restored, showing that the phenomenon of anaerobic antagonism is not a permanent one. Thus it seems that the 4-quinolones specifically need oxygen for them to be fully bactericidal and that they are only bacteriostatic when oxygen is absent.

Discussion

Although no significant inoculum size effect has been found with the 4-quinolones as determined by MIC studies (Chin and Neu 1983; Smith 1984) the bactericidal activities of ofloxacin and ciprofloxacin against both a Gram positive and Gram negative organism were found to be reduced by increasing the initial inoculum size from about 10^6 cfu ml^{-1} to about 10^8 cfu ml^{-1}. At an initial inoculum size of about 10^{10} cfu ml^{-1} the effect was not simply quantitative but qualitative as both 4-quinolones were only bacteriostatic against *E. coli*. A similar result occured with ciprofloxacin and *S. aureus* at this initial inoculum size but ofloxacin exhibited an ability to kill *S. aureus* at concentrations exceeding 15 μg ml^{-1} with a most bactericidal effect at 150 μg ml^{-1} when the initial inoculum size was about 10^{10} cfu ml^{-1}. Although this result may not be pertinent clinically at serum levels, it may have relevence for the treatment of urinary tract infections. It is not surprising that ofloxacin was more active than ciprofloxacin against *S. aureus* at an initial inoculum size of about 10^{10} cfu ml^{-1} because ciprofloxacin has been shown to have only one mechanism of action against *S. aureus* (mechanism A), while ofloxacin has been shown to have two mechanisms of action (A and B) against this species (Lewin and Smith 1988). This suggests that the possession of mechanism B causes death to occur at high ofloxacin concentrations with *S. aureus*. However, since both 4-quinolones possess mechanisms A and B against *E. coli* (Lewin and Smith 1988) the finding that they were essentially bacteriostatic at a high initial inoculum size over the whole drug concentration range tested with *E.coli* is puzzling.

When the cause of the inoculum size effect was investigated a lack of oxygen was implicated because bubbling air but not nitrogen restored the bactericidal activity of the 4-quinolones. Thus it would seem that when *E. coli* or *S. aureus* cultures contain about 10^{10} cfu ml^{-1} they consume so much oxygen that despite being exposed to air the cultural conditions remain essentially anaerobic. Furthermore, it would appear that oxygen is required for the full bactericidal activity of the 4-quinolones because under anaerobic conditions in Brewer's medium, even at an initial inoculum size as low as 10^6 cfu ml^{-1} only a bacteriostatic response was observed.

Clinically, this may explain why gut commensals, particularly the anaerobic flora, do not seem to be completely eliminated by oral 4-quinolone therapy (Brumfitt et al. 1984; Pequet et al. 1987) as conditions in the gut are well known to be anaerobic. This may provide an explanation as to why relatively few gastro-intestinal upsets have been reported in patients receiving oral 4-quinolone therapy (Ball 1986), despite the apparent potency of these drugs recorded in MIC tests which are traditionally done under aerobic conditions.

A second clinical implication of these results is that the 4-quinolones may only be bacteriostatic rather than bactericidal when bacterial numbers are very high so that conditions are essentially anaerobic. For example, in chest infections it is not unusual to find 10^9 cfu ml^{-1} in the sputum. Therefore before using 4-quinolones in such situations it should be considered whether bacteriostasis is sufficient because the bactericidal activity of these antibacterials may be absent. This is particularly pertinent in neutropenic patients where the bactericidal activity of the drug is especially relevant and in other conditions where a lack of oxygen may occur, for example, in the vicinity of prostheses or in some bone infections (Lesse et al. 1987). Indeed, osteomyelitis in patients with underlying peripheral arteriosclerotic disease (Gilbert et al. 1987), where a low oxygen tension could be anticipated, seems particularly difficult to treat with 4-quinolone therapy.

References

Ball P (1986) Ciprofloxacin: An overview of adverse experiences. J Antimicrob Chemother 18 suppl D:187-194

Bergan T, Thorsteinsson SB, Solberg R, Bjorskau L, Kolstad IM, Johnsen S (1987)   Pharmacokinetics of ciprofloxacin: intravenous and increasing oral doses. Am J Med 82 suppl 4a:97-102

Brumfitt W, Franklin I, Grady D, Hamilton-Miller JMT (1984) Changes in pharmacokinetics of ciprofloxacin and faecal flora during administration of a seven day course to human volunteers. Antimicrob Agents Chemother 26:757-761

Chin NX, Neu HC (1983) In-vitro activity of enoxacin, a quinolone carboxylic acid, compared to those of norfloxacin, new beta-lactams, aminoglycosides and trimethoprim. Antimicrob Agents Chemother 24:754-763

Gasser TC, Ebert SC, Graverson P, Madsen PO (1987) Pharmacokinetic study of ciprofloxacin in patients with impaired renal function. Am J Med 82 suppl 4a:139-141

Gilbert DN, Tice AD, Marsh PK, Craven PC, Preihem LC (1987) Oral ciprofloxacin therapy for chronic contiguous osteomyelitis caused by aerobic Gram-negative Bacilli. Am J Med 82 suppl 4a:254-258

Hane M, Wood T (1969) Escherichia coli K12 mutants resistant to nalidixic acid: Genetic mapping and dominance studies. J Bacteriol 99:238-241

Knox R, Smith JT (1961) The nature of penicillin resistance in Staphylococci. Lancet Sept 2:520-522

Lesse AJ, Freer C, Salata RA, Francis JB, Scheld WM (1987) Oral ciprofloxacin therapy for Gram-negative Bacillary osteomylitis. Am J Med 82 suppl 4a:247-253

Lewin CS, Smith JT. Have 4-quinolones an inoculum size effect? Congress on Bacterial and Parasitic Drug Resistance. Abstract no.84, p141

Lewin CS, Smith JT (1987) 4-quinolones need oxygen for kill. Abstracts of the 27th Interscience Conference on Antimicrobial Agents and Chemotherapy. Abstract no 473, p 179

Lewin CS, Smith JT (1988) Bactericidal mechanisms of ofloxacin. J Antmicrob Chemother 22 suppl C:1-8

Lewin CS, Morrissey I, Smith JT (1989) Role of oxygen in the bactericidal action of the 4-quinolones. Reviews of infectious diseases 11:S913-S914

Pequet S, Andrement A, Tancrede C (1987) Effect of oral ofloxacin on faecal bacteria in human volunteers. Antimicrob Agents Chemother 31:124-125
Ratcliffe NT, Smith JT (1983) Effects of magnesium on the activity of 4-quinolone antibacterial agents. J Pharm Pharmacol 35,p 61
Ratcliffe NT, Smith JT (1984) Ciprofloxacin's bactericidal and inhibitory activity in urine. Abstracts of the 4th Mediterranean Congress of Chemotherapy. Abstract no 608:347-348 and Chemotherapia 1985;4, suppl.2:385-386
Smith JT (1984) Awakening the slumbering potential of the 4-quinolone antibacterials. The Pharmaceutical Journal 233:299-304
Smith JT, Ratcliffe NT (1986) Einfluss von pH-wert und magnesium auf die antibakterielle activitat von chinolonpraparaten. Infection 14 suppl 1, 31-35

Discussion Summary

The apparent inability of 4-quinolone-treated cells to die under anaerobic condititons prompted detailed checking of the pre-treatment growth conditions of the cells in order to ascertain whether or not the drug was used to challenge cells which had already undergone the global switch to anaerobic metabolism. However, it was concluded by everyone that the presence of available oxygen represented a fundamentally important environmental component which facilitated the induced death process, and that the differences in the effects of the 4-quinolones under aerobic and anaerobic conditions could not be simply related to differences in growth rate.

Some consideration was given to the possibility that under oxygen-depleted conditions the cell may react in such a way that there is an induced (and reversible) change in the nature of the role of DNA gyrase, thus raising the possibility that the observed effects represented an alteration in the drug/target interaction rather than a change in the "death process". However, this was generally thought to be rather unlikely, and would be extremely difficult to investigate.

Perhaps the most fruitful area for future study of this phenomenon, which clearly must have clinical significance in the treatment of infections of anaerobic regions of the body, was the proposal that the lack of oxygen might reduce the uptake of the 4-quinolones. This area had not yet been investigated, but it was clearly a possible explanation which was readily amenable to investigation.

Not surprisingly it was again generally concluded that here again the 4-quinolones provided us with yet another unanswered question, the only worry however being that "no-one had actually bothered to look at this question over the preceding 25 years".

Chapter 4

Interactions of 4-Quinolones with Other Agents - The Importance in Assessing Practical Antibacterial Potency

C. S. Lewin and J. T. Smith

The 4-quinolones are broad spectrum antibacterials and therefore are generally used as single antimicrobial agents. However, in certain situations, they can be combined with a second antibacterial to provide better anti-staphylococcal and anti-streptococcal activity. For example, when empirically treating febrile episodes in immunocompromised patients (Smith *et al.* 1988). Furthermore, it has been suggested that the 4-quinolones should be used in conjunction with a second antibacterial in order to prevent the emergence of 4-quinolone-resistant organisms during therapy (Scully *et al.* 1986; Farrag *et al.* 1986).

If two antibacterials are to be used successfully in combination essentially no antagonism should occur. *In vitro* studies provide a method of assessing if any interaction, either negative or positive, occurs between two classes of drugs. The most commonly used method of assessing interactions between antibacterials are minimum inhibitory concentration (MIC) chequerboard studies. When the interaction of 4-quinolones with other drugs is assessed by this method little or no interaction between the 4-quinolones and other antibacterials can be detected (Neu and Labthavikul 1982; Van der Auwera 1985; Wolfson and Hooper 1985; Davies and Cohen 1985; Haller 1985; Smith *et al.* 1986; Neu 1989). It might therefore be concluded that there are few interactions between the 4-quinolones and other antibacterials. However, MIC's only provide information about the inhibition of bacterial multiplication by the 4-quinolones and provide no information on the bacterial killing process (Smith 1984; Smith and Lewin 1988). The bacteriostatic and the bactericidal activities of the 4-quinolones are quite distinct. For example, magnesium severly antagonises the bactericidal activity of 4-quinolones but has little effect on their bacteriostatic potencies (Ratcliffe and Smith 1983), while low pH antagonises the bacteriostatic potency of ciprofloxacin much more than its lethality (Ratcliffe and Smith 1985).

Bacterial protein and RNA synthesis are essential for the full bactericidal activity of the 4-quinolones (Dietz *et al*. 1966; Smith 1984; Smith and Lewin 1988). This is because protein and RNA synthesis are prerequisites for the primary bactericidal mechanism exerted by all 4-quinolones, termed mechanism A (Smith 1984; Smith and Lewin 1988). Hence, it is not surprising that both the protein synthesis inhibitor chloramphenicol and the RNA synthesis inhibitor rifampicin have been found to antagonise the bactericidal activity of the 4-quinolones against *Escherichia coli* KL16 in nutrient broth (Smith 1984).

Smith (1984) showed that bacteriostatic concentrations of rifampicin and chloramphenicol (160 µg/ml and 20 µg/ml respectively) totally abolished the killing activity of nalidixic acid and norfloxacin and significantly reduced the bactericidal activity of ciprofloxacin and ofloxacin. It could be argued that this antagonism might not be significant clinically because of the relatively high concentrations of chloramphenicol and rifampicin used in these experiments. However, subsequent experiments by Lewin and Smith (1989) showed that even sub-inhibitory concentrations of chloramphenicol and rifampicin antagonised the bactericidal activities of nalidixic acid, ciprofloxacin and ofloxacin against *E. coli* KL16 in nutrient broth. It would therefore seem that the antagonism of the bactericidal activity of the 4-quinolones by chloramphenicol and rifampicin could occur under clinical conditions.

The effect of sub-inhibitory concentrations of 14 other antibacterials on the bactericidal activities of ciprofloxacin, nalidixic acid and ofloxacin against *E. coli* KL16 in nutrient broth was also determined by Lewin and Smith (1989). Sub-inhibitory concentrations of the cell wall antagonists azlocillin, ceftazidime, mezlocillin and vancomycin all had no effect on the bactericidal activities of the 4-quinolones against *E coli* KL16 (Lewin and Smith 1989). On the other hand, rifampicin, tetracycline, chloramphenicol, clindamycin and erythromycin all antagonised the killing activity of the 4-quinolones (Lewin and Smith 1989). Seven aminoglycosides, including both D and L compounds, were also tested and, somewhat unexpectedly, all seven were found to potentiate the bactericidal activity of the 4-quinolones (Lewin and Smith 1989). Hence the aminoglycosides formed a distinct sub-group amongst the protein and RNA inhibitors as they potentiated the bactericidal activity of the 4-quinolones while the other drugs of this type antagonised their killing activity. It was therefore concluded that, with the notable exception of the aminoglycosides, the 4-quinolones should not be combined with protein or RNA synthesis inhibitors. They could, however, be used in combination with aminoglycosides or cell wall antagonists (Smith and Lewin 1988). Indeed clinical trials have shown that the combinations of ciprofloxacin with benzylpenicillin, vancomycin (cell wall antagonists) (Smith *et al* 1988;*) or netilmicin (aminoglycoside **) are successful in treating febrile episodes in neutropenic patients.

Unlike the other protein and RNA synthesis inhibitors tested by Lewin and Smith (1989), the aminoglycosides damage the bacterial membrane (Tanaka 1982; Davis 1987). This is a secondary effect related to protein synthesis inhibition as the addition of chloramphenicol has been shown to eliminate the membrane damage caused by streptomycin (Tanaka 1982; Tai and Davis 1985). It has been suggested that this damage caused to the bacterial membrane by the aminoglycosides explains why these drugs are bactericidal against *E. coli* in contrast to chloramphenicol, tetracycline and erythromycin which are merely bacteriostatic (Tanaka 1982; Tanaka *et al*. 1984). The interaction between the aminoglycosides and the bacterial membrane also provides a possible explanation as to why the aminoglycosides form a distinct sub-group amongst the protein synthesis inhibitors in their interaction with the 4-quinolones. The membrane damage caused by the aminoglycosides, which is thought to be due to the

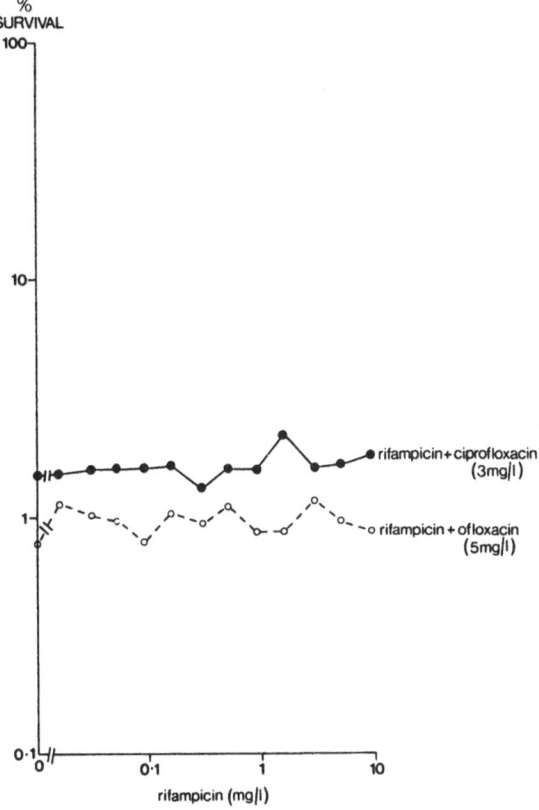

Fig. 4.1. Survival of *S. aureus* in nutrient broth after 3 h. *S. aureus* E3T was exposed to either 3 µg/ml ociprofloxacin or 5 µg/ml ofloxacin alone or plus varying concentrations of rifampicin in nutrient broth. After 3 h incubation at 37°C percentage survival was estimated.

incorporation of misread proteins into the membrane, increases the permeability of the membrane (Davis *et al.* 1986; Davis 1987). This membrane damage also affects the initiation step of DNA synthesis (Tanaka *et al.* 1984) probably by interfering with the formation of the *oriC*-membrane complex (Matsunaga *et al.* 1986). It would seem that the changes in membrane permeability caused by the aminoglycosides may increase 4-quinolone uptake. Alternatively, the aminoglycosides may potentiate 4-quinolone kill via their effect on the initiation of DNA replication as the 4-quinolones also affect the replication of DNA. The observation that sub-inhibitory concentrations of aminoglycosides enhance the killing activity of the 4-quinolones but not *vice versa* (Lewin 1987) suggests that the first hypothesis is more probable.

Because of our results (Smith 1984; Smith and Lewin 1988) we suggested rifampicin should not be combined with the 4-quinolones because the bactericidal activity of the 4-quinolone would be antagonised (Smith and Lewin 1988). Other workers have also demonstrated antagonism *in vitro* between rifampicin and ciprofloxacin (Haller 1985; Van der Auwera and Joly 1985; Hackbarth *et al.* 1986). However, clinical studies have reported that the combination of ciprofloxacin and rifampicin can be effective in treating infections caused by Staphylococci. For example, the combination of ciprofloxacin and rifampicin has been found to be highly effective in treating

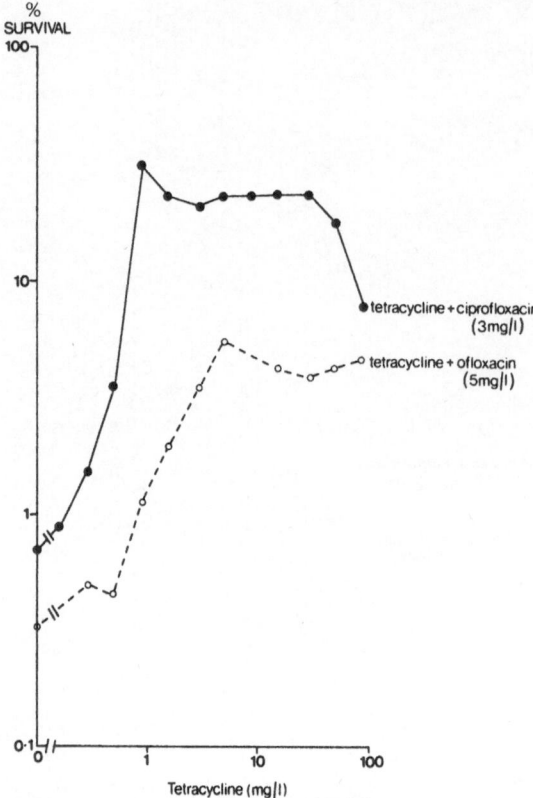

Fig. 4.2. Survival of *S. aureus* in nutrient broth after 3 h. *S. aureus* E3T was exposed to either 3 μg/ml ciprofloxacin or 5 μg/ml ofloxacin alone or plus varying concentrations of tetracycline in nutrient broth. After 3 h incubation at 37°C percentage survival was estimated.

staphylococcal endocarditis (Dworkin *et al.* 1989) and in treating experimental osteomylitis (Henry *et al.* 1986). Our conclusions (Smith and Lewin 1988) were based on experiments done with sub-inhibitory concentrations of the non-4-quinolone antibacterial against *E. coli*. Therefore the effect of a range of clinically achievable rifampicin concentrations on the bactericidal activity of ciprofloxacin and ofloxacin agains *S. aureus* or *S. warneri* in nutrient broth was examined.

S. aureus E3T was exposed to ciprofloxacin at its most (optimum) bactericidal concentration of 3 μg/ml along with increasing concentrations of rifampicin in nutrient broth for 3 h. Percent survival was then estimated by viable counting on nutrient agar as described by Lewin and Smith (1988). It can be seen in Fig. 4.1 that over the whole range of rifampicin concentrations tested the bactericidal activity was unaffected with kill comparable to ciprofloxacin on its own. Similarly, when *S. aureus* E3T was exposed to ofloxacin at its most bactericidal concentration of 5 μg/ml the bactericidal activity was not altered by increasing concentrations of rifampicin (Fig. 4.1). Hence somewhat unexpectedly rifampicin does not appear to antagonise the bactericidal action of ciprofloxacin or ofloxacin against *S. aureus*.

During our investigations rifampicin was found to be bactericidal against *S. aureus* but merely bacteriostatic against *E. coli*. The bactericidal activity of rifampicin against

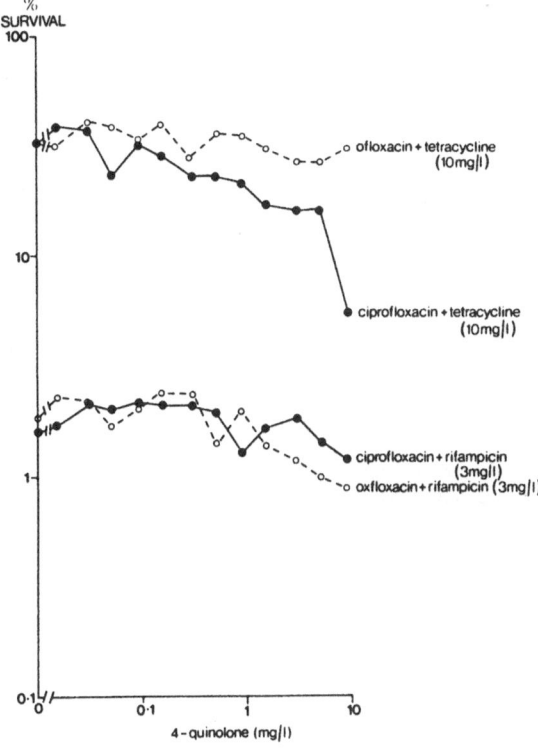

Fig. 4.3. Survival of *S. aureus* in nutrient broth after 3 h. *S. aureus* E3T was exposed to either 3 μg/ml rifampicin or 10 μg/ml tetracycline alone or plus varying concentrations of either ciprofloxacin or ofloxacin in nutrient broth. After 3 h incubation at 37°C percentage survival was estimated.

S. aureus might therefore compensate for any antagonism of the killing activiy of the 4-quinolones. This hypothesis was tested by investigating the effect of the protein synthesis inhibitor tetracycline (which is essentially bacteriostatic against *S. aureus* at clinically achievable concentrations) on the bactericidal activity of the 4-quinolones. It can be seen in Fig. 4.2 that as the concentration of tetracycline was increased, the bactericidal activity of the mixtures of tetracycline and ciprofloxacin, at its most bactericidal concentration against *S. aureus* E3T, decreased. Similar results were obtained when the experiment was repeated substituting ofloxacin (at its most bactericidal concentration) for ciprofloxacin (Fig. 4.2). Hence tetracycline, which is merely bacteriostatic, will reduce the amount of kill when added to a 4-quinolone. This suggests that the ability of rifampicin alone to kill *S. aureus* compensates for any antagonism of the bactericidal activity of the 4-quinolones.

Fig. 4.3 shows the results of an experiment where *S. aureus* was exposed to varying concentrations of ciprofloxacin or ofloxacin plus 3 μg/ml rifampicin for 3 h and percentage survival estimated. It can be seen that rifampicin 3 μg/ml is bactericidal against *S. aureus* on its own, and that the mixtures of either 4-quinolone plus rifampicin were bactericidal against *S. aureus* at all concentrations tested. Furthermore,

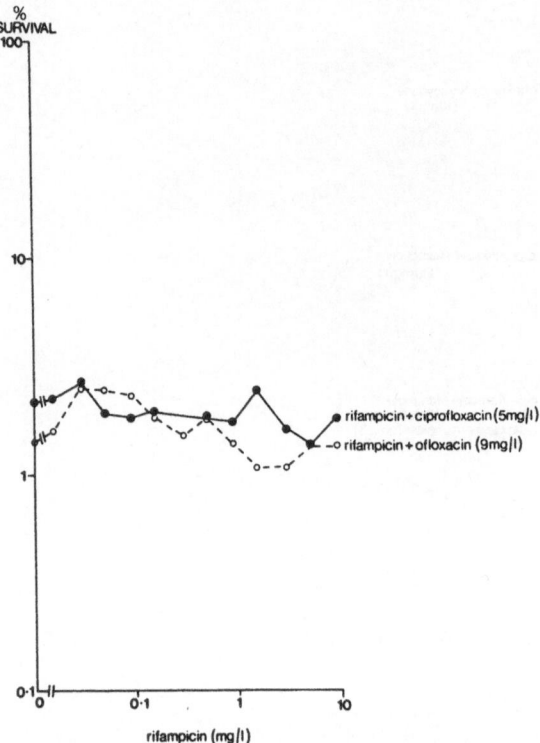

Fig. 4.4. Survival of *S. warneri* in nutrient broth after 3 h. *S. warneri* was exposed to either 5 µg/ml ciprofloxacin or 9 µg/ml ofloxacin alone or plus varying concentrations of rifampicin in nutrient broth. After 3 h incubation at 37°C percentage survival was estimated.

the bactericidal activities of the mixtures of rifampicin plus 4-quinolone were similar to those observed with either ofloxacin or ciprofloxacin alone at their optimum killing concentrations (Figs. 4.1 and 4.2). When the effect of tetracycline 10 µg/ml on the bactericidal activity of ofloxacin and ciprofloxacin was examined the protein synthesis inhibitor on its own was, as expected, not bactericidal (Fig. 4.3,) and the bactericidal activity of the two 4-quinolones was severely antagonised compared to the mixtures of rifampicin and the two 4-quinolones (Fig. 4.3).

The previous investigations were also carried out using *S. warneri*. The whole range of rifampicin concentrations tested in combination with ciprofloxacin and ofloxacin at their most bactericidal concentrations of 5 µg/ml and 9 µg/ml respectively, did not alter the bactericidal activity over 3 h (Fig. 4.4). Again it would appear that the bactericidal activity of rifampicin compensated for its antagonism of the killing activity of the two 4-quinolones against *S. warneri*.

In marked contrast this was not the case with tetracycline. It can be seen in Fig. 4.5 that as the tetracycline concentration was increased the amount of kill obtained by ciprofloxacin or ofloxacin at their most bactericidal concentrations against *S. warneri* decreased. Hence tetracycline, unlike rifampicin, appears to antagonise the bactericidal activity of the 4-quinolones against Staphylococci as well as *E. coli*.

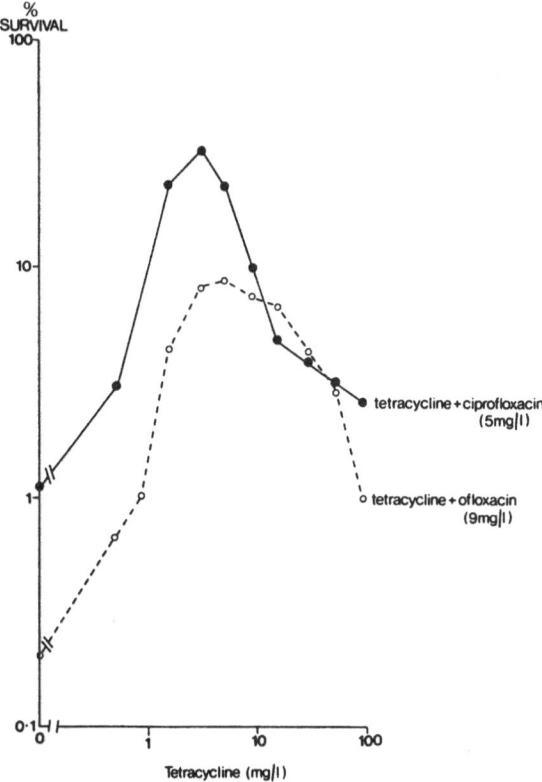

Fig. 4.5. Survival of *S. warneri* in nutrient broth after 3 h. *S. warneri* was exposed to either 5 μg/ml ciprofloxacin or 9 μg/ml ofloxacin alone or plus varying concentrations of tetracycline in nutrient broth. After 3 h incubation at 37°C percentage survival was estimated.

As with *S. aureus* E3T, rifampicin 3 μg/ml on its own was bactericidal against *S. warneri* (Fig. 4.6). The bactericidal activity of increasing concentrations ciprofloxacin or ofloxacin plus 3 μg/ml rifampicin was similar to that of the 4-quinolones alone at their most bactericidal concentrations (Figs. 4.4, 4.5 and 4.6). On the other hand, it can be seen in fig. 4.6 that 10 μg/ml tetracycline alone was not bactericidal against *S. warneri* in nutrient broth and severely antagonised the bactericidal activity of both ofloxacin and ciprofloxacin against this species.

It would therefore appear that the observation by Smith and Lewin (1988) that protein and RNA synthesis inhibitors (with the exception of the aminoglycosides) should not be combined with the 4-quinolones needs to be modified. Protein and RNA synthesis inhibitors which are bactericidal can be considered for combination with the 4-quinolones because their bactericidal activity can compensate for their antagonism of the killing activity of the 4-quinolones. However, protein and RNA synthesis inhibitors which can only inhibit bacterial multiplication but are unable to kill bacteria should not be combined with the 4-quinolones because a reduction in the bactericidal activity of the mixture compared to the 4-quinolones on their own would occur.

The gyrase B subunit antagonists and the 4-quinolones are two classes of antibacterials which inhibit the enzyme bacterial DNA gyrase, the only bacterial

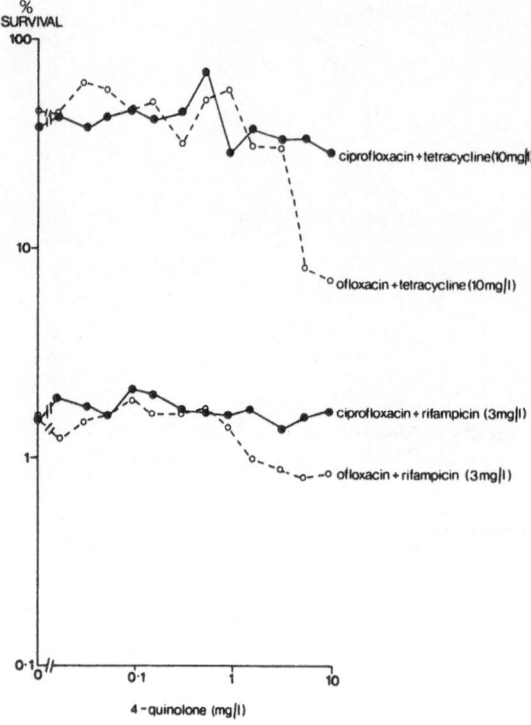

Fig. 4.6. Survival of *S. warneri* in nutrient broth after 3 h. *S. warneri* E3T was exposed to either 3 µg/ml rifampicin or 10 µg/ml tetracycline alone or plus varying concentrations of either ciprofloxacin or ofloxacin in nutrient broth. After 3 h incubation at 37°C percentage survival was estimated.

enzyme capable of introducing negative supercoils into DNA (Drlica and Franco 1988; Cullen *et al.* 1989). The mechanism of action of these two antibacterials is not identical as the gyrase B subunit antagonists act on the B subunit of gyrase while the 4-quinolones act primarily on the A subunit (Drlica and Franco 1988; Cullen *et al.* 1989). Fractional minimum inhibitory concentration (FIC) studies have shown that these two classes of drugs can act synergistically against both Gram negative and Gram positive bacteria including *S. aureus* (Chao 1978; Neu *et al.* 1984). However, *in vivo* studies have found that ciprofloxacin alone is more effective than ciprofloxacin combined with coumermycin (a gyrase B subunit antagonist) in treating *S. aureus* endocarditis in rats (Perronne *et al.* 1987).

As FICs are based on minimum inhibitory concentration (MIC) studies (which with the 4-quinolones only measure the inhibition of bacterial multiplication) the effect of two gyrase B subunit antagonists, novobiocin and coumermycin, on the bactericidal activities of ciprofloxacin and ofloxacin against *S. aureus* and *S. warneri* was investigated. If the killing activity of the 4-quinolones were to be antagonised by coumermycin then the results of Perronne *et al.* (1987) might be explained.

The effect of increasing concentrations of coumermycin on the bactericidal activity of ciprofloxacin against *S. aureus* E3T in nutrient broth at the most bactericidal

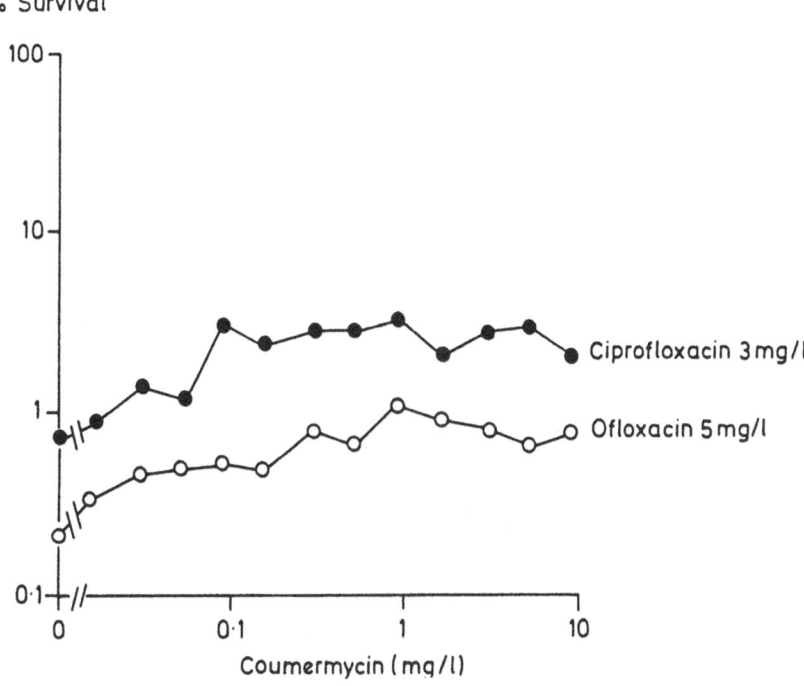

Fig. 4.7. Survival of *S. aureus* in nutrient broth after 3 h. *S. aureus* E3T was exposed to either 3 µg/ml ciprofloxacin or 5 µg/ml ofloxacin alone or plus varying concentrations of coumermycin in nutrient broth. After 3 h incubation at 37°C percentage survival was estimated.

concentration of this 4-quinolone (3 µg/ml) was investigated as described by Lewin and Smith (1988). It can be seen in Fig. 4.7 that as the concentration of coumermycin was increased the bactericidal activity of the mixtures of the two drugs against *S. aureus* E3T over a 3 h period was less than that of ciprofloxacin alone. 9 µg/ml of coumermycin (the highest coumermycin concentration tested) reducing ciprofloxacin's killing activity three-fold. When ofloxacin was investigated at its most bactericidal concentration (5 µg/ml) it can be seen in Fig. 4.7 that coumermycin also antagonised its bactericidal activity against *S. aureus* E3T. The bactericidal activity of the mixtures of the two drugs again decreased as the concentration of coumermycin was increased. The highest concentration of coumermycin tested (9 µg/ml) also causing a three-fold reduction in ofloxacin's bactericidal activity. The effects of a second gyrase B subunit antagonist, novobiocin, on the bactericidal activities of ciprofloxacin and of ofloxacin against *S. aureus* E3T in nutrient broth was also investigated. Novobiocin also appeared to antagonise the bactericidal activities of these two 4-quinolones at their most bactericidal concentrations (Fig. 4.8). The bactericidal activity of mixtures of either ciprofloxacin (3 µg/ml or ofloxacin 5 µg/ml) plus novobiocin decreased as the concentration of novobiocin in the mixture increased. Therefore both novobiocin and

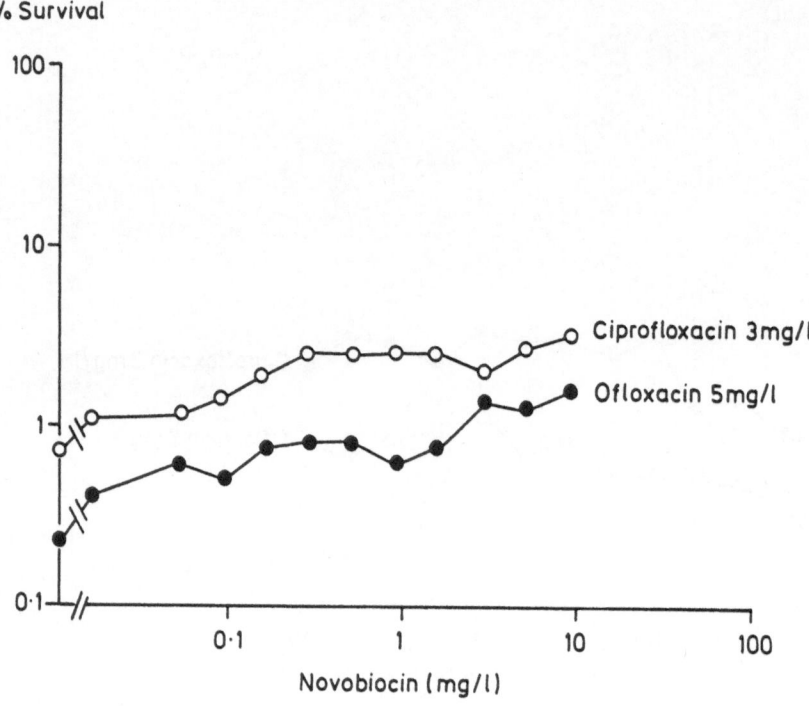

Fig. 4.8. Survival of *S. aureus* in nutrient broth after 3 h. *S. aureus* E3T was exposed to either 3 μg/ml ciprofloxacin or 5 μg/ml ofloxacin alone or plus varying concentrations of novobiocin in nutrient broth. After 3 h incubation at 37°C percentage survival was estimated.

coumermycin seem to antagonise the bactericidal activity of ciprofloxacin and ofloxacin against *S. aureus* E3T.

The effects of coumermycin and novobiocin on the bactericidal activities of ciprofloxacin and ofloxacin agains *S. warneri* were also investigated. It can be seen in Fig. 4.9 that as the coumermycin concentration was increased, the bactericidal activity of mixtures of coumermycin plus ciprofloxacin or ofloxacin (at their most bactericidal concentrations of 5 μg/ml and 9 μg/ml respectively) against *S. warneri* was significantly reduced. Novobiocin also antagonised the bactericidal activites of these two 4-quinolones as the bactericidal activity of the mixtures decreased with increasing novobiocin concentration (Fig. 4.10).

Both coumermycin and novobiocin were found to antagonise the bactericidal activity of ciprofloxacin or ofloxacin against *S. aureus* or *S. warneri* in nutrient broth. The bactericidal activity of either 4-quinolone at its most bactericidal concentration was progressively reduced by the addition of increasing concentrations of novobiocin or coumermycin. This antagonism of the killing activity of ciprofloxacin by coumermycin may explain why ciprofloxacin alone was found to be better than the combination of ciprofloxacin with coumermycin in treating *S. aureus* endocarditis in rats (Perronne *et al.* 1987) despite reports indicating synergy between these two classes

% Survival

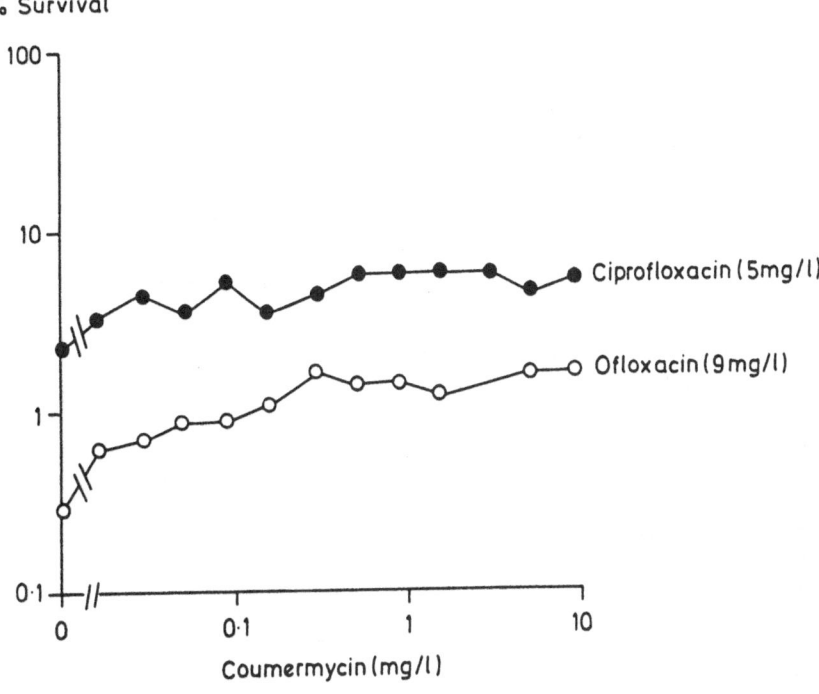

Fig. 4.9. Survival of *S. warneri* in nutrient broth after 3 h. *S. warneri* was exposed to either 5 μg/ml ciprofloxacin or 9 μg/ml ofloxacin alone or plus varying concentrations of coumermycin in nutrient broth. After 3 h incubation at 37°C percentage survival was estimated.

of drugs as determined by FIC against both Gram negative and Gram positive species (Chao 1978; Neu *et al.* 1983). Thus, in this case, the investigation of the bactericidal activity of mixtures of these drugs appears to provide a better prediction of the clinical outcome than provided by FIC studies.

In conclusion, *in vitro* investigations into interactions of the 4-quinolones with other antibacterials provide a method which helps determine the suitability of other antibacterials for combination with the 4-quinolones. However if accurate information is to be obtained from such investigations they must be undertaken with great care.

FIC studies need to be carried out to determine effect on inhibition of bacterial multiplication. However it is important to remember that FIC indices only provide information on the inhibition of bacterial multiplication. As changes in 4-quinolone lethality may affect the clinical outcome of combination therapy, just as much as changes in bacteriostatic potency, it would seem prudent to also investigate the bactericidal activity of combinations of the 4-quinolones with other antibacterials, as well as their bacteriostatic activity via MIC studies. In this manner the effects of combining 4-quinolones with a second antibacterial can be assessed more accurately.

It is not sufficient to investigate a single species when carrying out such studies as variations in interactions between the 4-quinolones and other antimicrobials can occur.

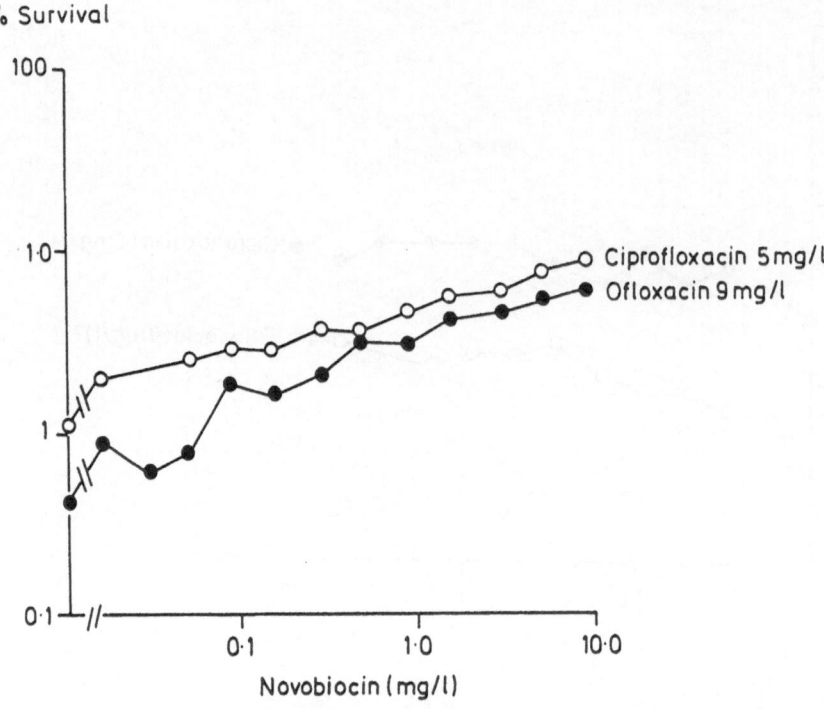

Fig. 4.10. Survival of *S. warneri* in nutrient broth after 3 h. *S. warneri* was exposed to either 5 μg/ml ciprofloxacin or 9 μg/ml ofloxacin alone or plus varying concentrations of novobiocin in nutrient broth. After 3 h incubation at 37°C percentage survival was estimated.

For example, combinations of some cell wall antagonists and ciprofloxacin have been found to be synergistic against *Ps. aeruginosa in vivo* as well as *in vitro* (Chalkley and Koornhof 1985; Giamarellou and Petrikos 1987; Fu *et al.* 1987; Bustamente *et al.* 1987) but not against Enterobacteriaceae (Chalkley and Koornhof 1985; Smith and Lewin 1988). Finally, it is important to see whether the predictions made from *in vitro* results are borne out by *in vivo* studies. This being the only way of assessing the accuracy of the predictions made using *in vitro* testing and can hopefully provide guidelines for future testing.

References

Bustamente CI, Drusano GL, Wharton RC, Wade JC (1987) Synergism of the combinations of imipenem plus ciprofloxacin and imipenem plus amikacin against *Pseudomonas aeruginosa*. Antimicrob Agents Chemother 31:632-634

Chalkley LJ, Koornhof HJ (1985) Antimicrobial activity of ciprofloxacin against *Pseudomonas aeruginosa, Escherichia coli* and *Staphylococcus aureus* determined by the killing curve method: antibiotic comparisons and synergistic interactions. Antimicrob Agents Chemother 28:331-342

Chao L (1978) An unusual interaction between the target of nalidixic acid and novobiocin. Nature 271:385-386

Cullen ME, Wyke AW, McEachern F, Austin CA, Fisher LM (1989) Inhibition of DNA gyrase: Bacterial sensitivity and clinical resistance to 4-quinolones. Current Topics in Infectious Diseases and Clinical Microbiology 2:41-47

Davies GSR, Cohen J (1985) *In vitro* study of the activity of ciprofloxacin against strains of *Pseudomonas aeruginosa* with multiple antibiotic resistance. J Antimicrob Chemother 16:713-717

Davis BD, Chen L, Tai PC (1986) Misread protein creates membrane channels: an essential step in the bactericidal action of the aminoglycosides. Procs Nat Acad Sci 83:6164-6168

Davis BD (1987) Mechanism of bactericidal action of aminoglycosides. Microbiol Revs 51:341-350

Dietz WH, Cook TM, Goss WA (1966) Mechanism of action of nalidixic acid on *Escherichia coli*: III conditions required for lethality. J Bacteriol 94:768-773

Drlica K, Franco RJ (1988) Inhibitors of DNA topoisomerases. Biochemistry 27:2254-2259

Dworkin RJ, Lee BL, Sande MA, Chambers HF (1989) Treatment of right-sided Staphylococcus aureus endocarditis in intravenous drug users with ciprofloxacin and rifampicin. The Lancet ii:1071-1073

Farrag NN, Bendig JWA, Talboys C, Azadian BS (1986) *In vitro* study of the activity of ciprofloxacin combined with amikacin or ceftazidine against *Pseudomonas aeruginosa*. J Antimicrob Chemother 18:770

Fu KP, Hetzel N, Gregory J, Hung PP (1987) Therapeutic efficacy of cefpirimide ciprofloxacin combination in experimental *Pseudomonas* infections in neutropenic mice. J Antimicrob Chemother 20:541-546

Giamerellou H, Petrikos G (1987) Ciprofloxacin interaction with imipenem and amikacin against multiresistant *Pseudomonas aeruginosa*. Antimicrob Agents Chemother 31:959-961

Hackbarth CJ, Chambers HF, Sande MA (1986) Serum bactericidal activity of rifampin in combinations with other antimicrobial agents against *Staphylococcus aureus*. Antimicrob Agents Chemother 29:611-613

Haller I (1985) Comprehensive evaluation of ciprofloxacin-aminoglycoside combination against enterobacteriaceae and *Pseudomonas aeruginosa*. Antimicrob Agents Chemother 28:663-666

Henry NK, Rouse MS, Whitesell AL, McConnell ME, Wilson WR (1987) Treatment of methicillin-resistant *Staphylococcus aureus* experimental osteomylitis with ciprofloxacin and vancomycin alone or in combination with rifampin. Am J Med 82 suppl 4A:73-75

Lewin CS (1987) Interactions of 4-quinolone antibacterials and antibiotics with Gram negative and Gram positive bacteria. PhD thesis, University of London

Lewin CS, Smith JT (1988) Bactericidal mechanisms of ofloxacin. J Antimicrob Chemother 22 suppl C:1-8

Lewin CS, Smith JT (1989) Interactions of the 4-quinolones with other antibacterials. J Med Microbiol 29:221-227

Matsunaga K, Yamaki H, Nishimuro T, Tanaka N (1986) Inhibition of DNA replication initiation by aminoglycoside antibiotics. Antimicrob Agents Chemother 30:468-474

Neu HC, Labthavikul P (1982) *In vitro* activity of norfloxacin, a quinolone carboxylic acid, compared with that of ß-lactams, aminoglycosides and trimethoprim. Antimicrob Agents Chemother 22:23-27

Neu HC, Chin N-X, Labthavikul P (1984) Antibacterial activity of coumermycin alone and in combination with other antibiotics. Antimicrob Agents Chemother 25:687-689

Neu HC (1989) Synergy of fluoroquinolones with other antimicrobial agents. Rev Infect Dis 11 suppl 5:S1025-1035

Perronne CM, Malinverni R, Glauser MP (1987) Treatment of *Staphylococcus aureus* endocarditis in rats with coumermycin A1 and ciprofloxacin, alone or in combination. Antimicrob Agents Chemother 31:539-543

Ratcliffe NT, Smith JT (1983) Effects of magnesium on the activity of 4-quinolone antibacterial agents. J Pharm Pharmacol 35:61p

Ratcliffe NT, Smith JT (1985) Ciprofloxacin's bactericidal and inhibitory activity in urine. Chemiotherapia 4 suppl 2:385-386

Scully BE, Parry MF, Neu HC (1986) Oral ciprofloxacin therapy for infections due to *Pseudomonas aeruginosa*. Lancet II:819-822

Smith GM, Leyland MJ, Farrell I, Geddes AM (1988) A clinical microbiological and pharmacokinetic study of ciprofloxacin plus vancomycin as initial therapy of febrile episodes in neutropenic patients. J Antimicrob Chemother 1:647-655

Smith JT (1984) Awakening the slumbering potential of the 4-quinolone antibiotics. Pharm J 233:299-305

Smith JT, Lewin CS (1988) Chemistry and mechanisms of action of the quinolone antibacterials. In: Andriole VT (ed) The quinolones, Academic Press, London, pp 23-82

Smith SM, Eng RHK, Berman E (1986) The effect of ciprofloxacin on methicillin-resistant *Staphylococcus aureus*. J Antimicrob Chemother 17: 287-295

Tai PC, Davis BD (1985) The actions of antibiotics on the ribosome. In: Greenwood D, O'Grady F (eds) The scientific basis of antimicrobial Chemotherapy, Cambridge University Press, Cambridge, pp 45-69

Tanaka N (1982) Mechanism of action of the aminoglycoside antibiotics. In: Umezawa H, Hooper IR (eds) Aminoglycoside antibiotics, Springer-Verlag, Berlin, Heidelberg, New York, pp 221-226

Tanaka N, Matsunaga K, Yamaki H, Nishimura TK (1984) Inhibition of initiation of DNA synthesis by aminoglycoside antibiotics. Biochem Biophys Res Comms 122:460-465

Van der Auwera P (1985) Interaction of gentamicin, dibekacin, netilmicin and amikacin with various penicillins, cephalosporins, minocycline and new fluoroquinolones against Enterobacteriaceae and *Pseudomonas aeruginosa*. J Antimicrob Chemother 16:581-587

Van der Auwera P, Jolly P (1985) Comparative *in vitro* activities of teicoplanin, vancomycin, coumermycin and ciprofloxacin alone and in combination with rifampicin or LM427 against *Staphylococcus aureus*. Abstracts of the 25th ICAAC:1058

Wolfson JS, Hooper DC(1985) The fluoroquinolones: Structures, mechanisms of action and resistance and spectra of activity *in vitro*. Antimicrob Agents Chemother 28:581-586

References added in proof

* Kelsey SM, Wood ME, Shaw E, Jenkins GC, Newland AC (1990) A comparative study of intravenous ciprofloxacin and benzylpenicillin versus netilmicin and piperacillin for the empirical treatment of fever in neutropenic patients. J Antimicrob Chemother 25, 149-157

**Chan CC, Oppenheim BA, Anderson H, Swindell R, Scarffe JH (1989) Randomized trial comparing ciprofloxacin plus netilmicin versus piperacillin plus netilmicin for empiric therapy in neutropenic patients. Antimicrob Agents Chemother 33: 87-91

Discussion Summary

It was evident from the discussion that the question of possible interactions between the 4-quinolones and other antibacterial agents which might be administered concurrently, represented a significant area for clinical concern. The concern was voiced with regard to both the effects upon the bactericidal effects of the 4-quinolones upon susceptible cells, and the possible effects upon the selection of resistant mutants during therapy. It was generally accepted that as yet very little is known about the possible effects upon resistance development, and that this was of lesser immediate concern than the possible interactive effects upon the direct antibacterial activity of the 4-quinolones.

The consequences of interactions had been shown to be demonstrable with regard to both the basic inhibition of the growth of cells (expressed as the MIC) and to the bactericidal consequences of inhibition. Discussion of methodology clearly established that whilst effects upon the MIC could be easily monitored, the consequences of interaction upon the bactericidal effects of the 4-quinolones represented the investment of a quite substantial effort. Not only would such studies, if carried out to investigate all of the likely clinical possible interactions, involve much effort, but also the interpretation of the data might be far from straightforward. This complexity being the result of trying to relate the *in vitro* studies to the *in vivo* situation where there are significant considerations of the differential pharmacology and pharmacokinetics of the individual agents. Some concern was expressed also about the question of whether two potentially interactive agents were applied as a simultaneous challenge, or whether a sequential challenge would yield similar results. It was also clear that there was some concern about the fact that all 4-quinolones might not behave in a common manner since it had already been shown that the effects of rifampicin in combination with ciprofloxacin and ofloxacin were clearly distinguishable from the simple antagonistic interaction with older 4-quinolones like nalidixic acid.

In mechanistic terms the potential for interactions could be divided into three principal areas; these were:

(a) the uptake of one or other agent (which is likely to affect the MIC);

(b) interactions which might interfere with the effects of the 4-quinolones upon DNA gyrase (also likely to affect the MIC); and

(c) interactions which perturb the "death processes" induced by the initial 4-quinolone challenge (which will only affect the bactericidal effects of the 4-quinolones and should have no significant influence upon the MIC).

Participants were however encouraged to some extent by the fact that all interactions were not simply antagonistic, some were also apparently synergistic. However, the fact that such potentially beneficial interactions might not be universal across the bacterial species clearly indicated that any potential clinical benefits could only be considered useful to the "thinking prescriber" who has full diagnostic laboratory support to give proper identification of the pathogenic organism.

Overall, it was generally considered that such interactions were more than just laboratory curiosities, they were of potential clinical significance (particularly with the hospitalised patient). It was also considered that here again was an aspect of the clinical use of the 4-quinolones which was in need of further study in order to elucidate both aspects of the mechanisms involved and the extent and degree of interactions with clinically significant species of pathogen.

Chapter 5

Molecular Effects of 4-Quinolones upon DNA Gyrase: DNA Systems

G. C. Crumplin

In 1963 it was shown that nalidixic acid inhibited DNA synthesis, but not until 1976 was the intracellular "target" protein identified as DNA gyrase. Until that time DNA gyrase had not even been identified as a cellular component, but subsequent studies have shown this enzyme to be an essential component of the bacterial cell and related type II DNA topoisomerases to be essential for all cellular systems.

Once DNA gyrase had been isolated and purified, the elegant simplicity of the *in vitro* insertion of negative superhelical turns into plasmid DNA molecules by DNA gyrase provided the basis of the classical inhibition of supercoiling by 4-quinolones. This inhibition of an ostensibly essential component of DNA replication seemed the obvious key to the mechanism of action of the 4-quinolones. Consequently this demonstrable *in vitro* reaction became considered as the basic component of the *in vivo* inhibition of DNA synthesis and subsequent death of cells treated with 4-quinolones. The initial adoption of this model of the mechanism of action of the 4-quinolones has become even more firmly established since it was shown that not only did mutations in the *gyrA* gene (the structural gene for the A subunit of DNA gyrase) confer resistance to the 4-quinolones, but also that DNA gyrase isolated from a cell harbouring such a mutation was resistant to the *in vitro* inhibition of DNA supercoiling by a 4-quinolone.

The demonstration of the inhibition of a fundamental biochemical reaction *in vitro* by a 4-quinolone was more than just a once-only research demonstration of the mechanism of action - it was also the practical basis for a convenient *in vitro* assay for the screening and comparison of new generations of 4-quinolone analogues derived from nalidixic acid. The application of this classical experiment to a comparison of the activity of the limited number of 4-quinolones based upon the chemical structure of nalidixic acid showed that for *Escherichia coli* (the standard laboratory model organism) there was a near linear correlation of the ability of the agents to block the supercoiling catalyzed by purified DNA gyrase and the Minimum Inhibitory Concentrations (MICs) of the agents against intact *E. coli* cells. This correlation was attainable with the use of a convenient "mail-order" plasmid like pBR322 which was relaxed with DNA topoisomerase I to generate a suitable substrate for supercoiling. Such a convenient correlation which seems to confirm the validity of the inhibition of supercoiling as the model for antibacterial activity contains only one inconvenience - the fact that the *in vitro* inhibition of DNA gyrase mediated supercoiling using the purified enzyme requires approximately ten-times the concentration of drug required to kill susceptible cells.

Examination of the mechanism of the inhibition of supercoiling by 4-quinolones subsequently showed that these drugs stabilized the binding of the enzyme at the site of a double-stranded break in the DNA to form the "cleavable complex". These cleavable complexes could be monitored by the generation DNA breakage under denaturing conditions (e.g. SDS or alkali-treatment). This generation of the cleavable complex was multiply useful since the phenomenon of double-strand breakage and strand-passaging by DNA gyrase could be conveniently monitored *in vitro* by:

(a)   the generation of linear fragments from circular or long linear DNA molecules;
(b)   the decatenation of circular catenanes;
(c)   the decatenation of kinetoplasts; and
(d)   the resolution of knotted DNA circles.

The generation of linear fragments after denaturation treatment of plasmid DNA in the presence of DNA gyrase, ATP and a 4-quinolone could be shown to be a drug-dose-related phenomenon in terms of the quantitative yield of linear fragments *in vitro* and represented an alternative assay for 4-quinolone activity. The fact that cleavable complexes could be identified in plasmid DNA isolated from intact *E. coli* cells challenged with a 4-quinolone also served to emphasize the probable role of the stabilization of the cleavable complex in the *in vivo* mechanism of action of the 4-quinolones.

In vitro assay systems which are technically very convenient have been developed using purified DNA gyrase from either *E. coli* or *Micrococcus luteus* as a routine screening in the development of new 4-quinolones, and as a basic criterion in structure-activity studies of this group of compounds. The same technology has also been used to study the susceptibility of DNA gyrases from other bacterial species like *Pseudomonas aeruginosa*, *Staphylococcus aureus*, and *Bacillus subtilis* as well as the resistance mechanism in laboratory and clinical isolates resistant to 4-quinolone challenge.

In setting up a new research programme to screen and monitor the activities of new generations of 4-quinolones, the data derived from mechanistic studies generates only a limited number of very basic questions which must be answered before embarking upon the experiments. These very basic questions are:-

(1)   Which DNA gyrase should be used?
(2)   Which DNA molecule should be used as a routine substrate?
(3)   Which reaction should be monitored - supercoiling inhibition or generation of the cleavable complex?

These are invitingly simple questions which seem to demand only simple choices, but they are extremely important questions which are based upon our perceptions of the mechanism of action of the 4-quinolones. If we attempt to answer these questions critically rather than by reflex or convenience considerations, then the derived answers are no longer as straightforward as we might wish.

Which DNA gyrase to use? In absolute terms we should assume that since DNA gyrase is an absolute requirement for viability it must be highly evolved and is thus likely to differ somewhat between bacterial species. Hence for the purist, if we are specifically interested in *E. coli* then we should use *E. coli* DNA gyrase. Similarly interest in *Pseudomonas* or *Staphylococci* should dictate the use of the DNA gyrase derived from the appropriate species, unless we can be assured that the enzyme from any one species represents a valid model for all potentially pathogenic species of bacteria. However, from the limited number of comparative studies that have been carried out there is evidence which indicates that there are significant differences in the susceptibility of different DNA gyrases to inhibition by 4-quinolones. From the purely academic standpoint, present knowledge would suggest that DNA gyrase(s) from species of specific interest should be selected for such studies. However, for a laboratory involved in the monitoring of a potentially large number of novel 4-quinolone molecules as potential clinical agents, the use of more than one or two enzymes is not really practicable or economic. Whilst we feel sure that DNA gyrase from different species functions in essentially the same manner, and that the basic interactions of a 4-quinolone with DNA gyrase and DNA *in vivo* should not differ significantly between species, it would be very presumptious to assume that the use of a single DNA gyrase will provide a "catch-all" screen. On the basis of observations of the spectra of activity of existing, well characterized 4-quinolones, it would seem reasonable to propose the use of a battery of 4-enzymes (when and if available) as an effective screen of novel compounds for potential clinical usefulness. Such a set of enzymes might consist of DNA gyrases from (i) *E. coli* (as a representative Gram negative organism); (ii) *S. aureus* (as a representative Gram positive organism of clinical significance); (iii) *Ps. aeruginosa* (included because of its particular clinical importance and specific differences to other Gram negative organisms) and (iv) an obligate anaerobic species (because of the relative lack of activity of existing 4-quinolones against this type of organism - remembering of course that the existing 4-quinolones have been identified and developed specifically upon the basis of their identifiable activity against aerobic organisms and against enzymes isolated from aerobic species).

Which DNA molecule should be chosen as a substrate? On the basis of generalized experience, DNA gyrase can utilise virtually any closed circular DNA molecule as a substrate for DNA supercoiling. Consequently first impressions suggest that a very practicable answer is the use of an easily isolated, high copy-number, amplifiable, small plasmid like pBR322. Such plasmid molecules are easily obtained and generally available, and seem to work equally well in different laboratories with DNA topoisomerase enzymes derived from a wide variety of sources.

Which reaction should be monitored? With this question we seem to be faced with a simple choice between the inhibition of the supercoiling reaction and the induction of the formation of the cleavable complex. All that we really need to know is which of these two represents the better indicator of potential antibacterial potency? A comparison of these two procedures has been carried out by Drs Elwell and Walton (see later in this volume) using *E. coli* DNA gyrase with pBR322 DNA as the common substrate and Fig. 5.1 shows the correlations between the activity of a number of

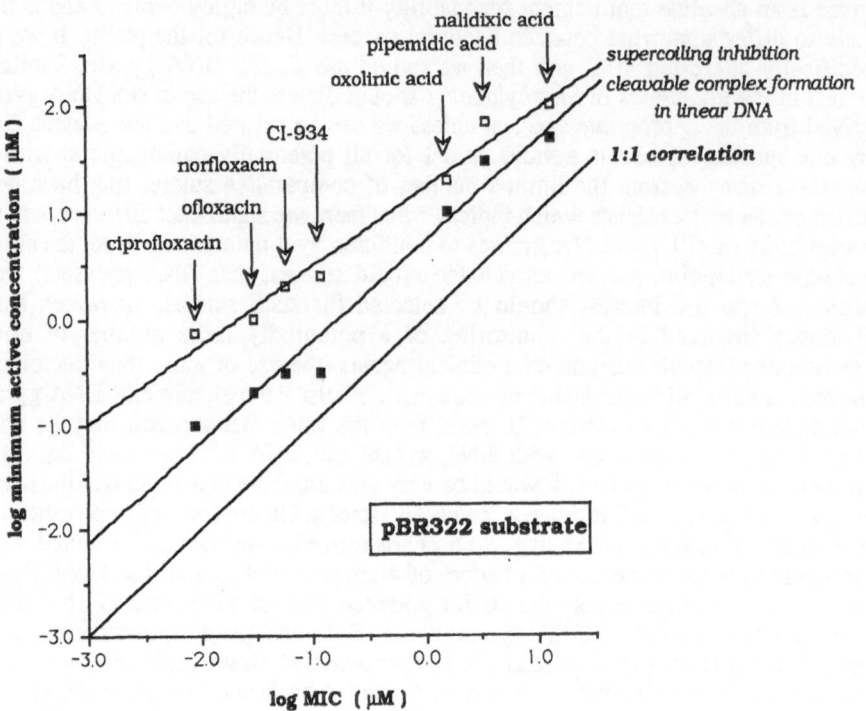

Fig. 5.1. The correlations between the inhibition of the supercoiling reaction and the formation of the cleavable complex with various 4-quinolones using pBR322 and *E. coli* DNA gyrase *in vitro*.

4-quinolones in these assays with the MIC of the agents against intact *E. coli* cells using data from my own laboratory along with data from Drs Elwell and Walton. In this figure it can be clearly seen that whilst both assay systems show a linear correlation with the MICs, and the formation of cleavable complexes in linear pBR322 giving a slightly closer correlation to the idealized 1:1 values, in both cases more drug is required to act upon the purified enzyme than is required to inhibit intact cells. In this figure the recorded values for the inhibition of supercoiling represent ID_{50} values for each agent, but better correlations with *in vivo* activity can be obtained if we use ID_{10} values instead. However, provided that we accept the differentials between the *in vitro* and the *in vivo* activities of the agents, both the supercoiling inhibition and generation of cleavable complexes yield acceptable linear correlations.

The differential between the *in vitro* and *in vivo* activities has been "explained away" by suggesting that *in vivo* the cells actually concentrate the 4-quinolones and yield higher intracellular levels of the agent than are administered. However, such a proposal does presuppose that all of the agents are concentrated to the same degree, which might seem a little unrealistic when extrapolating to species other than *E. coli* given the significant differences in cell-wall constitution between Gram negative and Gram positive organisms. In practical terms we might reasonably suggest that such assays do work in practice, and we can only really improve upon such systems by the

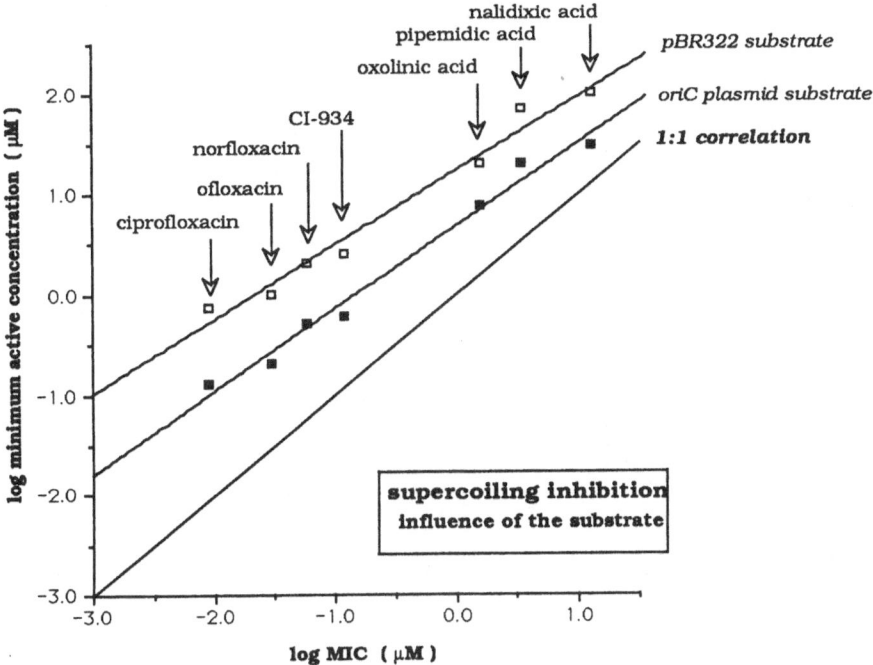

Fig. 5.2. The correlations between the inhibition of the supercoiling reaction with various 4-quinolones using *E. coli* DNA gyrase with pBR322 and an *oriC* plasmid *in vitro*.

introduction of similar assays using enzymes from other species where there are areas of specific clinical interest.

From the more esoteric standpoint of actually trying to understand the initial event in the inhibition of intact bacterial cells by the 4-quinolones, the inhibition of DNA supercoiling and the generation of cleavable complexes in plasmid DNA molecules have become accepted as the basic model. This is simply because it is aesthetically pleasing to accept that since DNA supercoiling requires the generation of double-strand breaks in DNA to facilitate strand-passaging, the inhibition of a sealing function of DNA gyrase by a 4-quinolone will, by definition, inhibit the completion of the introduction of superhelical turns. Since the A subunit of DNA gyrase (the *gyrA* gene-product) is known to have both a DNA nicking and DNA sealing function, and mutations in this gene confer resistance to the 4-quinolones, such a model for the inhibition of the sealing step seems obvious. In such a context a plasmid molecule is merely a "mini-chromosome" which is conveniently used to model the events which occur on the bacterial chromosome in treated intact cells. Acceptance of this basic model for the mechanism of action of the 4-quinolones indicates that all we really need to resolve before we can contemplate a purely rational designing of new 4-quinolones is the molecular chemistry of the interaction(s) between the 4-quinolone, DNA gyrase and/or DNA. However, we should not get carried away by the nearness of a final solution, because even the answer to the apparently simple question of whether or not

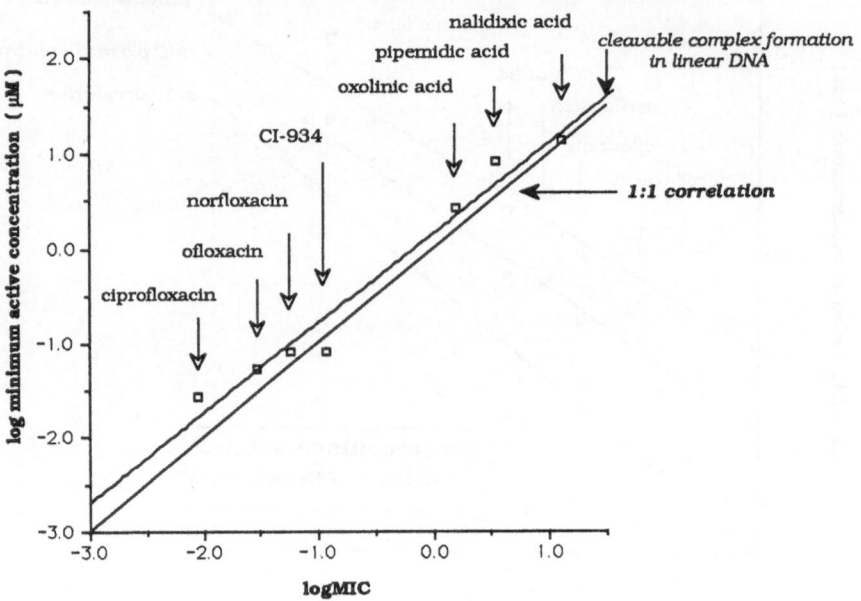

Fig. 5.3. The correlation between the formation of the cleavable complex with various 4-quinolones using an *oriC* plasmid with *E. coli* DNA gyrase *in vitro*.

4-quinolones bind to DNA has involved an enormous effort to attain even our present level of knowledge, which can still provoke disagreement between individual workers (see Chapter 10 by Dr Shen in this volume).

Before I try to review what we believe we know of the interaction(s) between 4-quinolones, DNA gyrase and DNA, it is worthwhile examining the results obtained in a variation of the basic assay system in which the DNA substrate is changed on the basis of some simplistic thinking. Inhibiting the replication of plasmids does not kill bacterial cells, but blocking chromosome replication is certain to be extremely deleterious. Consequently the assays have been carried out in the same way as before but with the replacement of pBR322 (which contains no intrinsic *E.coli* chromosomal DNA sequences) with an *E. coli oriC* plasmid containing the origin of chromosomal replication with the gene for TEM β-lactamase as the selectable marker. Using the inhibition of DNA supercoiling as the measureable event it can be seen in Fig. 5.2 that the reaction with the *oriC* plasmid is more sensitive to the effects of 4-quinolones than is the reaction using pBR322. Fig. 5.3 shows that if the experiment is repeated with the scoring of the formation of cleavable complexes the correlation of the *in vitro* activity with the MIC is very nearly 1:1. Such results simply suggest that the chromosome is more sensitive in some way to the effects of 4-quinolones than are plasmid molecules, even though this is contrary to the observations of Hill and Fangman (1973) who found that F' plasmid molecules were more susceptible than chromosomes to the formation of DNA strand breakages by cellular challenge with nalidixic acid.

The surprising closeness of the correlation of the results obtained for formation of cleavable complexes in the *oriC* plasmid with the MIC of the 4-quinolones is however not a solution to anything since in its own way *oriC* represents a possible artifactual situation. The *oriC* DNA is apparently unique since it is the only known case where challenge with a 4-quinolone leads to the release of DNA gyrase molecules from association with the DNA rather than stabilizing the formation of a covalent complex (Lother *et al*. 1984). Furthermore, we are also fairly certain that the activity of 4-quinolones is not focussed upon the initiation of DNA replication at the origin of replication. However, the clear differences between the results obtainable with pBR322 and an *oriC* plasmid may well be an important indicator of special considerations to be made with regard to the bacterial chromosome which is the specific intracellular target for the 4-quinolones as antibacterial agents.

4-quinolones, DNA gyrase and the bacterial chromosome: Whilst it is obvious to state that at bactericidal levels *in vivo*, 4-quinolones exert their principal effects upon the replicative apparatus of the bacterial chromosome, we have to remember that most of what we have learned about the effects of 4-quinolones has been derived from studies with plasmid DNA. Consequently, any attempt at clarifying our perception of the mechanism of action of the 4-quinolones should involve consideration of the following questions:-

(i)   Are there significant differences between chromosomes and plasmids?
(ii)   Do chromosomes and plasmids behave as identical substrates for interaction with DNA gyrase?
(iii)   Do the 4-quinolones do the same things to plasmid and chromosome systems?

In terms of the topological structure of chromosomes and plasmids, to the best of our knowledge plasmids represent single topological domains and have no higher-order structure whilst the chromosome of *E. coli* comprises 50-80 "domains of supercoiling" for which various models have been proposed (Worcel and Burgi 1972). Clearly there are evident structural differences beyond the simple size differentials between chromosomes and plasmids. Since DNA gyrase is a DNA topoisomerase, and such enzymes are by definition implicated in the regulation and/or maintenance of DNA structure, we should perhaps be wary of making too hasty an extrapolation from plasmid derived data to chromosome effects.

Any intuitive caution over extrapolating from chromosomes to plasmids is heightened by consideration of some recent results of the identification of cleavable complexes in the *gyrB* region of the *E. coli* chromosome when this chromosomal region is cloned into a plasmid molecule (Franco and Drlica 1988). In this case more than 20 quinolone-induced cleavage sites could be identified within the cloned region, thereby suggesting the likely presence of 10 000 such sites in the intact chromosome. This result contrasts markedly with the observation that bactericidal levels of nalidixic acid generate only 50-80 apparently evenly spaced cleavage sites in chromosomal DNA isolated from challenged intact cells (see Fig. 5.4) (Crumplin and Smith 1976). Clearly the location of a given DNA sequence is significant in considering the effects of the 4-quinolones, and it appears evident that more data is required to enable us to develop a model for the bactericidal effects of the 4-quinolones. We also need to address the question of whether or not the cleavable complexes we observe represent the inhibition of the sealing function within the overall strand-passaging reaction, or whether they

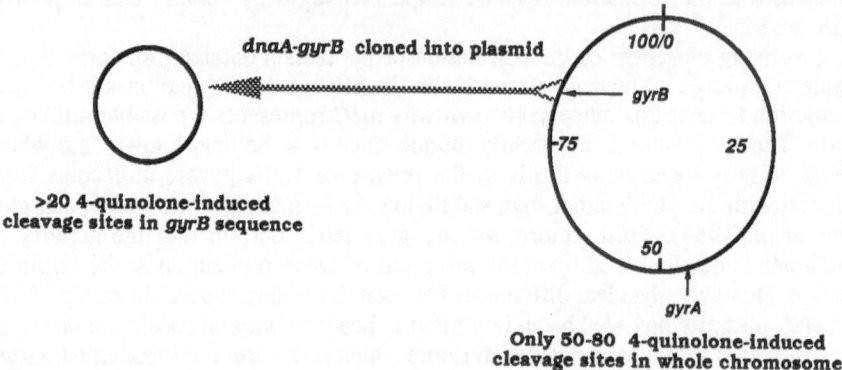

Fig. 5.4. Variation in the number of sites of the cleavable complex formation in chromosomal DNA related to the physical location of the DNA region.

represent specific induced, unsealable cleavages which might not be associated with the supercoiling reaction.

We can perhaps move towards some form of resolution by initially remembering that whilst DNA gyrase is exclusively a bacterial protein, it is still a type II DNA topoisomerase. The type II DNA topoisomerases have been widely studied since the first descriptions of DNA gyrase, and we have found that they are universally essential components of cell systems, and are significantly conserved, both compositionally and functionally, from T4 bacteriophage through bacteria to all eukaryotes so far studied. Consequently it is almost certainly worth us examining what we know of non-bacterial systems to see if there are data of value in our consideration of gyrase and bacteria.

Perhaps the first useful clue to be found outside the bacterial system *per se* is in the study of the replication of the linear DNA genome of bacteriophage T7. It has been shown that nalidixic acid is an effective inhibitor of T7 replication, thus indicating a role for DNA gyrase in the replication of the linear T7 DNA (Baird *et al.* 1972) and that in a *gyrA ts* background at the restrictive temperature T7 replication is relatively unimpaired (Kreuzer and Cozzarelli 1979; Steck and Drlica 1985). Since it is difficult to consider that DNA gyrase functions to insert negative superhelical turns into a linear DNA substrate, and since it is evident from the effects of the restrictive temperature upon T7 replication in a *gyrA ts* host that DNA gyrase subunit A is not required for competent replication, we can only conclude that the inhibitory effect of nalidixic acid is not the result of the inhibition of supercoiling. We can further conclude that the presence of nalidixic acid induces DNA gyrase to act upon the T7 DNA in a manner which is absent without the intervention of the drug.

This suggestion that 4-quinolones serve to induce DNA gyrase functions rather than inhibit them is not without precedent since it has also been shown that whilst nalidixic acid is an efficient inducer of the *recA* gene expression (Gudas and Pardee 1975) the exposure of a *gyrA ts* mutant to the restrictive temperature does not induce *recA* (Smith 1983). Furthermore, we have found that a *gyrA ts* mutant does not die at the restrictive temperature whilst 4-quinolone treatment of a susceptible cell induces cell death.

Since it is known that DNA gyrase is capable of nicking DNA strands and sealing DNA strand breaks, and that 4-quinolones stabilize double-strand breaks in DNA, then it is reasonable to propose that the 4-quinolones, if they induce a specific function, induce DNA cleavage rather than sealing (especially as the sealing function presupposes the existence of DNA strand breaks). As T7 DNA replication proceeds apparently uneventfully in the absence of the nicking function of DNA gyrase it would seem reasonable to presume that under normal conditions DNA gyrase is not required to nick and seal T7 DNA. Hence, if the presence of a 4-quinolone were to induce cleavage of the T7 DNA (leading to inhibition of the replication of the molecule) then the cleavage must be induced at a site on the DNA molecule where the enzyme interacts, but normally plays no catalytic role. The implications of this interpretation are two-fold:-

Firstly, the quinolone-induced gyrase cleavage sites mapped and sequenced upon plasmids like pBR322 represent specific quinolone-induced gyrase cleavage sites rather than sites for normal (drug-free) cleavage and sealing in the insertion of negative superhelical turns.

Secondly, in the context of the effects of 4-quinolones upon the chromosomes of treated cells the strand breakages induced *in vivo* by exposure to bactericidal levels of the drug represent induced cleavages at DNA gyrase binding sites where the enzyme has no routine catalytic function. On the basis of the evidence of the increased susceptibility of gyrase:DNA systems involving chromosomal DNA sequences (e.g. *oriC* plasmids) we can suggest that the chromosome contains DNA gyrase binding sites which are not routinely involved in the catalytic activity of the enzyme, and which on the basis of the evolution of both the intrinsic chromosome and the intrinsic gyrase are likely to be species specific. We can also suggest that these sites are particularly susceptible to the intervention of 4-quinolone molecules and that the inductive effect is manifested at lower concentrations than are required for the inhibition of the supercoiling reaction.

The demonstrable conservation of the composition and mechanistic functioning of the type II DNA topoisomerases from T4 bacteriophage through bacteria to *Drosophila melanogaster* (Moriya *et al.* 1985; Lynn *et al.* 1986; Huang *et al.* 1986; Heller *et al.* 1986; Uemura *et al.* 1986) suggests that useful data relevant to our consideration of the mechanism of antibacterial action of the 4-quinolones, might be found in studies of eukaryotic type II DNA topoisomerases. Recent studies of the functions of type II DNA topoisomerases in eukaryotic cells have shown that this protein has two essential functions, one of which is the familiar catalytic role of modulating the superhelical density of DNA, the other being a structural role as a major scaffold protein of chromatin; i.e. DNA topoisomerase II acts as a major component of the scaffold of eukaryotic chromosomes. It is also evident that in DNA replication and DNA transcription, the enzyme acts catalytically at very special sites on the chromosomal DNA which correspond to the sites where it interacts as a scaffold protein (Earnshaw and Heck 1985; Earnshaw *et al.* 1985; Gasser and Laemmli 1986). Furthermore treatment with epipodophyllotoxins, which induce and stabilize double-strand cleavages by type II topoisomerases, elicits chromosome breakages at the junctions of the radial loops at Scaffold Associated Region (SAR) sequences which are the junctions between individual radial loops of chromosomal DNA. These SAR sequences have also been shown to contain clusters of drug-induced topoisomerase II cleavage sites (8-15 sites per SAR sequence of approximately 250 base-pairs). If we simply extrapolate the concept of the chromosome scaffold (with type II DNA topoisomerase as a major structural component) to the prokaryotic cell and invoke the suggestion that the 4-quinolones induce DNA cleavage by DNA gyrase (bacterial topoisomerase II) at the equivalent of SAR sites on the chromosome, then we would expect 4-quinolone challenge of a susceptible cell to result in DNA cleavage at the junctions of the

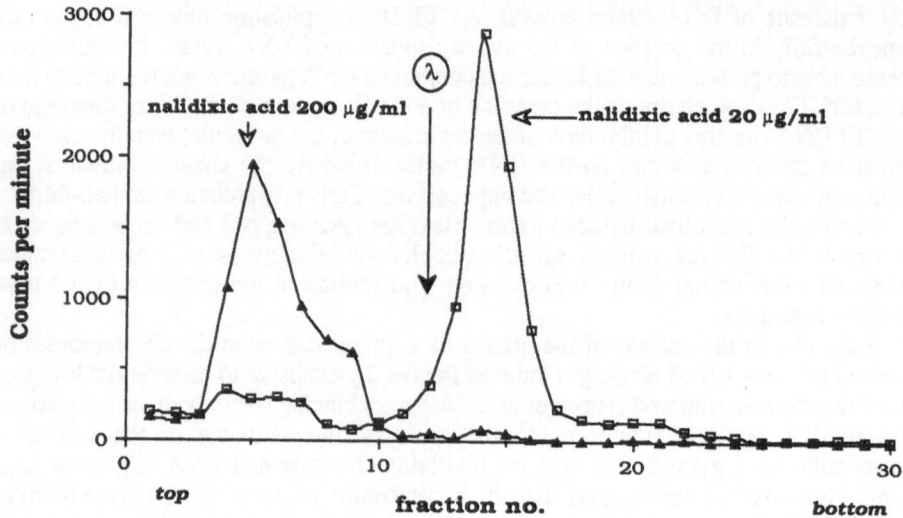

Alkaline-sucrose gradient profiles of chromosomal DNA

Fig. 5.5. 5%-20% alkaline sucrose gradient profiles of chromosomal DNA pulse-labelled with
³H-thymidine and isolated from *E. coli* strain KL16 treated for 20 min with nalidixic acid at either
20μg/ml or 200μg/ml.

equivalent of radial loops or domains of supercoiling. Since it has been shown that
both 4-quinolones and epipodophyllotoxins induce DNA cleavage *in vitro* by both
eukaryotic and prokaryotic topoisomerase II enzymes (Barrett *et al.* 1987), and that the
two enzymes are clearly related, it seems reasonable to draw analogies between the drug
induced effects in eukaryotic cells and those induced in prokaryotic cells.

Such an apparently plausible suggestion indicates that it is essential for us to
examine the effects of the 4-quinolones upon bacterial chromosomes. Such a study was
initially carried out several years ago (Crumplin and Smith 1976) when it was shown
that challenge of *E. coli* strain KL16 with 20μg/ml of nalidixic acid gave rise to an
anomalous accumulation of 38S single-stranded DNA fragments when the
chromosomal DNA was subjected to alkaline-sucrose gradient centrifugation (see Fig.
5.5). More recently we have repeated this study and subjected SDS treated chromosomal
DNA to pulse-field gel electrophoresis which shows that the challenge with a low,
bactericidal, dose of a 4-quinolone facilitates the isolation of large double-stranded DNA
fragments of approximately 100 kilobase-pairs. A simple arithmetic exercise indicates
that these fragments are each equivalent to approximately 0.015 - 0.02 of the entire
chromosome which equates to the apparent size of the previously demonstrated
"domains of supercoiling". It might thus be suggested that the chromosome fragments
generated by such 4-quinolone treatment represent the domains of supercoiling, and
that the DNA gyrase molecules presumably associated with the ends of these fragments
represent the residues of a chromosomal scaffold.

If we postulate that DNA gyrase in bacteria is a true analogue of the eukaryotic type
II DNA topoisomerase and thus has both a structural and a catalytic role in the bacterial
cell, then at "structural" sites on the chromosome DNA gyrase will bind and serve to

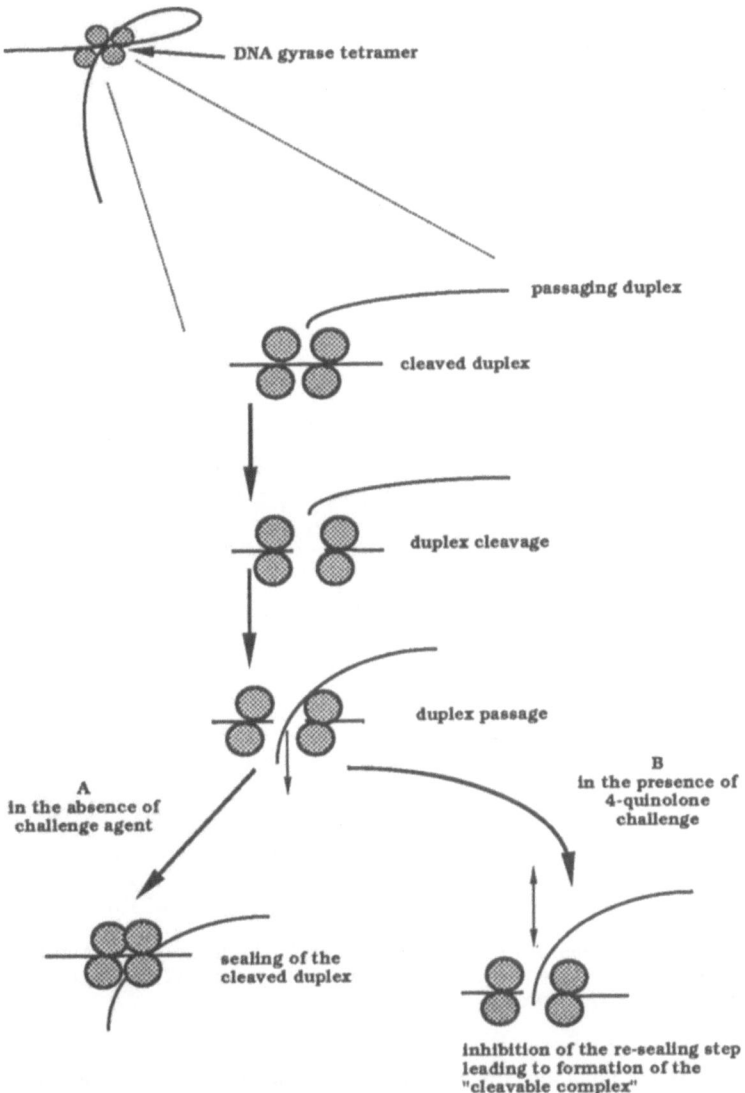

Fig. 5.6a. Model for the formation of the cleavable complex by the inhibition of the sealing step component of the strand-passaging reaction of DNA gyrase during the insertion of negative superhelical turns in DNA.

constrain the supercoiling within the domains of supercoiling. Hence it would seem reasonable to suggest further that at the structural association sites DNA gyrase does not routinely cleave the DNA and catalyze topological changes. Such a model now allows us to draw upon the data from bacteriophage T7 and suggest that challenge with a 4-quinolone induces an aberrant cleavage by structural DNA gyrase molecules at the

In a "structural" mode the type II
topoisomerase is bound to both the
"passaging" duplex and the "cleaved"
duplex but cleaves neither strand

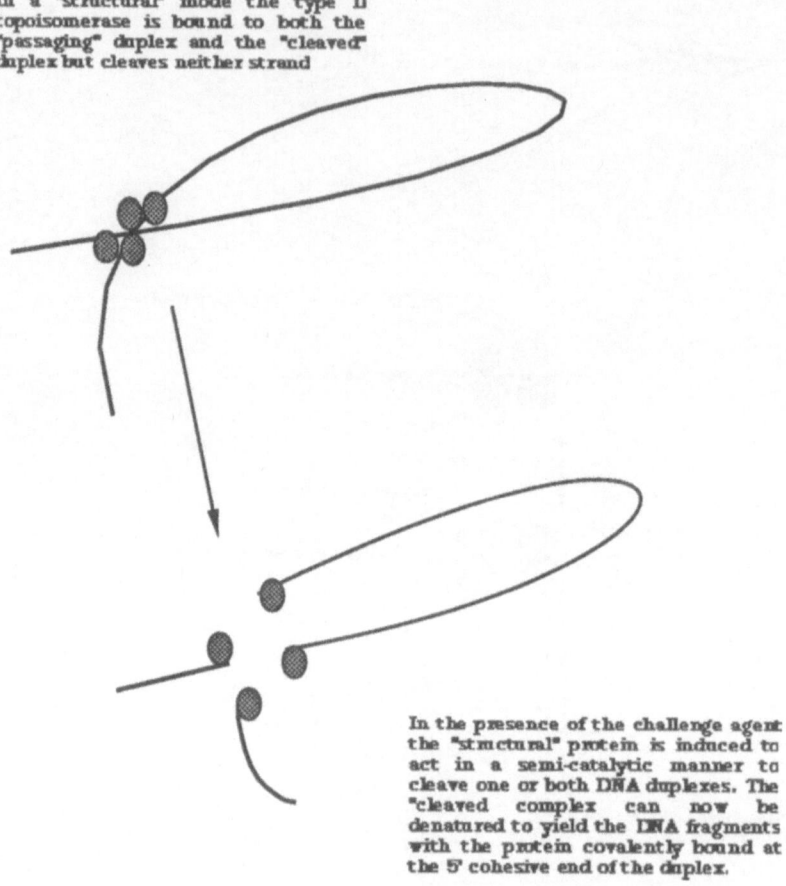

In the presence of the challenge agent
the "structural" protein is induced to
act in a semi-catalytic manner to
cleave one or both DNA duplexes. The
"cleaved complex can now be
denatured to yield the DNA fragments
with the protein covalently bound at
the 5' cohesive end of the duplex.

Fig. 5.6b. Model for the formation of the cleavable complex by the 4-quinolone induction of an anomalous double-strand breakage which cannot be sealed in the presence of the drug at a non-catalytic site of binding of DNA gyrase to DNA.

scaffold-association sites and leads to the formation of cleavable complexes at the junctions between individual domains of supercoiling.

This proposal of a chromosome scaffold function for DNA gyrase in bacterial cells implies that there may be two types of gyrase-DNA interaction on the chromosome with two forms of recognition/interaction nucleotide sequence. It further implies that the two types of gyrase-DNA complex show differential sensitivity to the 4-quinolones with the scaffold complex being the more susceptible to perturbation with a 4-quinolone. The consequence of such a proposal would then be that whilst at low, bactericidal, doses of a 4-quinolone the structural DNA gyrase molecules are targetted, and at higher doses of the drug the structural sites become saturated and the catalytic gyrase-DNA sites become targeted. Fig. 5.5 shows the results of a high-dose challenge with nalidixic acid at a dose of 200 μg/ml which is above the most-lethal concentration. In contrast to the large DNA fragments generated by treatment with the low dose of nalidixic acid, the high dose challenge generated much smaller fragments and no large fragments were apparent. These small fragments equate to something like 5000 cleavable complexes per chromosome, which is not too dissimilar to the 10 000 potential sites postulated by Franco and Drlica (1988).

This demonstration of what may represent two modes of 4-quinolone susceptibility of DNA gyrase on the bacterial chromosome raises the possibilities that these may equate with the nalidixic acid sensitive and resistant modes of DNA replication in *E. coli* (Dermody *et al.* 1974) and/or the two levels of action of DNA gyrase on the bacterial chromosome (Drlica *et al.* 1980). It might also help explain the observation that 4-quinolones will elicit demonstrable reductions in plasmid supercoiling *in vivo* at drug doses which exceed the most lethal concentration (N. Cozzarelli, personal communication; Fisher *et al.* 1986). Furthermore, since plasmids represent single topological domains, they are unlikely to contain sequences which would correspond to the possible scaffold association sites of the chromosome, hence when we study the effects of 4-quinolones and DNA gyrase upon plasmid substrates *in vitro* we can only examine the effects of the drug upon gyrase/DNA interaction sites which represent catalytic sites. The results of the studies on chromosomal DNA might indicate that catalytic sites are less susceptible to 4-quinolone induced perturbation, hence we should not be surprised when *in vitro* studies using plasmids indicate that the system is less susceptible to inhibition than are whole cells.

Such a postulated model for the effects of 4-quinolones upon DNA gyrase activity does however have further implications upon our perception of the generation of the cleavable complex. We have always been faced with the unresolvable question of whether the drugs induce an unsealable cleavage, or whether they inhibit the sealing of a "normally" generated cleavage for strand-passaging. With the possibility of two forms of gyrase-DNA interaction, it is quite feasible that at the catalytic sites of gyrase functioning the enzyme generates a normal strand breakage ready for strand-passaging and the presence of 4-quinolone molecules prevents the sealing of the strand-breakages (see Fig. 5.6a). In contrast, since structural DNA gyrase molecules might not routinely generate strand-breakages, the presence of 4-quinolone molecules might induce anomalous strand-breakages which cannot be sealed (see Fig. 5.6b). We are at present in the process of isolating and sequencing the sites of quinolone-induced cleavable complexes on the *E. coli* chromosome in order to compare these with the concensus cleavage sites which have been described for plasmid molecules and chromosomal fragments cloned into small plasmids and we hope to establish whether or not there are two identifiable types of sequence available for gyrase/DNA interaction.

Conclusions

Examination of the effects of 4-quinolones upon DNA gyrase/DNA systems associated with the bacterial chromosome has led to the proposal of a mechanism of action which involves the generation of a cleavable complex not associated with the supercoiling activity of the enzyme. In terms of the starting point of this communication, such a model clearly suggests that *in vitro* screens designed to indicate the potential antibacterial activity of 4-quinolone molecules should be based upon monitoring the formation of cleavable complexes rather than perturbation of the supercoiling reaction. It also suggests that the routine use of plasmid DNA substrate molecules only allows monitoring of the less susceptible mode of DNA gyrase/DNA association. However, in practice it seems evident that for most cases the relative susceptibilities to different

4-quinolones of the less susceptible (catalytic) DNA gyrase/DNA complex accurately reflects the relative susceptibilities of the postulated structural DNA gyrase/DNA complex on the chromosome. However there are exceptions to this equivalence which might only be identified if screening assays are developed using gyrases from different species and maybe appropriate substrates. For example a comparison of the antibacterial activities and activities against DNA gyrase *in vitro* of ciprofloxacin and the active (+) isomer of the experimental compound S-25930 shows that ciprofloxacin is approximately ten-times as active as S-25930 both *in vivo* and *in vitro* against intact *E. coli* and against the *E. coli* DNA gyrase whereas against *S. aureus* intact cells S-25930 is two - four times as active as ciprofloxacin. This differential relative activity might reflect differences in the susceptibilities of the two DNA gyrases or differences in the uptake of the drugs between the two bacterial species - it does not matter which is appropriate since the data is only cited here as a demonstration of the care we have to exercise in extrapolating conclusions from *in vitro* studies of only a single DNA gyrase.

It is evident from the standpoint of the basic science that there are significant limitations in the value of the *in vitro* assays we routinely use for the screening of new 4-quinolone molecules, but they do already serve as a useful guide. Perhaps greater care should be taken in extrapolating from convenient screens, which serve their purpose, to the modelling of mechanisms of action than is necessary in extrapolating from studies of the *E. coli* enzyme to possible activity against unrelated and clinically significant species of bacteria. It seems evident from both the scientific literature and the promotional material for new 4-quinolones that the mechanism of action of the 4-quinolones has been perceived as the inhibition of DNA supercoiling, possibly because this was the first definitive observed function after 17 years of searching for a mechanism of action for nalidixic acid. It is perhaps time that we re-appraised the data and maintain the option to change our perception of the mechanism of action of these compounds. This communication represents one such attempt and suggests that we might better perceive these compounds as inducers of the formation of the cleavable complex between DNA gyrase and DNA than as inhibitors of the supercoiling reaction catalyzed by DNA gyrase.

References

Baird JP, Bourguignon GJ and Sternglanz R (1972) Effects of Nalidixic acid on the growth of DNA bacteriophages. J Virol. 9:17-21

Barrett JF, Gootz TD, McGuirck PR, Farrell C, Sokolowski S and Frescura M (1987) Use of *in vitro* topoisomerase II assays in studying nalidixic acid derivatives. XXV ICAAC (New York)

Crumplin GC and Smith JT (1976) Nalidixic acid and bacterial chromosome replication. Nature 260: 643-645

Dermody JJ, Bourguignon GJ, Foglesong PD and Sternglanz R (1974) Nalidixic acid-sensitive and resistant modes of DNA replication in *Escherichia coli*. Biochem Biophys Res Comm 61:1340-1347

Drlica K, Engle EE and Manes SH (1980) DNA gyrase on the bacterial chromosome: Possibility of two levels of action. Proc Natl Acad Sci 77:6879-6883

Earnshaw WC and Heck MMS (1985) Localization of topoisomerase II in mitotic chromosomes. J Cell Biology 100:1716-1725

Earnshaw WC, Halligan B, Cooke CA, Heck MMS and Liu LF (1985) Topoisomerase II is a structural component of mitotic chromosome scaffolds. J Cell Biology 100:1706-1715.

Fisher LM, Barot HA, Cullen ME and Hulton CS (1986) DNA breakage by *E. coli* DNA gyrase *in vitro* and *in vivo*. Biochem Soc Transact 14: 493-496

Franco RJ and Drlica K (1988) DNA gyrase on the bacterial chromosome. Oxolinic acid-induced DNA cleavage in the *dnaA-gyrB* region. J Molec Biol 201:229-233

Gasser SM and Laemmli UK (1986) Cohabitation of scaffold binding regions with upstream/enhancer elements of the developmentally regulated genes of *D. melanogaster*. Cell 46:521-530

Gudas LJ and Pardee AB (1975) Model for the regulation of DNA repair functions.Proc Natl Acad Sci 72:2330-2334

Heller RA, Fairman R, Philip M and Brutlag DL (1986) Cloning and characterization of the *Drosophila* DNA topoisomerase II gene and its expression during development. Cold Spring Harbor Symposium (Sept)

Hill WE and Fangman WL (1973) Single-strand breaks in deoxyribonucleic acid and viability loss during deoxyribonucleic acid synthesis in *Escherichia coli*. J Bacteriol 116:1329-1335

Huang WM, Nicholson G, Fang M and Gibson A (1986) Structure and function of T4 DNA topoisomerase subunits. Cold Spring Harbor Symposium (Sept)

Kreuzer KN and Cozzarelli NR (1979) *Escherichia coli* mutants thermosensitive for DNA gyrase subunit A: Effect on DNA replication, transcription and bacteriophage growth. J Bacteriol 140:424-435

Lother H, Lurz R and Orr E (1984) DNA binding and antigenic specificities of DNA gyrase. Nucleic Acids Research 12:901-913

Lynn R, Giaever G, Swanberg SL and Wang JC (1986) Tandem regions of yeast DNA topoisomerase II share homology with different subunits of bacterial gyrase. Science 233:647-649

Moriya S, Ogasawara N and Yoshikawa H (1985) Structure and function of the region of the replication origin of the *Bacillus subtilis* chromosome : III Nucleotide sequence of some 10 000 base pairs in the origin region. Nucleic Acids Res 13:2251-2265

Smith CL (1983) *recF* -dependent induction of *recA* synthesis by coumermycin, a specific inhibitor of the B subunit of DNA gyrase. Proc Natl Acad Sci 80:2510-2513

Steck TR and Drlica K (1985) Involvement of DNA gyrase in bacteriophage T7 growth. J Virol 53:296-298

Uemura T and Yanagida M (1986) Functional domains in DNA topoisomerase II. Cold Spring Harbor Symposium (Sept)

Worcel A and Burgi E (1972) On the structure of the folded chromosome of *Escherichia coli*. J Molec Biol 71:127-147

Discussion Summary

The proposal that the formation of the cleavable complex induced by the 4-quinolones represented a more valid model for the mechanism of action of these compounds than did the inhibition of supercoiling was seemingly accepted as quite straightforward. However, the suggestion that the use of plasmid molecules, like pBR322, as substrates for *in vitro* assays, as a potentially invalid model system, was considered rather surprising. The fact that the presence of an *oriC* sequence on a substrate plasmid radically altered the sensitivity of the system to perturbation by 4-quinolones was seemingly quite unexpected. The niceties of the use of a proper model system for *in vitro* screening, along with the use of more than one DNA gyrase, each with a specific individual substrate, was generally considered to be rather impractical in the context of the commercial screening of a potentially large number of new synthetic entities. It was generally appreciated that the use of a single standardized DNA gyrase did provide for the risk of "missing" the identification of new molecules which might be particularly active against individual species of bacteria presenting particular clinical problems.

The demonstration of a very limited number of cleavage sites of the *E. coli* chromosome, particularly when compared to cloned regions of the chromosome on a plasmid showing far more apparent cleavage sites in the same piece of DNA, was not new information, but no-one was able to provide an explanation of these findings. It was suggested that the specific sites of 4-quinolone activity on the chromosome might correspond to the *rep* sequences which have recently been identified. However, in the absence of any proper knowledge of the function of the *rep* sequences, and in the absence of available sequence data on the 4-quinolone cleavage sites, it was not reasonable to launch into another bout of speculation on the mechanism of action of the 4-quinolones.

The demonstration of different size classes of DNA fragments from the chromosome resulting from the treatment of intact cells with different doses of a 4-quinolone was considered to be potentially significant, but in general this was not considered to be too surprising a finding in the light of the well established concept of these drugs displaying two dose-related mechanisms of action. However, it was appreciated that this fact did represent a complication in any attempt to try and elucidate the mechanism of action of the 4-quinolones.

The raising of several, apparently unanswerable, questions (in the light of present knowledge) was believed to be par for the course of 4-quinolone research. It re-affirmed the overall feeling of the meeting which had repeatedly shown by recent research that what we believe we know of these agents is still very much open to question.

Chapter 6

Mechanisms of Killing of Bacteria by 4-Quinolones

J. S. Wolfson and D. C. Hooper

Introduction

The 4-quinolones are a major new class of potent orally-absorbed antibacterial agents. These drugs, derivatives of the earlier marketed drugs nalidixic acid and oxolinic acid, include norfloxacin, ciprofloxacin, ofloxacin, fleroxacin, lomefloxacin, temafloxacin, spafloxacin, and others. A favorable property of the 4-quinolones is rapid killing of bacteria. Mechanisms of killing, however, are not well understood (Smith and Lewin 1988; Wolfson *et al.* 1989d), despite considerable knowledge of the effects of these drugs on the intracellular target, the enzyme DNA gyrase, and on cellular metabolic processes (Cozzarelli 1980; Gellert 1981; Drlica 1984; Wang 1987). Information on mechanisms of killing of bacteria by 4-quinolones will be reviewed, discussing definitions and methodology, the killing phenomenon, studies of mechanisms, differences among 4-quinolones, and bacterial mutants exhibiting reduced killing.

Definitions and Methodology

It is important to consider definitions and methodology used in the study of bacterial killing, to minimize confusion related to inexact terminology and experimental techniques which do not adequately differentiate among resistance, inhibition, and killing.

Exposure to an antibacterial agent at sufficient concentrations might have three effects on a population of growing bacteria. Firstly, growth might continue, in which case bacteria are resistant. Secondly, bacterial growth might stop but the viable cell number remain constant, in which case the growth of bacteria is inhibited. Or thirdly, viability might decrease, in which case bacteria are killed. Thus "inhibition" signifies a bacteriostatic and not a bactericidal (killing) effect.

Minimal inhibitory concentration (MIC) is defined as the lowest drug concentration of two-fold increments which blocks visible growth of a known inoculum of cells during incubation for 18 h. Minimal bactericidal concentration (MBC) is defined as the lowest drug concentration of two-fold increments which reduces the viable count 1000-fold or more during incubation for 18 h. It is important to note that methods which measure inhibition, such as determination of an MIC in broth, or on agar, or efficiency of plating assays using drug-containing agar, do not differentiate between the bacteriostatic and bactericidal effects of an antimicrobial agent. To differentiate between inhibitory and killing activity, the ability of cells to grow on drug-free agar subsequent to drug exposure must be determined.

Exposure of a population of cells to a bactericidal drug may reveal a subpopulation of cells which are inhibited but which are either slowly killed or not killed at all. This subpopulation does not represent resistant mutants, because the cells do not grow in the presence of drug. The cells also are not mutants genotypically refractory to killing, because exposure of descendants of these cells to the same concentration of drug results in a similar killing curve (Moyed 1983; Wolfson et al. 1989c). Such a subpopulation of cells are called persisters, and the phenomenon is called persistence (Bigger 1944; Moyed 1983; Greenwood 1985).

Tolerance, which may be related to persistence, describes a phenomenon in which a bacterial strain is inhibited but not killed by a bactericidal agent (Sabath et al. 1977; Handwerger and Tomasz 1985; Woolfrey et al. 1985). Tolerant strains are killed more slowly than non-tolerant strains in time-killing studies, and the MBC of a tolerant strain may greatly exceed its MIC. The term tolerance has also been used to indicate reduced antagonism by a drug by any mechanism, including resistance as defined above. Definition of tolerance seems better restricted, however, to signify selective reduction in killing by a bactericidal agent.

A "gold standard" assay for measuring bacterial viability is the ability of a cell to form a colony on agar, quantitated as colony forming units (CFUs). More rapid and easier assays of viability exist but may be misleading. For example, for 4-quinolones, increases in mass measured by optical density or light scattering may occur with either increases in cell number or filamentation of cells unable to form colonies (Smith 1984; Elliott et al. 1987). Also, determination of cell number by particle count with a Coulter counter, Klett colorimeter, or counting chamber does not measure the viability of the particles (Crumplin et al. 1984).

Determination of CFUs of bacteria growing in broth may be inaccurate as well. Cells may attach to glass above the meniscus, survive due to lack of drug exposure, be reintroduced into broth with mixing, and form colonies when spread on agar (Handwerger and Tomasz 1985). This problem may be circumvented by care not to deposit bacteria above the meniscus. Another approach is to avoid use of liquid media altogether. Cells may be spread on nitrocellulose filters placed on the surface of drug-containing agar, with subsequent removal of filters to drug-free agar to allow growth of survivors. For norfloxacin, the kinetics of killing of *E. coli* KL16 on filters (Fig. 6.1) are identical to those in broth (Wolfson et al. 1989c).

Many killing studies have been carried out by measurement of MICs and MBCs. Similarity of MIC and MBC values for a bacterial strain, however, may also be misleading, because differences present early after drug exposure may no longer exist by 18 h when viability is determined. This occurrence has been documented for ß-lactams

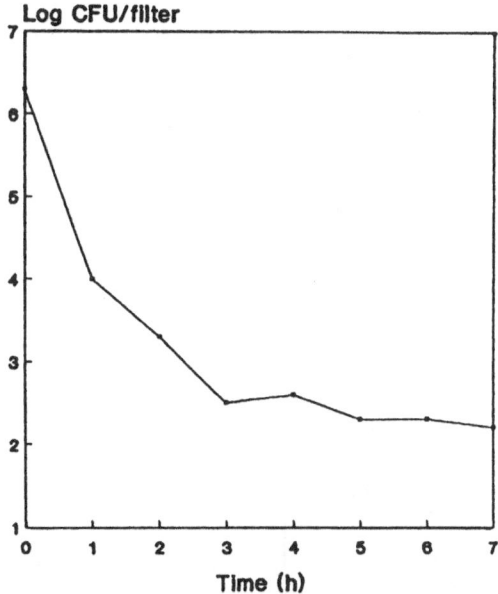

Fig. 6.1. Killing of *Escherichia coli* KL16 by norfloxacin at 37°C. Cells spread on nitrocellulose filters were exposed to drug by placement of filters on the surface of drug-containing agar for indicated times. Filters were removed to drug-free agar to allow growth of survivors (for details, see Wolfson *et al.* 1989c).

Handwerger and Tomasz 1985) and for coumermycin (an antagonist of the B subunit of DNA gyrase) (Wolfson *et al.* 1989c).

   The growth phase of the cells, the inoculum size, and oxygenation should be standardized, as these parameters affect the bactericidal activity of 4-quinolones (discussed below).

   It is important that studies comparing killing of strains by drugs be performed using drug concentrations equivalent multiples above MICs or at a range of concentrations above the MICs of the strains being evaluated. These approaches optimize the possibility that differences observed between strains or drugs relate to the killing phenomenon itself and not simply to differences in drug susceptibility. For example, norfloxacin kills *E. coli* KL16 more effectively than a similar molar concentration of nalidixic acid, but killing activities are similar when compared at equivalent multiples above their respective MICs (Smith 1984).

Kinetics and Characteristics of Killing

Exposure of *E. coli* KL16 to a 4-quinolone results in initial rapid killing which then slows or reaches a plateau (Goss *et al.* 1964; Cook *et al.* 1966; Deitz *et al.* 1966; Smith 1984; Crumplin *et al.* 1984; Diver and Wise 1986; Elliott *et al.* 1987; Wolfson *et al.* 1989c) (Fig. 6.1). These kinetics suggest at least two components to killing. The less rapidly killed cells represent persisters and not resistant or less effectively killed mutants (Wolfson *et al.* 1989c). The mechanism of persistence for 4-quinolones

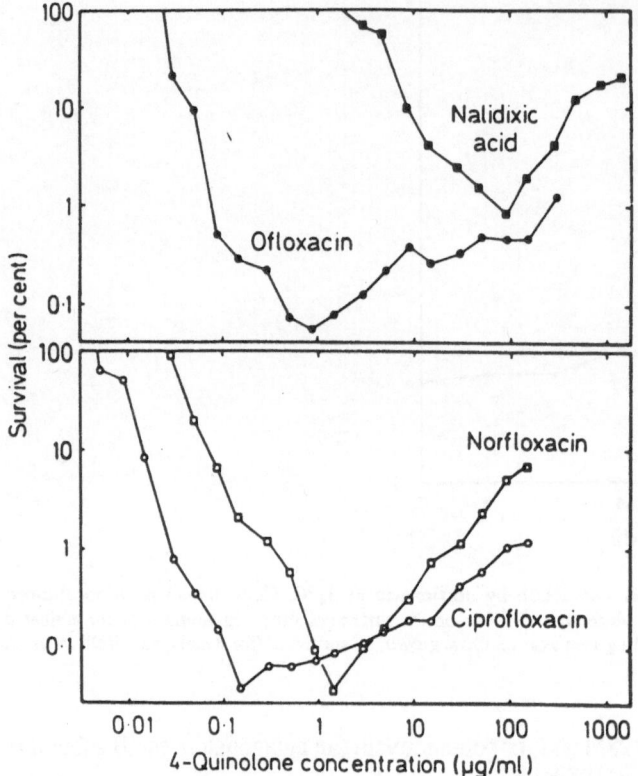

Fig. 6.2. The paradoxical (biphasic) response of *Escherichia coli* KL16 exposed for 3 h to different concentrations of 4-quinolones (from Smith 1984, with permission).

is unknown. It also has not been determined whether persistence represents exposure of a subpopulation to 4-quinolones during a refractory period of the cell cycle, as has been documented for ß-lactams for *E. coli* (Mathison 1968; Hoffmann *et al.* 1972) and *S. aureus* (Holzhoffer *et al.* 1985).

The rate of bacterial killing by a 4-quinolone increases as drug concentrations increase but then either approaches a plateau or decreases at higher drug concentrations (Winshell and Rosenkranz 1970; Crumplin and Smith 1975; Smith 1984; Diver and Wise 1986) (Fig. 6.2). Thus killing may lessen at higher drug concentrations, a "paradoxical effect" initially described for killing of certain Gram-positive bacterial strains by penicillin (Eagle and Musselman 1948). The paradoxical effect allows determination of a "most (maximal) bactericidal concentration" (Smith 1984) or "optimal bactericidal concentration" (Phillips 1987) for 4-quinolones, which tends to be 30- to 60-fold above the MIC, for killing during exposure to varied concentrations of drug for 3 h.

Agents Decreasing Killing and Differences Among 4-Quinolones

Agents Decreasing Killing by 4-Quinolones

Killing of *E. coli* by a 4-quinolone is substantially diminished or completely blocked by chloramphenicol, rifampin, erythromycin, tetracycline, clindamycin, dinitrophenol, or nutrient starvation (Deitz *et al.* 1966; Winshell and Rosenkrantz 1970; Crumplin and Smith 1975; Stevens 1980; Crumplin *et al.* 1984; Smith 1984; Zeiler 1985; Diver and Wise 1986; Smith and Lewin 1988). These treatments have in common inhibition of protein synthesis, suggesting that killing requires active synthesis of proteins.

Studies of bacterial DNA, RNA, and protein synthesis during exposure to different concentrations of 4-quinolones also suggest a requirement for protein synthesis for killing. Exposure to a 4-quinolone at concentrations below the most bactericidal concentration inhibits DNA synthesis only, concentrations at the most bactericidal concentration inhibit RNA and protein synthesis about 20% also (Stevens 1980), and concentrations above the most bactericidal concentration are associated with progressively greater inhibition of RNA and protein synthesis (Winshell and Rosenkranz 1970; Crumplin and Smith 1975; Stevens 1980). These findings in addition suggest that inhibition of protein synthesis at high drug concentrations might be responsible for the paradoxical effect. The association of decreased killing with increased inhibition of protein synthesis is further illustrated by studies with the 4-quinolone derivative 17607. This agent inhibits, but does not kill, *E. coli* KL16 and inhibits RNA and protein synthesis at concentrations close to those inhibiting DNA synthesis (Stevens 1980). Whether the requirement for active protein synthesis for killing represents need for a specific protein(s) or the requirement for bacterial growth in general is unknown.

While variation in inoculum by up to 10 000-fold alters inhibition by 4-quinolones little, marked decreases in killing of *E. coli* and *S. aureus* may occur (Lewin and Smith 1988a; Lewin *et al.* 1989b). This decreased killing results from oxygen starvation. Killing at low inoculum appears to have an absolute requirement for oxygen.

Non-dividing cells starved by incubation in phosphate-buffered saline are less effectively killed by 4-quinolones than cells in exponential growth (Zeiler and Grohe 1984; Ratcliffe and Smith 1985; Zeiler 1985). With starvation, *E. coli* Neumann is more effectively killed by ciprofloxacin than by norfloxacin.

Increased concentrations of magnesium ions antagonise the bactericidal activity of the 4-quinolones to a greater extent than the bacteriostatic activity (Ratcliffe and Smith 1984).

Differences in Killing Among 4-Quinolones

Three differences in killing activity among 4-quinolones have been identified (Smith 1984). Firstly, addition of chloramphenicol or rifampin completely blocks killing of *E. coli* KL16 by nalidixic acid and norfloxacin but only reduces killing by ciprofloxacin and ofloxacin (Fig. 6.3). Secondly, the time necessary to kill 90% of a population of *E. coli* KL16 by exposure to the most bactericidal concentration is less

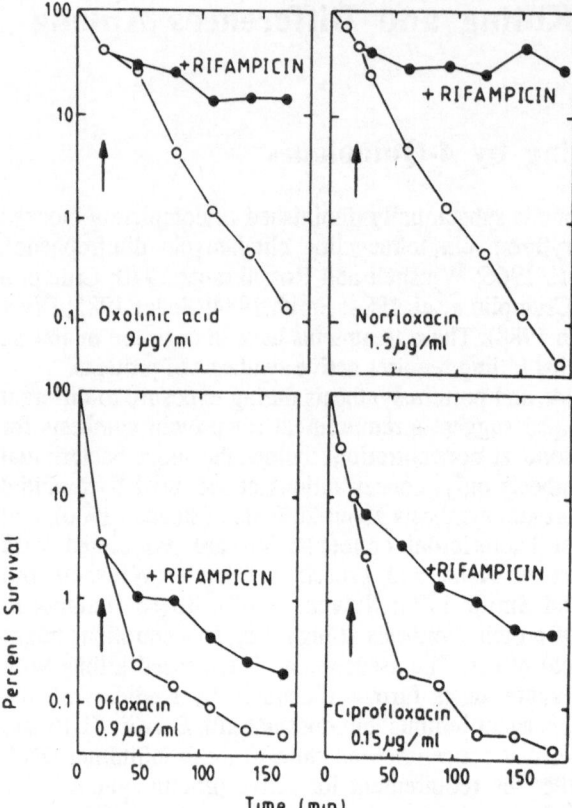

Fig. 6.3. Effect of rifampin on killing of *E. coli* KL16 by 4-quinolones (from Smith 1984, with permission).

for ciprofloxacin and ofloxacin (19 min) than for nalidixic acid and norfloxacin (44-52 min). And thirdly, the paradoxical effect for *E. coli* KL16 is less marked for ciprofloxacin and ofloxacin than for nalidixic acid and norfloxacin (Fig. 6.2).

These differences have been interpreted to define two mechanisms of killing of *E. coli* by 4-quinolones: mechanism A present for all the drugs and an additional mechanism B present for ciprofloxacin and ofloxacin (Smith 1984). Mechanism B is thought to act on either dividing or non-dividing bacteria treated with ciprofloxacin (Zeiler and Grohe 1984; Chalkley and Koornhof 1985). An important recent observation is that killing by ciprofloxacin of *E. coli* KL16 carrying a *gyrA* Nal[R] mutation is completely blocked by chloramphenicol, indicating involvement of the gyrase A subunit (encoded by the *gyrA* gene located at 48 min on the *E. coli* genetic map) in mechanism B (Lewin and Smith 1989b).

A third mechanism, mechanism C, has been defined based on killing of non-dividing bacteria by norfloxacin. Killing by mechanism C is abolished by either chloramphenicol or rifampin but, unlike mechanism B, occurs in phosphate-buffered saline in which bacteria are unable to divide (Ratcliffe and Smith 1985).

Ciprofloxacin and ofloxacin additionally differ in their ability to kill isolates of the Gram-positive bacteria *S. aureus* and *S. warneri* (Lewin and Smith 1987, 1988, 1989a,b; Smith and Lewin 1988). Chloramphenicol completely blocks killing by ciprofloxacin but only reduces killing by ofloxacin. Thus ciprofloxacin kills staphylococci by mechanism A only, while ofloxacin kills by both mechanisms A and B.

Mechanisms of Killing: DNA Gyrase, DNA Synthesis, DNA Damage

Antagonism of DNA Gyrase

Three observations indicate involvement of antagonism of the A subunit of DNA gyrase in bacterial killing by 4-quinolones. Firstly, mutations in the *gyrA* gene causing resistance to 4-quinolones usually result in proportional increases in drug concentrations required to inhibit and kill bacteria (Stevens 1980; Chow *et al*. 1988). Secondly, some mutations in the *gyrA* gene causing 4-quinolone resistance can result in a more selective marked decrease in killing by 4-quinolones, as has been documented for nalidixic acid and *E. coli* MH5 (*gyrA*, Nal^R) (Stevens 1980; Crumplin *et al*. 1984). Interestingly, DNA, RNA, and protein synthesis in MH5 are inhibited by similar drug concentrations, in contrast to the parent KL16, in which 4-quinolones preferentially inhibit of DNA synthesis (Stevens 1980). And thirdly, mutation in the *gyrA* gene eliminates killing mechanism B in *E. coli* KL16 exposed to ciprofloxacin (Lewin and Smith 1989b).

It is important to note that these observations, while indicating involvement of DNA gyrase in killing by 4-quinolones, do not differentiate between direct involvement of the enzyme (e.g. non-repairable cleavage of DNA) or indirect involvement (e.g. as an initial trigger of a cascade of events ultimately leading to bacterial death).

Inhibition of DNA Synthesis

Exposure of bacteria to bactericidal concentrations of 4-quinolones results in preferential inhibition of DNA synthesis (discussed above). Inhibition is rapid (Crumplin *et al*. 1984; Drlica 1984; Courtright *et al*. 1988), indicating direct antagonism of DNA synthesis in the replication fork. Concentrations of 4-quinolones that inhibit DNA synthesis approximate the MBC (Chow *et al*. 1988), suggesting that antagonism of DNA synthesis may be related to the bactericidal effect. In the presence of chloramphenicol concentrations that block killing of bacteria by 4-quinolones, however, nalidixic acid still inhibits DNA synthesis (Deitz *et al*. 1966; Winshell and Rosenkranz 1970), suggesting that inhibition of DNA synthesis alone is not sufficient to produce the bactericidal activity of 4-quinolones.

DNA Damage

Antagonism of purified and likely intracellular DNA gyrase by 4-quinolones stimulates enzyme-mediated cleavage of double-stranded DNA under certain conditions (Gellert 1981; Drlica 1984; Wang 1987). In addition, exposure of *E. coli* to 4-quinolones results in damage to the bacterial chromosome (Deitz *et al*. 1966; Crumplin and Smith 1976; Drlica 1984; Courtright *et al*. 1988), with preferential degradation of newly synthesized DNA in the replication fork (Ramareddy and Reiter 1969; Drlica and Franco 1988). An association between increased killing and increased DNA degradation has been noted (Cook *et al.* 1966). These observations suggest that killing by 4-quinolones might represent introduction of a non-repairable (and thus lethal) lesion into the bacterial chromosome by DNA gyrase, perhaps in the region of the replication fork. In recent studies, however, DNA degradation has been found to occur in the presence of concentrations of chloramphenicol and rifampin that block bactericidal activity, suggesting that DNA degradation alone is not sufficient to explain killing by 4-quinolones (Lewin and Smith 1988b, 1989c).

E. coli recA SOS DNA Repair System

4-quinolones are potent inducers of the *recA* SOS DNA repair system of *E. coli* confirming the DNA-damaging nature of these drugs (Gudas and Pardee 1976; Phillips *et al. 1987*; Piddock and Wise 1987; Courtright *et al.* 1988). Induction of the SOS system by 4-quinolones requires a replication fork (Gudas and Pardee 1976) and the helicase (unwinding) activity of exonuclease V, the product of the *recBC* gene (Chaudhury and Smith 1985). *RecA13* and *recB21 E. coli* mutants, which are unable to induce either the recombination-repair or SOS DNA repair functions of the SOS system, are inhibited by lower concentrations of 4-quinolones than are wild-type isogenic strains (McDaniel *et al.* 1978; Lewin *et al.* 1989a).

The most-bactericidal-concentrations of 4-quinolones are similar to concentrations that maximally induce the SOS system (Phillips *et al.* 1987), suggesting involvement of the SOS system in bacterial killing (Diver and Wise 1986; Phillips *et al.* 1987; Chow *et al.* 1988). An alternative explanation, however, is that induction of SOS proteins decreases at higher drug concentrations because of inhibition of protein synthesis (Lewin *et al.* 1989a).

Recent studies exclude involvement of the SOS system as a mediator of bacterial killing by 4-quinolones. Wild-type cells and a *lexA3* mutant (in which the SOS system is uninducible by 4-quinolones) are killed to similar extents by nalidixic acid (Lewin *et al.* 1989a). In addition, wild-type cells and a *recA430* mutant (which is recombination-repair but not SOS repair proficient) are similarly killed by nalidixic acid, demonstrating involvement of recombination-repair in repair of 4-quinolone-induced DNA damage. These data indicate that the SOS response repairs DNA damage caused by exposure to 4-quinolones rather than contributing to killing directly (Drlica 1984; Lewin *et al.* 1989a).

Mechanisms of Killing: 4-Quinolones and ß-lactams

Mutant Strains Exhibiting Decreased Killing

One approach to dissecting mechanisms of killing by 4-quinolones has been isolation and characterization of bacterial mutants selectively deficient in killing. Mutant genes and strains identified include the *gyrA* allele of *E. coli* MH5 (Steven 1980; Crumplin *et al.* 1984; Sato *et al.* 1986), the *gyrA* (*nalA*) allele of *E. coli* KL16 (Lewin and Smith 1989b), the *oxr-1* (*ox*olinic acid-resistant) and *oxr-2* mutations of *Bacillus subtilis*

(Vazquez-Ramos and Mandelstam 1981), the DS1 and KL500 mutants of *E. coli* KL16 (Wolfson *et al.* 1989a,b,c), and the *hipA* (high persistence) mutants of *E. coli* K-12 (Moyed and Bertrand 1983; Moyed and Broderick 1986; Scherrer and Moyed 1988). *E. coli* MH5 and the *nalA* mutant of *E. coli* KL16 (discussed above) document involvement of the A subunit of DNA gyrase in killing. The other mutations are not located in the *gyrA* gene and will now be considered.

B. subtilis oxr-1 and oxr-2 mutations

The *oxr-1* and *oxr-2* mutations of *B. subtilis* NiI were selected by cycled exposure to oxolinic acid either without, or after, mutagenesis with ethyl methanesulfonate respectively (Vazquez-Ramos and Mandelstam 1981). The two mutations confer low-level resistance to oxolinic acid, are located close to each other on the bacterial chromosome, and are near but not alleles of the *gyrA* and *gyrB* genes of *B. subtilis*. The mechanism of decreased killing by these mutant loci is unknown and warrants further study.

E. coli DS1 and KL500

E. coli DS1 (Wolfson *et al.* 1989a,b,c) is a mutant of *E. coli* KL16 isolated by mutagenesis with nitrosoguanidine followed by cycled exposure to norfloxacin. DS1 is two-fold more resistant to norfloxacin than KL16 but is 1000-fold less effectively killed by exposure to a wide range of drug concentrations. In time-killing studies, exposure of DS1 and KL16 to norfloxacin at 10-fold above their respective MICs results in similar initial rapid killing followed by a slowed phase of killing with DS1 at a titer 1000-fold higher than KL16 (Fig. 6.4). This pattern of kinetics of killing demonstrates that a population of DS1 contains a 1000-fold higher number of persisters than the wild-type strain KL16, and thus DS1 is a *hip* mutant (Moyed and Bertrand 1983). DS1 grows at a rate similar to that of its parent (Fig. 6.4).

DS1 also exhibits reduced killing by other 4-quinolones (ofloxacin and ciprofloxacin) and by antagonists of the B subunit of DNA gyrase (novobiocin and coumermycin). In contrast, DS1 and KL16 are killed to a similar extent by the aminoglycoside gentamicin and the RNA polymerase antagonist rifampin.

An unexpected finding was that DS1 exhibits reduced killing by ß-lactam agents, including ampicillin, with a pattern of high persistence, like 4-quinolones. Decreased killing extends to ß-lactams with preferred binding to different penicilling-binding proteins (PBPs), including cefsulodin (PBP 1a) (Tuomanen and Schwartz 1987), piperacillin (PBP 3) (Botta and Park 1981), and ampicillin, cefoxitin, and imipenem (multiple PBPs) (Spratt 1977; Baron *et al.* 1984; Tuomanen and Schwartz 1987). Thus reduced killing of DS1 by ß-lactams is not associated with binding to a specific PBP.

Because isolation of DS1 involved mutagenesis with nitrosoguanidine, the possibility exists that reduced killing by DNA gyrase antagonists and ß-lactams represents effects of two unrelated mutations. Characterization of strain DS1-2, however, makes this possibility unlikely. DS1-2 is a spontaneous derivative of DS1 killed to an intermediate level between DS1 and KL16 by norfloxacin. DS1-2 is also killed to an intermediate level by ampicillin, thus suggesting the same mutation(s) is altering killing by these two different drugs.

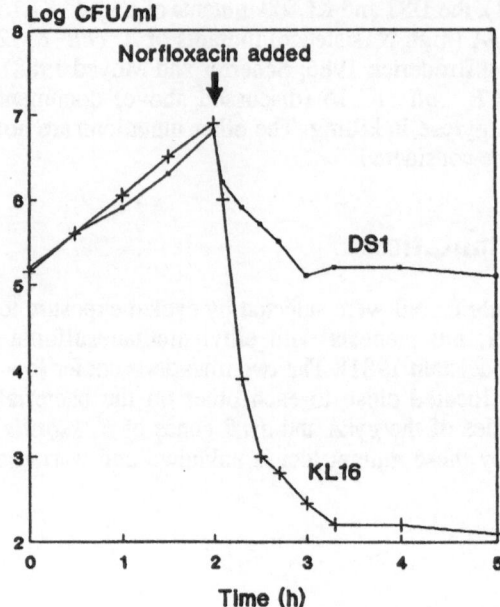

Fig. 6.4. Time-killing study of *Escherichia coli* DS1 and KL16 exposed to 3.0 μg and 1.5 μg norfloxacin/ml at 37°C (ten-fold the respective MICs) (for details, see Wolfson *et al.* 1989c).

Genetic studies were undertaken to identify mutant loci in DS1 responsible for reduced killing by 4-quinolones and ß-lactams. P1 transduction of wild-type DNA into the 2 min region of the DS1 chromosome resulted in increased killing by norfloxacin to or near to wild-type levels in five of 18 transductants (23%), indicating the presence of a mutation(s) in this region. Two transductants exhibiting increased killing by norfloxacin also exhibited increased killing by ampicillin, while two transductants still exhibiting reduced killing by norfloxacin also exhibited reduced killing by ampicillin, confirming an overlap in mechanisms of killing by 4-quinolones and ß-lactams.

E. coli KL500 is a strain isolated without mutagenesis by cycled exposure of *E. coli* KL16 (*dapE*) to norfloxacin and selected for loss of growth requirement for diaminopimelic acid (Wolfson *et al.* 1989b). KL500 exhibits reduced killing by norfloxacin and by ampicillin. KL500 thus represents a spontaneously occurring derivative selected for reduced killing by norfloxacin which also acquired reduced killing by ampicillin, again confirming a relationship between mechanisms of killing of *E. coli* by 4-quinolones and ß-lactams.

E. coli hip A (High Persistence) Mutants

hipA mutants of *E. coli* K-12 were isolated by mutagenesis with ethyl methanesulfonate followed by cycled exposure to ampicillin (Moyed and Bertrand 1983; Moyed and Broderick 1986; Scherrer and Moyed 1988). *hipA* mutants exhibit decreased killing by ampicillin and other inhibitors of murein synthesis (phosphomycin, cycloserine, and starvation for diaminopimelic acid), with kinetics revealing a high persistence pattern. The *hipA* gene is located at 34 min on the *E. coli* map, within the region of the terminus of DNA replication of the *E. coli* chromosome. The *hipA*[+] gene encodes a 50 000 kilodalton protein and is dominant over mutant alleles.

hipA mutants also exhibit decreased killing by exposure to nalidixic acid (Scherrer and Moyed 1988) and to norfloxacin and ofloxacin (Wolfson *et al.* 1989a). *hipA* mutants thus represent an additional example of the close relationship of killing of *E. coli* by 4-quinolones and ß-lactams.

hipA mutants additionally exhibit a reversible cold-sensitive block in cell division and reduced killing by thymine starvation and by elevated temperature in strains containing thermosensitive mutations in DNA replication genes (Scherrer and Moyed 1988). These pleiotropic effects suggest involvement of the *hipA* protein in a cell division step which, when mutant, results in a bacteriostatic rather than bactericidal effect when peptidoglycan or DNA synthesis is interrupted (Scherrer and Moyed 1988).

E. coli Cell Division, 4-quinolones, and ß-lactams

Reduced killing of DS1, KL500, and mutants by both 4-quinolones and ß-lactams indicates overlap in a mechanism(s) of killing by these two classes of drugs (Scherrer and Moyed 1988; Wolfson *et al.* 1989a,b). Genetics of *E. coli*, morphologic effects of drug exposure (discussed below), and characteristics of *hipA* mutants mentioned suggest cell division as a candidate common site (or pathway) for killing (Donachie and Robinson 1987; Scherrer and Moyed 1988), as follows.

Exposure of *E. coli* to 4-quinolones damages DNA, resulting in exonuclease V-dependent activation of the *recA* protein, which inactivates the *lexA* repressor (Gottesman 1984; Walker 1984). A consequence is induction of the *sulA* (*sfiA*) gene (21 min), the product of which antagonizes the product of the *sulB* (*sfiB*) gene (2 min), which is required for an early step in septum formation (Donachie and Robinson 1987). These events are responsible at least in part for growth of *E. coli* in aseptate filaments in the presence of a 4-quinolone.

Exposure of *E. coli* to ß-lactams results in the binding of drug to PBPs, including PBP 3, which is encoded by the *ftsI* (Filamentation Temperature-Sensitive) gene. *ftsI* is one of the 15 genes in the "major morphogene cluster" at 2 min on the *E. coli* chromosome (Donachie and Robinson 1987; Donachie *et al.* 1984). This cluster also includes *ftsQ*, *ftsA*, and *ftsZ*, the functions of which are required for septation (Donachie and Robinson 1987).

Importantly, *ftsZ* and *sulB* are alleles of the same gene (Lutkenhaus 1983). *ftsZ* (*sulB*) thus represents a common point in two sets of otherwise unrelated genes, one set affected by 4-quinolones and the other by ß-lactams, raising the possibility of involvement of the *ftsZ* gene product in some way in killing by these two classes of drugs. Whether the mutation(s) in the 2 min region of the DS1 chromosome involves a known cell division gene is as yet unknown.

Morphological Changes Associated with Killing

Exposure of *E. coli* to a 4-quinolone induces formation of filaments (Goss 1964; Crumplin *et al.* 1984; Smith 1984; Dougherty and Saukkonen 1985; Voigt and Zeiler 1985; Diver and Wise 1986; Elliott *et al.* 1987) which lack septae (Walker and Pardee 1968; Walker *et al.* 1973; Begg and Donachie 1985). For norfloxacin or ciprofloxacin, filamentation begins by 30 min and continues for two to four hours, followed by formation of vacuoles predominantly at the poles of cells, breaks in the cell wall at these sites, and lysis (Elliott *et al*. 1987). Under these conditions, decrease in viability occurs primarily during the first 30 to 90 min of drug exposure. Similar findings have been documented for exposure of *E. coli* in pooled active human serum to ciprofloxacin (Voigt and Zeiler 1985).

Exposure of *E. coli* to a ß-lactam such as piperacillin which binds preferentially to PBP 3 (Botta and Park 1981) also causes formation of aseptate filaments, indicating overlap in morphological effects of ß-lactams and 4-quinolones on bacterial cells. Exposure of *E. coli* to the most bactericidal concentration of ciprofloxacin, however, results in less filamention than at lower, less bactericidal drug concentrations (Diver and Wise 1986). Thus the exact relationship of filamentation to the bactericidal effect of quinolones remains unclear and warrants further study.

Conclusions

Summary

Despite a large body of information on the effects of 4-quinolones on bacteria and on DNA gyrase, relatively little is known about the mechanisms by which these agents kill bacteria. Time-killing studies reveal that cell death is rapid and extensive but not complete, with a population of persisters remaining. Killing can be paradoxically less at high drug concentrations and requires oxygen. Bactericidal activity is blocked or reduced by treatments which inhibit protein synthesis, and differences in the extent of antagonism of killing by chloramphenicol have been used to deduce existence of three mechanisms of killing. The roles of DNA damage, inhibition of DNA synthesis, and induction of the SOS system in cell death have not been demonstrated convincingly and are controversial.

Recent insight into mechanisms has come from studies of mutant cells exhibiting reduced killing by 4-quinolones. Several mutations have been described in the *gyrA* gene, indicating involvement of DNA gyrase in the killing phenomenon. It is unknown, however, whether antagonism of DNA gyrase directly kills cells, represents an early step in a cascade of events which ultimately results in cell death, or both.

Mutants of *E. coli* have also been isolated which exhibit reduced killing by both 4-quinolones and ß-lactam agents, indicating overlap either in a pathway of killing by these two classes of agents or alternatively in an otherwise undefined protection system. Studies of these mutants and knowledge of the genetics of *E. coli* suggest cell division as a possible common target site. Whether autolysins are involved in the killing mechanism is unknown.

Hypotheses for Mechanisms for Killing by 4-Quinolone

On the basis of the observations presented in this review, three general hypotheses for mechanisms of bacterial killing by 4-quinolones have been proposed. One hypothesis is that DNA gyrase introduces a lethal lesion, such as a non-repairable break, into DNA. Thus DNA gyrase itself could be the protein whose synthesis is required for killing, an intriguing possibility in that exposure of cells to nalidixic acid induces synthesis of the enzyme (Menzel and Gellert 1983). A related hypothesis is that 4-quinolones cause DNA gyrase to damage DNA and that another protein, such as a DNA repair enzyme, is synthesized and introduces a lethal lesion in attempting to repair the damage (Cook 1966; Drlica 1984; Smith 1984; Diver and Wise 1986; Phillips *et al.* 1987; Piddock *et al.* 1987). This possibility seems unlikely, at least for SOS repair enzymes, because data indicate involvement of SOS functions in repair of damage and rather than contributing to killing (Drlica 1984; Lewin *et al.* 1989a).

A second hypothesis is that inhibition of DNA synthesis is responsible for killing by 4-quinolones. This possibility is in question, however, because nalidixic acid inhibits DNA synthesis in the presence of concentrations of chloramphenicol which block bactericidal activity of the 4-quinolone (Deitz *et al.* 1966; Winshell and Rosenkrantz 1970).

A third hypothesis is that 4-quinolones kill bacteria by activation of a generalized killing system. Mutants exhibiting decreased killing by 4-quinolones and ß-lactam agents might represent mutants altered in such a system. Whether such a system involves interference with cell division, induction of autolysins, or other processes is unknown.

Thus the exact mechanisms of killing of bacteria by 4-quinolones are not well understood, and studies are needed to define which of the above or other mechanisms are responsible for the potent bactericidal action of these drugs.

A Comment on Clinical Importance

The clinical importance of the bactericidal activity of 4-quinolones currently is an unanswered and important question. For ß-lactams, data from clinical and animal studies strongly suggest that decreased killing is associated with unfavorable outcome for certain streptococcal infections and perhaps staphylococcal infections (Handwerger and Tomasz 1985; Sherris 1986; Tuomanen *et al.* 1986). For 4-quinolones, the data are sparse. Reduced killing has been raised as possibly contributing to therapeutic failures in treatment of urinary tract infections caused by *S. saprophyticus* (Garlando *et al.* 1987). Also ciprofloxacin-resistant isolates of *E. coli* have been documented to be deficient in killing mechanism B (Lewin and Smith 1989b). It would seem worthwhile to evaluate non-resistant bacterial isolates from patients failing therapy with 4-quinolones for reduced killing *in vitro*.

Acknowledgments

Experiments presented in Figs. 6.1 and 6.4 were done by David J. Shih and by Gail L. McHugh respectively. Studies with DS1 and KL500 were supported in part by U.S. Public Health Service grant AI23988 from the National Institutes of Health and a grant from Ortho Pharmaceutical Corporation.

References

Baron P, Michelot D, Masson JM, Labia R (1984) Comparison of three ^{125}I-radiolabelled penicillins in PBP studies. Drug Exptl Clin Res 10:1-4

Begg KJ, Donachie WD (1985) Cell shape and division in *Escherichia coli*: experiments with shape and division mutants. J Bacteriol 163:615-622

Bigger JW (1944) Treatment of staphylococcal infections with penicillin by intermittent sterilisation. Lancet ii:497-500

Botta GA, Park JT (1981) Evidence for involvement of penicillin-binding protein 3 in murein synthesis during septation but not during cell elongation. J Bacteriol 145:333-340

Chalkley LJ, Koornhof HJ (1985) Antimicrobial activity of ciprofloxacin against *Pseudomonas aeruginosa*, *Escherichia coli* and *Staphylococcus aureus* determined by the killing curve method: antibiotic comparisons and synergistic interactions. Antimicrob Agents Chemother 28:331-342

Chaudhury AM, Smith GR (1985) Role of *Escherichia coli RecBC* enzyme in SOS induction. Mol General Genetics 201:525-528

Chow RT, Dougherty TJ, Fraimow HS, Bellin EY, Miller MH (1988) Association between early inhibition of DNA synthesis and the MICs and MBCs of carboxyquinolone antimicrobial agents for wild-type and mutant [*gyrA nfxB (ompF*) *acrA*] *Escherichia coli* K-12. Antimicrob Agents Chemother 32:1113-1118

Cook TM, Deitz WH, Goss WA (1966) Mechanism of action of nalidixic acid on *Escherichia coli*. IV. Effects on the stability of cellular constituents. J Bacteriol 91:774-779

Courtright JB, Turowski DA, Sonstein SA (1988) Alteration of bacterial DNA structure, gene expression, and plasmid encoded antibiotic resistance following exposure to enoxacin. J Antimicrob Chemother 21(Suppl B):1-18

Cozzarelli NR (1980) DNA gyrase and supercoiling of DNA. Science 207:953-960

Crumplin GC, Smith JT (1975) Nalidixic acid: an antibacterial paradox. Antimicrob Agents Chemother 8:251-261

Crumplin GC, Smith JT (1976) Nalidixic acid and bacterial chromosome replication. Nature 260:643-645

Crumplin GC, Kenwright M, Hirst T (1984) Investigation into the mechanism of action of the antibacterial agent norfloxacin. J Antimicrob Chemother 13(Suppl B):9-23

Deitz WH, Cook TM, Goss WA (1966) Mechanism of action of nalidixic acid on *Escherichia coli*. III. Conditions required for lethality. J Bacteriol 91:768-773

Diver JM, Wise R (1986) Morphological and biochemical changes in *Escherichia coli* after exposure to ciprofloxacin. J Antimicrob Chemother 18(Suppl D):31-41

Donachie W, Robinson A (1987) Cell division: parameter values and the process. In: Neidhardt F (ed), *Escherichia coli* and *Salmonella typhimurium*, American Society for Microbiology, Washington, DC, pp 1578-1593

Donachie WD, Begg JK, Sullivan NF (1984) Morphogenes of *Escherichia coli*. In: Losick R, Shapiro L (ed), Microbial development, Cold Spring Harbor Laboratory, Cold Spring Harbor, NY, pp 27-62

Dougherty TJ, Saukkonen JJ (1985) Membrane permeability changes associated with DNA gyrase inhibitors in *Escherichia coli*. Antimicrob Agents Chemother 28:200-206

Drlica K (1984) Biology of bacterial DNA topoisomerases. Microbiol Rev 48:273-289

Drlica K, Franco RJ (1988) Inhibitors of DNA topoisomerases. Biochemistry 27:2253-2259

Eagle H, Musselman AD (1948) The rate of bactericidal action of penicillin *in vitro* as a function of its concentration, and its paradoxically reduced activity at high concentrations against certain organisms. J Exp Med 88:99-131

Elliot TSJ, Shelton A, Greenwood D (1987) The response of *Escherichia coli* to ciprofloxacin and norfloxacin. J Med Microbiol 23:83-88

Garlando F, Rietiker S, Tauber MG, Flepp M, Meier B, Luthy R (1987) Single-dose ciprofloxacin at 100 versus 250 mg for treatment of uncomplicated urinary tract infections in women. Antimicrob Agents Chemother 31:354-356

Gellert M (1981) DNA topoisomerases. Ann Rev Biochem 50:879-910

Goss W, Deitz WH, Cook TM (1964) Mechanism of action of nalidixic acid on *Escherichia coli*. J Bacteriol 88:1112-1118

Gottesman S (1984) Bacterial regulation: global regulatory networks. Annu Rev Genetics 18:415-441

Greenwood D (1985) Phenotypic resistance to antimicrobial agents. J Antimicrob Chemother 15:653-658 (leading article)

Gudas LJ, Pardee AB (1976) DNA synthesis inhibition and the induction of protein X in *Escherichia coli*. J Molec Biol 101:459-477

Handwerger S, Tomasz A (1985) Antibiotic tolerance among clinical isolates of bacteria. Rev Infect Dis 7:368-386

Hoffmann B, Messer W, Schwarz U (1972) Regulation of polar cap formation in the life cycle of *Escherichia coli*. J Supramolecular Structure 1:29-37

Holzhoffer S, Sussmuth R, Haag R (1985) Oscillating tolerance in synchronized cultures of *Staphylococcus aureus*. Antimicrob Agents Chemother 28:456-457

Lewin CS, Smith JT (1987) Ciprofloxacin does not exhibit mechanism B against *Staphylococcus albus*. J Pharmacy Pharmacology 39:21P

Lewin CS, Smith JT (1988a) Bactericidal mechanisms of ofloxacin. J Antimicrob Chemother 22(Suppl C):1-8

Lewin CS, Smith JT (1988b) DNA breakdown and its significance. 28th Interscience Conference on Antimicrobial Agents and Chemotherapy, American Society for Microbiology, Los Angeles, California, Abstract #541

Lewin CS, Smith JT (1989a) Bactericidal activity of ciprofloxacin against *Staphylococcus aureus*. J Antimicrob Chemother 24:77-78 (letter)

Lewin CS, Smith JT (1989b) Loss of ciprofloxacin's second killing action by mutation. 29th Interscience Conference on Antimicrobial Agents and Chemotherapy, American Society for Microbiology, Houston, Texas, Abstract #142

Lewin CS, Smith JT (1989c) DNA breakdown by the 4-quinolones and its significance. J Med Microbiol, in press

Lewin CS, Howard BMA, Ratcliffe NT, Smith JT (1989a) 4-Quinolones and the SOS response. J Med Microbiol 29:139-144

Lewin CS, Morrissey I, Smith JT (1989b) Role of oxygen in bactericidal action of the 4-quinolones. Rev Infect Dis 11(Suppl 5):S913-S914 (extended abstract)

Lutkenhaus JF (1983) Coupling of DNA replication and cell division: *sulB* is an allele of *ftsZ*. J Bacteriol 154:1339-1346

Mathison GE (1968) Kinetics of death induced by penicillin and chloramphenicol in synchronous cultures of *Escherichia coli*. Nature 219:405-407

McDaniel LS, Rogers LH, Hill WE. Survival of recombination-deficient mutants of *Escherichia coli* during incubation with nalidixic acid. J Bacteriol 134:1195-1198

Menzel R, Gellert M (1983) Regulation of the genes of *E. coli* DNA gyrase: homeostatic control of DNA supercoiling. Cell 34:105-113

Moyed HS, Bertrand KP (1983) *hipA*, a newly recognized gene of *Escherichia coli* K-12 that affects frequency of persistence after inhibition of murein synthesis. J Bacteriol 155:768-77

Moyed HS, Broderick SH (1986) Molecular cloning and expression of *hipA*, a gene of *Escherichia coli* K-12 that affects frequency of persistence after inhibition of murein synthesis. J Bacteriol 166:399-403

Phillips I, Culebras E, Moreno F, Baquero F (1987) Induction of the SOS response by new 4-quinolones. J Antimicrob Chemother 20:631-638

Piddock LJV, Wise R (1987) Induction of the SOS response in *Escherichia coli* by 4-quinolone antimicrobial agents. FEMS Microbiol Lett 41:289-294

Ramareddy G, Reiter H (1969) Specific loss of newly replicated DNA in nalidixic acid-treated *Bacillus subtilis* 168. J Bacteriol 100:724-729

Ratcliffe NT, Smith JT (1984) The mechanism of reduced activity of 4-quinolone agents in urine. In: Adam D, Stille W, Ruckdeschel G, Knothe H, Lode H, Eikenberg H-V (eds) Gyrase Hammer, Forschritte der antimikrobiellen und antineoplastischen Chemotherapie FAC 3-5, Futuramed Verlage, Munchen, pp 563-569

Ratcliffe NT, Smith JT (1985) Norfloxacin has a novel bactericidal mechanism unrelated to that of other 4-quinolones. J Pharmacy Pharmacology 37:92P

Sabath LD, Wheeler N, Laverdiere M, Blazevic D, Wilkinson BJ (1979) A new type of pencillin resistance in *Staphylococcus aureus*. Lancet i:443-447

Sato K, Inoue Y, Fujii T, Aoyama H, Mistuhashi S (1986) Antibacterial activity of ofloxacin and its mode of action. Infection 14 (Suppl 4):S226-S230

Scherrer R, Moyed HS (1988) Conditional impairment of cell division and altered lethality in *hipA* mutants of *Escherichia coli* K-12. J Bacteriol 170:3321-3326

Sherris JC (1986) Problems in *in vitro* determination of antibiotic tolerance in clinical isolates. Antimicrob Agents Chemother 30:633-637

Smith JT (1984) Awakening the slumbering potential of the 4-quinolone antibacterials. Pharmaceutical J 233:299-305

Smith JT, Lewin CS (1988) Chemistry and mechanisms of action of the quinolone antibacterials. In: Andriole VT (ed) The quinolones, Academic Press, London, England, pp 23-82

Spratt BG (1977) Properties of penicillin-binding proteins of *Escherichia coli* K-12. Eur J Biochem 72:341-352

Stevens PJE (1980) Bactericidal effect against *Escherichia coli* of nalidixic acid and four structurally related compounds. J Antimicrob Chemother 6:535-542

Tuomanen E, Schwartz J (1987) Penicillin-binding protein 7 and its relationship to lysis of nongrowing *Escherichia coli*. J Bacteriol 169:4912-4915

Tuomanen E, Durack DT, Tomasz A (1986) Antibiotic tolerance among clinical isolates of bacteria. Antimicrob Agents Chemother 30:521-527

Vazquez-Ramos JM, Mandelstam J (1981) Oxolinic acid-resistant mutations of *Bacillus subtilis*. J General Microbiol 127:1-9

Voigt W-H, Zeiler H-J (1985) Influence of ciprofloxacin on the ultrastructure of Gram-negative and Gram-positive bacteria. Arzneim-Forsch 35:1601-1603

Walker GC (1984) Mutagenesis and inducible responses to DNA damage in *Escherichia coli*. Microbiol Rev 48:60-93

Walker JR, Pardee AB (1968) Evidence for a relationship between DNA metabolism and septum formation in *Escherichia coli*. J Bacteriol 95:123-131

Walker JR, Ussery CL, Allen JS (1973) Bacterial cell division regulation: lysogenization of conditional cell division *lon* mutants of *Escherichia coli* by bacteriophage λ. J Bacteriol 113:1326-1332

Wang JC (1987) Recent studies of DNA topoisomerases. Biochim Biophys Acta 909:1-9

Winshell EB, Rosenkranz HS (1970) Nalidixic acid and the metabolism of *Escherichia coli*. J Bacteriol 104:1168-1175

Wolfson JS, Hooper DC, McHugh GL, Bozza MA, Shih DJ, Swartz MN (1989a) Co-tolerance of *Escherichia coli* mutants to quinolone and ß-lactam agents. Annual Meeting of the American Society for Microbiology, New Orleans, Louisiana, Abstract #A82

Wolfson JS, Hooper DC, McHugh GL, Bozza MA, Swartz MN (1989b) Genetic studies of killing of *Escherichia coli* by quinolones and ß-lactams. 29th Interscience Conference on Antimicrobial Agents and Chemotherapy, American Society for Microbiology, Houston, Texas, Abstract #141

Wolfson JS, Hooper DC, Shih DJ, McHugh GL, Swartz MN (1989c) Isolation and characterization of an *Escherichia coli* strain exhibiting partial tolerance to quinolones. Antimicrob Agents Chemother 33:705-709

Wolfson JS, Hooper DC, Swartz MN (1989d) Mechanisms of action of, and resistance to, quinolone antimicrobial agents. In: Wolfson JS, Hooper DC (eds) The quinolone antimicrobial agents. American Society for Microbiology, Washington, DC, pp 5-34

Woolfrey BE, Lally RT, Ederer MN (1985) Evaluation of oxacillin tolerance in *Staphylococcus aureus* by a novel method. Antimicrob Agents Chemother 28:381-388

Zeiler H-J (1985) Evaluation of the *in vitro* bactericidal action of ciprofloxacin on cells of *Escherichia coli* in the logarithmic and stationary phases of growth. Antimicrob Agents Chemother 28:524-527

Zeiler H-J, Grohe K (1984) The *in vitro* and *in vivo* activity of ciprofloxacin. Eur J Clin Microbiol 3:339-343

Discussion Summary

The attempted selection of drug-tolerant mutants as a means of analysing the mechanism of killing of susceptible cells by the 4-quinolones was regarded as an ideal approach to what is a long-standing problem. However, it was evident from the comments and questions from a number of participants that the definition of a non-viable (or dead) cell was of prime importance in dealing with this whole area of the effects of 4-quinolones. It was understandably difficult to accept that a cell which was non-viable, in terms of its ability to reproduce and generate a colony on subsequent cultivation, was able to continue growing and actively producing proteins. Equally problematical was the attempt to distinguish between drug-tolerant cells and true "persisters". Here the general concensus seemed to be that whilst persisters represented a minority population of cells, which did not manage to die during the course of the drug challenge (but were intrinsically capable of doing so), tolerant cells represented populations which intrinsically had a reduced ability to complete the death processes induced by a drug challenge.

Comments indicated that this presentation also represented a significant step towards altering our perception of the killing of susceptible bacterial cells by the 4-quinolones in that it could be accepted that we are unable at present to determine the specific lethal event. It was concluded that in its own way the mechanism of killing, consequent upon the initial effects upon DNA gyrase and DNA, represented an area which is likely to prove as problematical to the scientist as was the initial search for the intracellular target of the 4-quinolones (DNA gyrase).

Chapter 7

Use of an In Vitro DNA Strand-Breakage Assay to Monitor Compound Interactions with DNA Gyrase

L. P. Elwell, L. M. Walton, J. M. Besterman and A. Hudson

Introduction

Topoisomerases are enzymes that regulate the superhelical density of DNA by transiently nicking either one (type I) or both (type II) strands of the DNA helix (Drlica 1984; Wang 1985; Osheroff 1989). DNA gyrase is a type II topoisomerase which catalyzes the ATP-dependent negative supercoiling of closed circular duplex DNA and is an essential enzyme in *Escherichia coli* (Gellert 1981). Gyrase actively maintains the supercoiled state of bacterial DNA and is involved in DNA replication, transcription, and recombination (Cozzarelli 1980). *Escherichia coli* DNA gyrase is an A_2B_2 tetramer; the A subunits mediate DNA breakage and rejoining, while the B subunits bind ATP and participate in energy transduction (Gellert 1981). The subunit A protein is the target of the quinolone family of antibacterial agents, and the newer fluoroquinolones, such as norfloxacin and ciprofloxacin, strongly inhibit the catalytic (strand-passing) activity of DNA gyrase. In the presence of oxolinic acid (as well as the newer 4-quinolone antibacterial agents), DNA gyrase forms a complex with DNA, which can be activated by treatment with detergent (e.g. SDS) to produce double-strand breaks in DNA. Detailed analyses of the broken complex have revealed that the break is a four-base stagger and that a *gyrA* subunit is covalently linked to each 5' protruding end of the break (Morrison and Cozzarelli 1979; Fisher *et al.* 1981). The formation of this complex *in vivo* in the presence of drugs like norfloxacin presumably poisons DNA gyrase on the DNA template and thus inhibits cell growth (Higgins and Cozzarelli 1982). Why drug-trapping of these presumed abortive topoisomerase II:DNA complexes leads to rapid cell death and low incidence of drug resistance is far from clear (Wolfson and Hooper 1985).

An analogous situation exists in the case of certain antitumor drugs and the eukaryotic counterpart of DNA gyrase; DNA topoisomerase II. Using purified mammalian topoisomerase II, it was found that intercalative antitumor drugs like m-AMSA, adriamycin, ellipticine and the nonintercalative epipodophyllotoxins, (e.g. teniposide and etoposide), interfere with the breakage and rejoining reaction of topoisomerase II due to the reversible formation of the so-called cleavable-complex (Nelson et al. 1984; Chen et al. 1984; Ross et al. 1984; Rowe et al. 1985). Structure-activity studies of drugs from the same chemical class showed a strong correlation between cytotoxicity and the ability of the drug to induce the in vitro cleavable-complex (Rowe et al. 1985; Long 1984).

Inhibition of Catalytic (Strand-passing) Activity

The quinolone antibacterial agents perturb several gyrase-mediated functions, including supercoiling, decatenation and unknotting (Wang 1985; Wolfson and Hooper 1985; Wang 1987). Therefore, one of the most frequently used experimental approaches to monitor the effects of quinolones on their intracellular target, DNA gyrase, has been the supercoiling-inhibition assay (Fu et al. 1986; Sato et al. 1986; Zweerink and Edison 1986; Imamura et al. 1987; Inoue et al. 1987; Takahata and Nishino 1988; Hoshino et al. 1989). Not surprisingly, the potency of the quinolones as DNA gyrase inhibitors does not always correlate with antimicrobial potency. Zweerink and Edison (1986), for example, found that ofloxacin and amifloxacin differ only two fold in their ability to inhibit Micrococcus luteus gyrase-mediated supercoiling, yet ofloxacin was 60-fold more potent than amifloxacin against intact cells. Furthermore, cell permeability differences were not sufficiently different to account for the observed discrepancy. The imperfect correlation between antimicrobial potency and DNA gyrase inhibitory activity among the 4-quinolones has its eukaryotic counterpart. For example, the aminoacridine isomers o-AMSA and m-AMSA are virtually equipotent with respect to inhibition of catalytic activity, yet m-AMSA is far more active with respect to stimulating DNA cleavage and producing cytotoxicity than is the ortho isomer (Nelson et al. 1984). Ethidium bromide is another example. This avid DNA intercalator does not induce the topoisomerase II-mediated cleavable complex, but it does inhibit the catalytic (strand-passing) activity of the enzyme (Tewey et al. 1984b). In addition, ethidium bromide is much less potent with respect to cytotoxicity than other DNA intercalators.

In addition to generating the in vitro cleavable-complex, biologically active epipodophyllotoxins (e.g. teniposide and etoposide) inhibit the catalytic activity of mammalian topoisomerase II in vitro (Chen et al. 1984). However, Rowe et al. (1985) screened a series of epipodophyllotoxin analogs and found that the ability to generate the cleavable-complex correlates better with cytotoxicity than does the ability to inhibit the strand-passing activity of purified enzyme. Hence, these workers consider cleavable-complex formation to be the "primary" event in the antitumor effect of active epipodophyllotoxin congeners.

DNA Breakage as an Indicator of Quinolone Potency

If inhibition of the catalytic activity of DNA gyrase is not a particularly good indicator of a compound's antibacterial potency, what is? Zweerink and Edison (1986) suggest

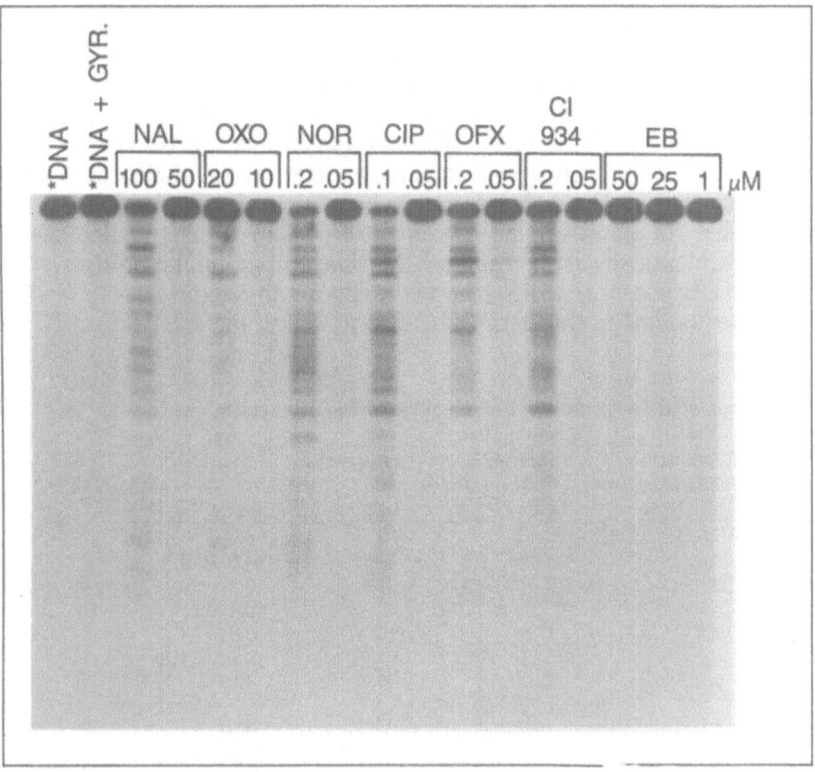

Fig. 7.1. DNA breakage by various quinolones mediated by *E. coli* DNA gyrase. Assays for DNA gyrase-mediated DNA cleavage were performed as described by Walton and Elwell (1988). Lanes *DNA and *DNA + GYR do not contain drug. Incubation of DNA gyrase plus ^{32}P-end-labeled pBR322 DNA was done in the presence of various quinolones at the concentrations (μM) indicated. NAL, nalidixic acid; OXO, oxolinic acid; NOR, norfloxacin; CIP, ciprofloxacin; OFX, ofloxacin; CI 934; EB, ethidium bromide.

that other assays, e.g. DNA replication, phage replication or DNA strand-breakage might provide a more reliable indication of the effects of these drugs on critical intracellular events.

Domagala *et al.* (1986) included a nonradiolabeled DNA-cleavage assay in their systematic analysis of 60 quinolone antibacterials; some novel and some well characterized. In a similar vein, Barrett *et al.* (1989) studied drug-induced DNA cleavage in a quantitative non-radiolabeled cleavage assay to determine the extent to which various quinolones interacted with eukaryotic topoisomerase II. Walton and Elwell (1988) examined six quinolones in an *in vitro* cleavable-complex assay using uniquely ^{32}P-end-labeled, linear DNA as substrate. The results of this analysis as well as the results of a more extensive survey of some novel heterocyclic carboxylic acids are presented below.

Figure 7.1. depicts the results of a representative *E. coli* DNA gyrase-mediated cleavable-complex assay and shows the extent of DNA breakage induced by various quinolones. Each drug is represented by two lanes; the concentration in the left lane represents the lowest concentration of each drug that generated a cleavable complex, whereas the concentration on the right is a no-effect concentration for that particular drug. Ethidium bromide is unable to generate the cleavable-complex in the presence of mammalian topoisomerase II (Tewey *et al.* 1984b) and is equally inactive in the gyrase assay (Fig. 7.1). The DNA fragment patterns elicited by the six quinolones are similar if not identical. This is consistent with previous studies on the interaction of certain intercalative antitumor drugs with mammalian DNA topoisomerase II in which drugs of the same chemical class were shown to stimulate cleavage at similar sites while drugs of different chemical classes showed strikingly different cleavage patterns (Tewey *et al.* 1984a). Table 7.1 shows that the rank-order of potency for these quinolones in the cleavable-complex assay is the same as the supercoiling-inhibition assay. The primary difference is one of sensitivity in that it appears to require two- to ten-fold lower drug concentrations to fragment DNA than to inhibit the catalytic activity of gyrase.

Table 7.1 Comparison of DNA gyrase results with growth inhibition against six clinical isolates of *E. coli* .

Compound/Drug	Supercoiling Inhibition Assay (µM)	Cleavable Complex Assay (µM)	Avg. MIC[a] (µM)	Avg. MIC[a] (µg ml^{-1})
Nalidixic Acid	100	50	13	3.0 (1.0-10.0)[b]
Oxolinic Acid	20	10	1.3	0.3 (0.1-0.6)
CI-934	2.0	0.2	0.10	0.04 (0.03-0.10)
Ofloxacin	2.0	0.2	0.08	0.03
Norfloxacin	1.0	0.2	0.06	0.02 (0.01-0.03)
Ciprofloxacin	1.0	0.1	0.009	0.003 (0.001.0.004)

[a] MIC results for 6 *E. coli* clinical isolates were simply averaged.
[b] Range of MICs for the 6 *E. coli* clinical strains.

Table 7.1 compares the minimum drug concentrations required to elicit an effect in the gyrase assays versus their MICs against three clinical strains of *E. coli* . Of the six quinolones tested, ciprofloxacin and norfloxacin were the most active in the *in vitro* enzyme assays and were the most potent inhibitors of cell growth. This generality appears to hold throughout this series of drugs. However, the values are not always directly proportional; for example, norfloxacin and ciprofloxacin were essentially equipotent in the gyrase assays and yet ciprofloxacin was about seven fold more active against intact cells. This nonproportionality has been noted by others (Domagala *et al.* 1986; Zweerink and Edison 1986) and although differences in cell permeation has been invoked as a possible explanation, this is clearly not the reason in every situation (Zweerink and Edison1986).

The most notable nonproportionality can be seen between the inhibitory concentrations for gyrase activity and cell growth with some of the fluoroquinolone antibacterial agents tested. For example, there are 25- and 100-fold-greater concentration requirements for gyrase inhibition than for cell growth inhibition for ofloxacin and ciprofloxacin, respectively (Table 7.1). Previous studies have shown that the concentration of quinolones required to inhibit the catalytic activity of gyrase is substantially higher than that required to inhibit the growth of the organism from which the gyrase was purified, including *Pseudomonas* spp. (Miller and Scurlock1983),

M. luteus (Fu *et al.* 1986; Zweerink and Edison 1986) and *E. coli* (Sugino *et al.* 1977; Sato *et al.* 1986). In addition, Gellert *et al.* (1977) reported that it required 50 times the concentration of oxolinic acid to inhibit the DNA gyrase-mediated supercoiling reaction than to inhibit cell growth. Thus, Gellert *et al.* (1977) suggested the possibility of a more subtle interaction between these inhibitors and DNA gyrase than mere enzyme inhibition/titration. The possible nature of the drug:gyrase interaction will be discussed in more detail later.

Table 7.2. Compounds tested for their activities against DNA gyrase and three clinical isolates of *E. coli* .

Compound	ChemicalName
Norfl-oxacin	1-Ethyl-6-fluoro-1,4-dihydro-4-oxo-7-(1-piperazinyl)-quinoline-3-carboxylic acid
A	1,4-Dihydro-4-oxo-5-phenyl-1-(4-pyridyl) pyridine-3-carboxylic acid
B	4,10-Dihydro-4-oxo-pyrimido [1,2-a] benzimidazole-3-carboxylic acid
C	1-(4-Dimethylaminophenyl)-4-oxo-6-phenyl-1,4,5,6-tetrahydronicotinic acid
E	1-(2,4-Dimethoxyphenyl)-4-oxo-6-phenyl-1,4,5,6-tetrahydronicotinic acid
G	6-(4-Dimethylaminophenyl)-1-(4-methoxy-2-methylphenyl)-4-oxo-1, 4, 5,6-tetrahydronicotinic acid
H	1-(2,4-Dichlorophenyl)-4-oxo-6-phenyl-1,4,5,6-tetrahydronicotinic acid
I	1-(4-Hydroxyphenyl)-4-oxo-6-phenyl-1,4,5,6-tetrahydronicotinic acid
K	6-(4-Dimethylaminophenyl)-1-(3-hydroxy-2-methylphenyl)-4-oxo-1, 4,5,6-tetrahydronicotinic acid
M	1-(4-Hydroxy-2-methylphenyl)-5-methyl-4-oxo-6-phenyl-1,4,5,6- tetrahydronicotinic acid
N	1-(4-Hydroxy-2-methylphenyl)-5-methyl-4-oxo-6-phenyl-1H-nicotinicacid
O	6-(4-Dimethylaminophenyl)-1-(4-hydroxy-2-methoxyphenyl)-4-oxo-1, 4,5,6-tetrahydronicotinic acid
P	6-(4-Dimethylaminophenyl)-1-(3-hydroxy-6-methylphenyl)-4-oxo-1, 4,5,6-tetrahydronicotinic acid
Q	6-(4-Dimethylaminophenyl)-1-(2,6-dimethyl-4-hydroxyphenyl)-4-oxo- 1,4,5,6-tetrahydronicotinic acid
R	6-(4-Dimethylaminophenyl)-1-(2-fluoro-4-hydroxyphenyl)-4-oxo-1,4,5,6-tetrahydronicotinic acid
S	1-(4-Hydroxy-2-methylphenyl)-4-oxo-6-phenyl-1,4,5,6-tetrahydro- nicotinic acid
T	6-(4-Dimethylaminophenyl)-1-(2-ethyl-4-hydroxyphenyl)-4-oxo-1,4,5, 6-tetrahydronicotinic acid
U	1-(4-Hydroxy-2-methylphenyl)-6-phenyl-4 (1H)-pyridinone-3- carboxylic acid
V	6-(4-Dimethylaminophenyl)-1-(4-hydroxy-2-methylphenyl)-4-oxo-1,4, 5,6-tetrahydronicotinic acid
W	1-(2-Chloro-4-hydroxyphenyl)-6-(4-dimethylaminophenyl)-4-oxo-1,4, 5,6-tetrahydronicotinic acid
X	6-(4-Dimethylaminophenyl)-1-(4-hydroxy-2-trifluoromethylphenyl)-4- oxo-1,4,5,6 tetrahydronicotinic acid
Y	1-(2-Bromo-4-hydroxyphenyl)-6-(4-dimethylaminophenyl)-4-oxo-1,4, 5,6-tetrahydronicotinic acid
Z	1-(2-Chloro-4-hydroxyphenyl)-4-oxo-6-phenyl-1,4,5,6-tetrahydro-nicotinic acid

One cautionary note with respect to the data presented in Table 7.1 is in order; this comparison includes three substantially different assays. One consists of purified gyrase and uniquely 3'-end-labeled linear DNA, the second assay employs purified enzyme and relaxed, covalently closed, circular plasmid DNA and the third assay involves living cells (inhibition of growth).

Although the rank-orders of potency of the six quinolones tested in this study were essentially the same in all three assays, it was somewhat disappointing to find that the cleavable-complex test was only two- to ten-fold more sensitive than the enzyme-inhibition assay. In addition, the non-proportionately of the MIC results between norfloxacin and ciprofloxacin is not reflected in the DNA-cleavage assay. One possible explanation for the increased potency of one quinolone over another is that the more active drug binds more avidly to a critical cellular target(s). Assuming, for the moment, that the primary target here is the gyrase:DNA binary complex (rather than gyrase alone or DNA alone), then one might expect ciprofloxacin to be significantly more active than norfloxacin in the cleavable-complex assay as a consequence. This does not appear to be the case under these experimental conditions.

Anti-gyrase Activity of Selected Heterocyclic Carboxylic Acids

Having established the fact that well characterized 4-quinolone antimicrobials could be ranked in order of potency in the cleavable-complex assay, we undertook a similar analysis of 22 heterocyclic carboxylic acids. These were mainly tetrahydronicotinic acids (table 7.2) and were novel with the exception of compounds "B" (Dunwell and Evans 1973) and "U" (Narita et al. 1983). No gyrase data have been reported for these compounds although "U" is known to have potent antibacterial properties (Narita et al. 1983). We also found this compound (for structure see Fig. 7.2) to be active against a wide range of microorganisms and to be a modest inhibitor of two clinical isolates of Pseudomonas aeruginosa (Table 7.3). Furthermore, compound "U" is inactive against a gyr A, Nal[r] mutant of E. coli (data not shown).

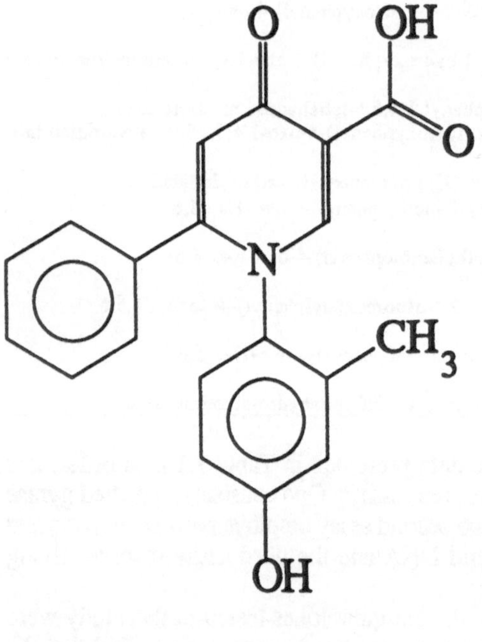

Fig. 7.2. Structure of compound "U".

Table 7.3. MICs of Compound "U" against a variety of bacterial strains.

Organisms (no.of isolates)	MIC ($\mu g\ ml^{-1}$)
St. pyogenes (1)	0.1
St. faecalis (1)	0.1
S. aureus (2)	0.1
V. cholerae (1)	0.1
Past. multocida (1	0.1
Salm. typhosa (1)	0.1
Salm. typhimurium (1)	0.1
Shig. flexneri (1)	0.1
E. coli (4)	0.1
S. marcescens (2)	1.0
C. freundii (1)	0.1
P. vulgaris (1)	0.3
P. mirabilis (1)	0.3
Ps. aeruginosa (2)	3.0
C. albicans (1)	>100.0

Table 7.4. Comparison of DNA gyrase results with growth inhibition against three clinical isolates of E. coli.

Compound or Drug	Inhibitory conc. (μM)		Avg MIC[a]	
	Supercoiling-inhibition assay	Cleavable-complex assay	μM	$\mu g\ ml^{-1}$ (range)
Norfloxacin	1.0	0.2	0.031	(0.01)
A.	>100	>1000	310	100
B	>100	>1000	340	>100
C	>100	>1000	>300	>100
E	>100	>1000	>280	>100
G	>100	500	>260	>100
H	>100	50	>270	>100
I	>100	50	>320	>100
K	88	10	>270	>100
M	50	5	6.3	2.1 (0.3-3.0)
N	50	5	6.1	2.1 (0.3-3.0)
O	50	5	196	77 (30-100)
P	50	5	20	7.7 (3.0-10)
Q	50	5	61.3	23 (10-30)
R	5	5	2.0	0.8 (0.3-1.0)
S	10	1	6.1	2.1 (0.3-3.0)
T	5	1	1.4	0.5 (0.3-3.0)
U	5	0.5	0.10	0.03 (0.03)
V	5	0.5	0.7	0.3 (0.3)
W	1.0	0.5	0.08	0.03 (0.03)
X	10	0.5	3.3	1.4 (0.3-3.0)
Y	5	0.5	0.1	0.05 (0.03-0.1)
Z	10	0.5	0.6	0.2 (0.1-0.3)

[a] MIC results for three E. coli strains were simply averaged.

Table 7.4 summarizes the results of this analysis. Compounds (with the exception of the control drug, norfloxacin) are ranked in ascending order of activity in the cleavable-complex assay. Within this group of 22 compounds, there is generally a good correlation in compound activity among the three assays employed as well as in the overall rank-order of potency. None of these compounds were as active as norfloxacin but compounds 'W' and 'U' are very close (Table 7.4). In this regard, Takahata et al. (1988) recently described a pyridone carboxylic acid, T-3262, with a broad spectrum of potent antibacterial activity against Gram-positive and Gram-negative bacteria.

Compounds "M", "N", "O", "P", "Q" and "R" all generate the cleavable-complex at a minimum concentration of 5 μM, however, the MICs among these six compounds range from 2.0 to 196 μM; a relatively large range of values. Similarly, compounds "U", "V", "W", "X", "Y" and "Z" all induce the cleavable-complex at a minimum concentration of 0.5 μM and have an MIC range, as a group, of 0.1 to 3.3 μM. Compound "O" appears to be somewhat of an outlier, in that it was equipotent to compounds 'M' and 'N' in the supercoiling-inhibition and cleavable-complex assays and yet was approximately 30-fold less active against intact *E. coli*. It is tempting to explain this discrepancy in terms of cell permeation, however, based on structural considerations, there is no obvious reason why this compound should have trouble penetrating Gram-negative bacteria. It is worth noting that the uptake of quinolone antibacterials by *E. coli* occurs by more than one pathway. Quinolones are known to penetrate the outer membrane of this organism through the *OmpF* and *OmpC* porins; this has been demonstrated by both the use of defined porin-deficient strains (Hirai *et al.* 1986) and the characterization of quinolone-resistant mutants that lack porins (Hooper *et al.* 1986). In addition, Chapman and Georgopapadakou (1988) propose that quinolones interact with the outer membrane as chelating agents thereby destabilizing the lipid bilayer and allowing diffusion of the drug through these exposed lipid domains. Outer membrane perturbations typical of those seen with gentamicin and ethylene diaminetetracetic acid were seen in quinolone-treated cells. Regardless of the preferred pathway(s) for the penetration of active hetercyclic carboxylic acids, careful uptake studies are required before one can conclude, with certainty, that compound "O" has difficulty penetrating cells.

In addition, there are some notable differences in the concentration of compound required to inhibit strand-passing activity versus that required to inhibit the growth of intact cells. In the case of compound "U", there is a 50-fold greater concentration needed for gyrase inhibition than required to inhibit *E. coli*. Similar discrepancies have been reported for quinolone anti-microbials and were discussed at length earlier in this chapter.

As was the case in the quinolone analysis (Table 7.1), the potencies of these heterocyclic carboxylic acids appear to encounter a "floor" of activity (0.5 μM) with respect to the induction of the cleavable-complex. We had hoped that a strand-breakage assay might provide a more direct and, therefore, more *sensitive* indication of the effect of these drugs. Possible refinements to this experimental approach will be discussed in the final section of this chapter.

Anti-gyrase Activity of the Diuretic Amiloride

Amiloride is a potassium sparing diuretic capable of inhibiting the progression of the cell cycle of animal cells both *in vitro* and *in vivo* (Pieri *et al.* 1983; Besterman *et al.* 1984). In addition to its diverse effects on intact animals, perfused organs, cultured tissues and cells (Benos 1982), amiloride has intrinsic antibacterial activity. It is bacteriostatic against certain Gram-positive bacteria at concentrations of 25 to 1300 μg ml[-1] (Giunta *et al.* 1984) and bactericidal against selected strains of hemolytic *Streptococci* (Giunta *et al.* 1985). In addition, Cohn *et al.* (1988) have shown the combination of amiloride plus tobramycin is synergistic against many clinical strains of *Pseudomonas cepacia*.

The mechanism by which this drug inhibits susceptible bacteria is far from clear. Giunta *et al.* (1986) found that millimolar concentrations of amiloride caused marked changes in the growth-dependent intracellular balance of Na^+ and K^+ in *Streptococcus faecalis*. These workers suggested that amiloride's effect on bacterial electrolytes may lead to growth inhibition. Related to these observations, Cohn *et al.* (1988) have speculated that amiloride affects transcellular Na^+ flux in *P. cepacia* , which, in turn enhances tobramycin uptake hence, the observed synergy of the combination amiloride plus tobramycin against this particular organism.

The possibility that DNA topoisomerase II is an important intracellular target for this pyrazine diuretic was suggested by two lines of evidence. First, work from several laboratories (Besterman *et al.* 1984; L'Allemain *et al.* 1984) indicates that inhibition of mitogen-induced DNA synthesis by amiloride could not be entirely accounted for by the ability of amiloride to inhibit mitogen-stimulate Na^+/H^+ exchange. Second, a great deal of evidence suggests that DNA topoisomerase II activity is probably required for DNA synthesis by mammalian cells in culture (Mattern and Painter1979; Colwill and Sheinin 1983; Nelson *et al.* 1986). On the strength of these independent observations, we tested the effect of amiloride on purified mammalian (calf thymus) topoisomerase II *in vitro* and on DNA topoisomerase II-dependent cell functions *in vivo* (Besterman *et al.* 1987). Amiloride displayed inhibitory effects in both systems. Furthermore, we found that this drug was a DNA intercalator in that it was able to

1) shift the thermal denaturation profile of DNA,
2) increase the viscosity of linear DNA and
3) unwind circular, covalently closed DNA (Besterman *et al.* 1987).

The fact that amiloride intercalates into DNA is not that surprising when one considers the conformation of this drug as it has been observed in crystal structures. Studies by Smith *et al.* (1979) using NMR in conjunction with quantum mechanics computations suggest that the ground-state tautomer of amiloride is the planar, hydrogen-bonded, tricyclic acylamino tautomer (Fig. 7.3). This conformation of amiloride could easily intercalate into DNA: it provides a planar ring system approximately 10 Å across with

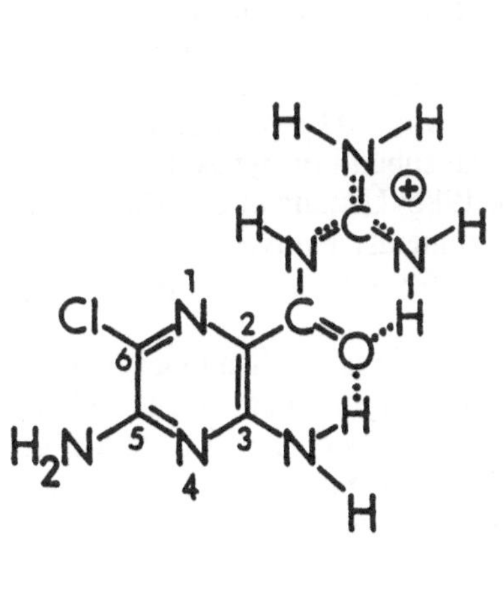

Fig. 7.3. Proposed planar, hydrogen-bonded, tricyclic conformation of amiloride (Smith *et al.* 1979).

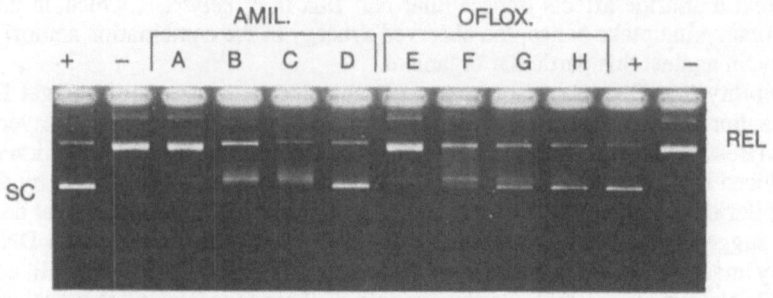

Fig. 7.4. Agarose gel electrophoresis assay of the inhibition of *Micrococcus luteus* DNA gyrase supercoiling activity. Positions of relaxed (REL) and supercoiled (SC) pBR322 DNA are indicated. Lane (+) was the enzyme control (gyrase plus relaxed pBR322 DNA, no drug). Lane (-) was the DNA control (relaxed pBR322 DNA, no drug, no enzyme). Lanes A through D; incubation of gyrase plus relaxed DNA in the presence of various concentrations of amiloride at 1.0, 0.5, 0.25 and 0.125 mM respectively. Lanes E through H; incubation in the presence of various concentrations of ofloxacin at 1.0, 0.5, 0.25 and 0.125 mM respectively.

a set of hydrogen bond-donating groups on either end of the system (Besterman *et al.* 1987).

To determine whether the ability of amiloride to intercalate into DNA and to inhibit topoisomerase II was dependent upon its ability to assume a planar, cyclized conformation, we studied the structure-activity relationship for 12 amiloride analogs (Besterman *et al.* 1989). Results indicated that only those analogs capable of cyclization could intercalate into DNA and inhibit the catalytic activity of purified topoisomerase II as well as topoisomerase II- mediated cell functions *in vivo*.

In evolutionary terms, DNA topoisomerase II appears to be a highly conserved enzyme. Thus, the N-terminal region of eukaryotic topoisomerase II is homologous to *Gyr B* (the subunit of bacterial gyrase which interacts with ATP), whereas the middle portion of the molecule is homologous to *Gyr A* (the subunit of gyrase that cleaves and religates DNA) (Lynn *et al.* 1986; Giaver *et al.* 1986; Uemura *et al.* 1986). It has been postulated, therefore, that human DNA topoisomerase evolved from bacterial topoisomerase II by fusion of the two gyrase subunits into a single polypeptide (Pflugfelder *et al.* 1988).

The fact that amiloride inhibited the catalytic activity of mammalian topoisomerase II and the collateral facts that the eukaryotic and procaryotic enzymes are highly conserved, (at least in their functionally "active" portions) encouraged us to test the effect of amiloride on DNA gyrase. Figure 7.4 compares the potency of ofloxacin and amiloride in terms of their ability to inhibit the supercoiling activity of *M. luteus* DNA gyrase. Both drugs completely inhibited this enzyme at a concentration of 1.0 mM. Amiloride and ofloxacin had estimated IC_{50} values of 0.25 mM and 0.5 mM, respectively, against *M. luteus* gyrase (Table 7.5). On the other hand, amiloride was 1000-fold less active than ofloxacin against *E. coli* gyrase with an estimated IC_{50} of 2.5 mM. Thus, amiloride is approximately ten fold more active against DNA gyrase originating from a Gram-positive organism compared to enzyme isolated from a Gram-negative bacteria. These data are consistent with the fact that the bacteriostatic and bacteriocidal effects of amiloride are confined to selected Gram-positive microorganisms (Giunta *et al.* 1984; Giunta *et al.* 1985). We have tested this drug, *in vitro*, against a wide variety of both Gram-positive and Gram-negative bacteria. In our hands, the only organisms inhibited by amiloride were isolates of *Streptococcus pyogenes* (MICs ranging from 31-62 μg ml^{-1}) and *M. luteus* (MICs ranging from 50-100 μg ml^{-1}).

We also tested amiloride for its ability to generate the DNA gyrase-mediated cleavable-complex. Whereas ofloxacin induced the cleavable-complex in the presence of both *E. coli* and *M. luteus* DNA gyrase (at concentrations ranging from 0.2 - 5 mM), amiloride was totally inactive over the concentration range of 10 - 100 mM, regardless of enzyme source (data not shown). Earlier, Besterman *et al.* (1987) reported that amiloride did not stimulate cleavable-complex formation in the presence of mammalian (calf thymus) topoisomerase II.

Table 7.5. Inhibition of the catalytic activity of *M. luteus*-and *E. coli* - derived DNA gyrase by amiloride and ofloxacin.

	Enzyme	
	M. luteus gyrase	*E. coli* gyrase
Drug	IC$_{50}$[a]	IC$_{50}$[a]
Amiloride	0.25 mM	2.5 mM
Ofloxacin	0.50 mM	2.5 μM

[a]Drug concentrations required to inhibit approximately 50% of the activity of the enzymes were estimated by visually examining the gels.

Thus, amiloride inhibits the strand-passing activity of both eukaryotic and procaryotic DNA topoisomerase II without being able to generate the cleavable-complex. Such a dichotomy of properties is not without precedent. Ethidium bromide, for example, is an avid DNA intercalator, inhibits the catalytic activity of topoisomerase II, but does not generate the cleavable complex. Furthermore, the antibiotics novobiocin and coumermycin both inhibit DNA gyrase and are inactive in the cleavable-complex assay. However, unlike novobiocin and coumermycin, amiloride does not appear to inhibit topoisomerase II by competing with ATP (Besterman *et al.* 1987). It is possible, of course, that amiloride interacts in some other manner with the gyrase B subunit and as a consequence inhibits catalytic activity. Clearly, definition of the mechanism of topoisomerase II-inhibition by amiloride awaits further research.

How does one assess the association (if any) between amiloride's rather selective antibacterial activity and its anti-gyrase properties? Unfortunately, there are no well characterized *gyrA*, Nal^r mutant strains of *St. pyogenes* or *M. luteus* . It would be helpful to know whether mutants of this kind were cross-resistant to amiloride. Also, no one, to our knowledge, has reported the isolation of DNA gyrase from hemolytic *Streptococci*. Since certain strains of *S. pyogenes* are particularly susceptible to amiloride one wonders whether purified *Streptococcal* DNA gyrase might be hypersusceptible? Finally, Figgitt *et al.* (1989) have reported amiloride to be a more potent inhibitor of the catalytic activity of purified *Saccharomyces cerevisae* topoisomerase II activity than either *m*-AMSA or etoposide. However, amiloride lacks intrinsic antifungal activity which these authors suggest is the result of poor cell penetration (Figgitt *et al.* 1989). These workers did not test whether *Saccharomyces* topoisomerase II induced DNA strand breaks *in vitro* in the presence of amiloride.

In conclusion, the mechanism of amiloride's antibacterial activity, either alone or in combination with known antimicrobial agents against susceptible *Pseudomonas,*

Streptococcal and *Micrococcal* species is far from clear. It has been postulated that amiloride blocks cellular Na^+/H^+ transport activity which ultimately leads to inhibition of bacterial cell growth and, in some cases, cell death (Giunta *et al.* 1986; Cohn *et al.* 1988). This explanation is certainly consistent with the well characterized actions of this pyrazine diuretic on eukaryotic cells (Benos 1982). The discovery, however, that amiloride is more potent than ofloxacin as an inhibitor of *Micrococcal* DNA gyrase and more potent than *m*-AMSA as an inhibitor of *Saccharomyces* topoisomerase II suggests an alternative explanation for its inherent antibacterial activity.

Conclusion

Although a considerable body of evidence indicates that DNA gyrase is the target of the 4-quinolone antibacterials, the precise mechanism of inhibition remains a mystery. This question is particularly pertinent in light of the controversy over whether quinolones bind directly to DNA, to gyrase or preferentially to the DNA:gyrase binary complex? (Shen and Pernet 1985; Hussy *et al.* 1986; Forsgren *et al.* 1987; Palu *et al.* 1988). Recent evidence by Shen *et al.* (1989) suggests that norfloxacin binds nonspecifically to relaxed DNA and binds specifically to gyrase at sites on the enzyme that appear only *after* gyrase:DNA complex formation.

Equally unclear is how the inhibition of DNA gyrase by the quinolones ultimately leads to cell death. In experiments with bacteriophage T7, Kreuzer and Cozzarelli (1979) showed that the action of nalidixic acid is not strictly equivalent to inhibition of gyrase. Thus, the burst size of phage T7 was little affected by thermal inactivation of a thermosensitive gyrase A subunit, a finding indicating that for phage maturation a requirement for bacterial gyrase is minimal. However, nalidixic acid still markedly inhibited phage burst size when infection occurred at a permissive (but not restrictive) temperature. In addition, Crumplin *et al.* (1984) have shown that the bactericidal effect of norfloxacin require competent RNA and protein synthesis, suggesting an active *process* rather than an irreversible inhibition of a vital cell enzyme. Finally Chow *et al.* (1988) showed that there is an excellent correlation between quinolone-induced inhibition of early (i.e. 10 to 15 min) DNA synthesis and the corresponding MICs for all strains and carboxyquinolines tested. These authors suggest that the initial event in the cascade leading to quinolone-induced cell death may be gyrase-mediated, supercoiling-independent early inhibition of DNA synthesis.

Results of these and other studies are consistent with the hypothesis that 4-quinolones act within bacteria primarily to form drug:enzyme:DNA complexes that function as poisons, rather than as simple inhibitors of enzymatic strand-passing activity. In bacteria, one of the active processes induced by the 4-quinolones (Phillips *et al.* 1987) is expression from the network of genes that collectively comprise the so-called SOS response (Little and Mount 1982). When chromosome replication in *E. coli* is affected by DNA lesions, an "SOS" signal is produced that invokes the activation of *RecA* protein. This, in turn, promotes the cleavage of *Lex A* and prophage repressors, resulting in derepression of cellular SOS genes and prophages. Might not the trapped gyrase:quinolone:DNA complex provide the necessary SOS "signal" and, as a consequence, one (or more) of the normal products of the SOS gene network, cause misrepair of the lesions and thereby kill the cell? (Chow *et al.* 1988; Drlica and Franco 1988; Hooper and Wolfson1988).

Assuming, for the moment, that all this speculation is true, one might expect that an *in vitro* screening procedure that measures the minimum concentration of an agent required to generate the SOS signal (i.e. cleavable-complex) might be a very sensitive indicator of drug potency. Based upon the data shown in Tables 7.1 and 7.4, the cleavable-complex assay correlates well with antibacterial activity but it is not especially sensitive. For example, compounds "U" through "Z" encounter a "floor" of 0.5 μM with respect to the concentration required to generate gyrase-mediated double strand DNA breakage, yet they vary considerably in antibacterial potency (Table 7.4). This argument is based, of course, on one bacterial species, (*E. coli*), and discounts cell permeability as a major contributing factor to the observed discrepancies.

To date, we have interpreted the results of the cleavable-complex assay by visually examining autoradiograms and making a subjective judgement as to the extent of the DNA cleavage that has occurred. Instrumentation is now available (e.g. Molecular Dynamics Phospho Imager system and Betagen's 603 Blot Analyzer) that has the capability to objectively quantitate the degree of DNA-cleavage with more than 50-times the dynamic range of conventional X-ray film. Perhaps this automated approach will enhance the sensitivity of the cleavable-complex assay and thus make it a more powerful tool with which to study the extremely subtle and complex interaction between active compound and procaryotic DNA topoisomerase II.

Acknowledgement

We could not have done the experiments discussed in this chapter without the generosity and advice of Drs. Leroy Liu, Martin Gellert and David Knowles, for which we are most grateful. In addition, we thank Dr. Michael Cory for his help, both practical and theoretical, regarding the rather unexpected interactions of amiloride with DNA.

References

Barrett JF, Gootz TD, McGuirk PR, Farrell CA, Sokolowski SA (1989) Use of *in-vitro* topoisomerase II assays for studying quinolone antibacterial agents. Antimicrob Agents Chemother 33:1697-1703

Benos DJ (1982) Amiloride: a molecular probe of sodium transport in tissues and cells. Am J Physiol 242:C131-C145

Besterman JM, Elwell LP, Blanchard SG, and Cory M (1987) Amiloride intercalates into DNA and inhibits DNA topoisomerase II. J Biol Chem 262:13352-13358

Besterman JM, Elwell LP, Cragoe EF, Andrews CW, Cory M (1989) DNA intercalation and inhibition of topoisomerase II: Structure-activity relationships for a series of amiloride analogs. J Biol Chem 265:2324-2330

Besterman JM, Tyrey SJ, Cragoe Jr EJ, Cuatrecasas P (1984) Inhibition of epidermal growth facter-induced mitogenesis by amiloride and an analog: Evidence against a requirement for Na⁺/H⁺ exchange. Proc Natl Acad Sci USA 81:6762-6766

Chapman JS, Georgopapadakou NH, (1988) Routes of quinolone permeation in *Escherichia coli*. Antimicrob Agents Chemother 32:438-442

Chen GL, Yang L, Rowe TC, Halligan BD, Tewey KM, Liu LF (1984) Nonintercalative antitumor drugs interfere with the breakage-reunion reaction of mammalian DNA topoisomerase II. J Biol Chem 259:13560-13566

Chow RT, Dougherty TJ, Fraimow HS, Bellin EY, Miller MH (1988) Association between early inhibition of DNA synthesis and the MICs and MBCs of carboxyquinolone anti-microbial agents for wild-type and mutant [gyrA nfxB (ompF) acr A] E. coli K-12. Antimicrob Agents Chemother 32:1113-1118

Cohn RC, Jacobs M, Aronoff SC (1988) *In vitro* activity of amiloride combined with tobramycin against *Pseudomonas* isolates from patients with cystic fibrosis. Antimicrob Agents Chemother 32:395-396

Colwill RW, Sheinin R (1983) *ts* AlS9 locus in mouse L cells may encode a novobiocin binding protein that is required for DNA topoisomerase II activity. Proc Natl Acad Sci USA 80:4644-4648

Cozzarelli NR (1980) DNA gyrase and the supercoiling of DNA. Science 207:953-960

Crumplin GC, Kenwright M, Hirst T (1984) Investigations into the mechanism of action of the antibacterial agent norfloxacin. J Antimicrob Chemother 13(Suppl. B):9-23

Domagala JM, Hanna LD, Heifetz CL, Hutt MP, Mich TF, Sanchez JP, Solomon M (1986) New structure-activity relationships of the quinolone antibacterials using the target enzyme. The development and application of a DNA gyrase assay. J Med Chem 29:394-404

Drlica K (1984) Biology of bacterial deoxyribonucleic acid topoisomerases. Microbiol Rev 48:273-289

Drlica K, Franco RJ (1988) Inhibitors of DNA topoisomerases. Biochem 27:2253-2259

Dunwell DW, Evans D (1973) Reactions of 2-aminobenzimidazoles with propiolic esters and diethylethoxymethylenemalonate. J Chem Soc Perk 1:1588-1590

Figgitt DP, Denyer SP, Dewick PM, Jackson DE, Williams P (1989) Topoisomerase II: A potential target for novel antifungal agents. Biochem Biophys Res Commun 160:257-262

Fisher LM, Mizuuchi K, O'Dea MH, Ohmori H, Gellert M (1981) Site-specific interaction of DNA gyrase with DNA. Proc Natl Acad Sci USA 78:4165-4169

Forsgren A, Bredberg A, Pardee AB, Schlossman SF, Tedder TF (1987) Effects of ciprofloxacin on eukaryotic pyrimidine nucleotide biosynthesis and cell growth. Antimicrob Agent Chemother 31:774-779

Fu KP, Grace ME, McCloud SJ, Gregory FJ, Hung PP (1986) Discrepancy between the antibacterial activities and the inhibitory effect on *Micrococcus luteus* DNA gyrase of 13 quinolones. Chemother 32:494-498

Gellert M (1981) DNA topoisomerases. Ann Rev Biochem 50:879-910

Gellert M, Mizuuchi K, O'Dea MH, Itoh T, Tomizawa JT (1977) Nalidixic acid resistance, a second genetic character involved in DNA gyrase activity. Proc Natl Acad Sci USA 74:4772-4776

Giaever G, Lynn R, Goto T, Wang JC (1986) The complete nucleotide sequence of the structural gene TOP2 of yeast DNA topoisomerase II. J Biol Chem 261:12448-12454

Giunta S, Galeazzi L, Turchetti G, Sampaoli G, Groppa G (1985) *In vitro* antistreptococcal activity of the potassium-sparing diuretics amiloride and triamterene. Antimicrob Agents Chemother 28:419-420

Giunta S, Galeazzi L, Turcheti G, Sampaoli G, Groppa G (1986) Effect of amiloride on the intracelluar sodium and potassium content of intact *Streptococcus faecalis* cells *in vitro*. Antimicrob Agents Chemother 29:958-959

Giunta S, Pieri C, Groppa G (1984) Amiloride, a diuretic with *in vitro* antimicrobial activity. Pharmacol Res Commun 16:821-829

Higgins NP, Cozzarelli NR (1982) The binding to gyrase to DNA: analysis by retention by nitrocellulose filters. Nucleic Acids Res 10:6833-6847

Hirai K, Aoyama T, Irikura T, Iyobe S, Mitsuhashi S (1986) Differences in susceptibility to quinolones of outer membrane mutants of *Salmonella typhimurium* and *Escherichi coli* . Antimicrob Agents Chemother 29:535-538

Hooper DC, Wolfson J, Souza K, Tung C, McHugh G, Swartz M (1986) Genetic and biochemical characterization of norfloxacin resistance in *Escherichia coli* . Antimicrob Agents Chemother 29:639-644

Hooper DC, Wolfson J (1988) Mode of action of the quinolone antimicrobial agents. Rev Inf Dis 10:(Suppl 1)S14-S21

Hoshino K, Sato K, Une T, Osada Y (1989) Inhibitory effects of quinolones on DNA gyrase of *Escherichia coli* and topoisomerase II of fetal calf thymus. Antimicrob Agents Chemother 33:1816-1818

Hussy P, Maass G, Tummler B, Grosse F, Schomburg U (1986) Effect of 4-quinolones and novobiocin on calf thymus DNA polymerase primase, topoisomerases I and II, and growth of mammalian lymphoblasts. Antimicrob Agents Chemother 29:1073-1078

Imamura M, Shibamura S, Hayakawa I, Osada Y (1987) Inhibition of DNA gyrase by optically active ofloxacin. Antimicrob Agents Chemother 31:325-327

Inoue Y, Sato K, Fujii T, Hirai K, Inoue M, Iyobe S, Mitsuhashi S (1987) Some properties of subunits of DNA gyrase from *Pseudomonas aeruginosa* PAO1 and its nalidixic acid-resistant mutant. J Bacteriol 169:2322-2325

Kreuzer KN, Cozzarelli NR (1979) *Escherichia coli* mutants thermosensitive for deoxyribonucleic acid gyrase subunit A: effects on DNA replication, transcription and bacteriophage growth. J Bacteriol 140:424-435

L'Allemain G, Franchi A, Cragoe E, Pouyssegur J (1984) Blockade of the N^+/H^+ antiport abolishes growth factor-induced DNA synthesis in fibroblasts. J Biol Chem 259: 4313-4319

Little JW, Mount DW (1982) The SOS regulatory system of *Escherichia coli*. Cell 29:11-22.

Long BH, Musial ST, Brattain MG (1984) Comparison of cytotoxicity and DNA breakage activity of congeners of podophyllotoxin including VP16 and VM26: a quantative structure-activity relationship. Biochem 23:1183-1188.

Lynn R, Giaever G, Swanberg SL, Wang J (1986) Tandem regions of yeast DNA topoisomerase II share homology with different subunits of bacterial gyrase. Science 233:647-649

Mattern MR, Painter RB (1979) Dependence of mammalian DNA replication on DNA supercoiling. II Effects of novobiocin on DNA synthesis in CHO cells. Biochemica Biophys Acta 563:306-312

Miller RV, Scurlock TR (1983) DNA gyrase (topoisomerase II) from *Pseudomonas aeruginosa*. Biochem Biophys Res Commun 110:694-700

Morrison A, Cozzarelli NR (1979) Site-specific cleavage of DNA by *E. coli* DNA gyrase. Cell 17:175-184

Narita H, Konishi Y, Nitta J, Misumi S, Nagaki H, Nagai Y, Watanabe Y, Matsubara N, Minami S, Saikawa I (1983) Novel 4-oxo-1, 4-dihydronicotinic acid derivatives, salts thereof and antibacterial agents containing the same. UK Patent GB 2 130 580 B

Nelson WG, Liu LF, Coffey DS (1986) Newly replicated DNA is associated with DNA topoisomerase II in cultured rat prostatic adenocarcinoma cells. Nature 322: 187-189

Nelson EM, Tewey KM, Liu LF (1984) Mechanism of antitumor drug action: poisoning of mammalian DNA topoisomerase II on DNA by 4'-(9-acridinylamino) methane sulfon-*m*-anisidide. Proc Natl Acad Sci USA 81:1361-1365

Osheroff N (1989) Biochemical basis for the interactions of type I and type II topoisomerases with DNA. Pharmac Ther 41:223-241

Palu G, Valisena S, Peracchi M, Palumbo M (1988) Do quinolones bind to DNA? Biochem Pharmacol 37:1887-1888.

Pflugfelder MT, Liu LF, Liu AA, Tewey KM, Whang-Peng J, Knutsen T, Huebner K, CM, Wang JC (1988) Cloning and sequencing of cDNA encoding human DNA topo-isomerase II and localization of the gene to chromosome region 17q21-22. Proc Natl Acad Sci USA 85:7177-7181.

Phillips I, Culebras E, Moreno F, Baquero F (1987) Induction of the SOS response by new 4-quinolones. J Antimicrob Chemother 20:631-638

Pieri C, Giunta S, Giuli C, Bertoni-Freddari C, Muzzioli M (1983) *In vitro* block of murine L 1210 leukemia cell growth by amiloride, an inhibitor of passive sodium influx. Life Sci 32:1779-1784

Rowe TC, Chen GL, Hsiang Y, Liu LF (1986) DNA damage by antitumor acridines mediated by DNA topoisomerase II. Cancer Res 46:2021-2026

Rowe T, Kupfer G, Ross W (1985) Inhibition of epipodophyllotoxin cytotoxicity by interference with topoisomerase-mediated DNA cleavage. Biochem Pharm 34:2483-2487

Sato K, Inoue Y, Fujii T, Aoyama H, Inoue M, Mitsuhashi S (1986) Purification and properties of DNA gyrase from a fluoroquinolone-resistant strain of *Escherichia coli*. Antimicrob Agents Chemother 30:777-780

Shen L, Kohlbrenner WE, Weigl D, Baranowski J (1989) Mechanism of quinolone inhibition of DNA gyrase. Proc Natl Acad Sci USA 264:2973-2978

Shen L, Pernet AG (1985) Mechanism of inhibition of DNA gyrase by analogues of nalidixic acid: the target of the drug is DNA. Proc Natl Acad Sci USA 82:307-311

Smith RL, Cochran DW, Gund P, Cragoe EJ (1979) Proton, carbon-13 and nitrogen-15 nuclear magnetic resonance and CNDO/2 studies on the tautomerism and conformation of amiloride, a novel acylguanidine. J Am Chem Soc 101:191-201

Sugino A, Peebles CL, Kreuzer KN, Cozzarelli NR (1977) Mechanism of action of nalidixic acid: purification of *Escherichia coli Nal A* gene product and its relationship to DNA gyrase and a novel nicking-closing enzyme. Proc Natl Acad Sci USA 74:4767-4771

Takahata M, Nishino T (1988) DNA gyrase of *Staphylococcus aureus* and inhibitory effect of quinolones on its activity. Antimicrobial Agents Chemother 32:1192-1195

Takahata M, Otsuki M, Nishino T (1988) *In vitro* and *in vivo* activities of T-3262, a new pyridone carboxylic acid. J Antimicrob Chemother 22:143-154

Tewey KM, Chen GL, Nelson EM, Liu LF (1984a) Intercalative antitumor drugs interfere with the breakage-reunion reaction of mammalian DNA topoisomerase II. J Biol Chem 259:9182-9187

Tewey KM, Rowe TC, Yang L, Halligan BD, Liu LF (1984b) Adriamycin-induced DNA damage mediated by mammalian DNA topoisomerase II. Science 226:466-468

Uemura T, Morikawa K, Yanagida M (1986) The nucleotide sequence of the fission yeast DNA topoisomerase II gene: structural and functional relationships to other DNA topoisomerases. EMBO J 5:2355-2361

Walton L, Elwell LP (1988) *In vitro* cleavable-complex assay to monitor antimicrobial potency of
   quinolones. Antimicrob Agents Chemother 32:1086-1089
Wang JC (1985) DNA topoisomerases. Ann Rev Biochem 54:665-697
Wang JC (1987) DNA topoisomerases: nature's solution to the topological ramifications of the double-
   helix structure of DNA. Harvey Lect 81:93-110
Wolfson JS, Hooper DC (1985) The fluoroquinolones: structures, mechanisms of action and resistance,
   and spectra of activity *in vitro*. Antimicrob Agents Chemother 28:581-586
Zweerink MM, Edison A (1986) Inhibition of *Micrococcus luteus* DNA gyrase by norfloxacin and 10
   other quinolone carboxylic acids. Antimicrob Agents Chemother 29:598-601

Discussion Summary

The simple comparison of the sensitivity of the strand-breakage and supercoiling assays
as the means of assessing the potential potencies of new 4-quinolones was generally
regarded as a significant comparison. It was generally considered that the use of either
different enzymes from various bacterial species, and the use of differing assay
protocols, in different laboratories, made it difficult to draw valid comparisons between
different 4-quinolones. However, it was also realized that whatever method was used in
an individual laboratory, both systems did give an acceptable measurement of the
relative potencies of 4-quinolones. Until a satisfactory alternative assay system was
available either of these assays could be regarded as satisfactory provided that the study
incorporated sufficient control compounds to permit the drawing of a reasonably
accurate correlation line.

   Some concern was expressed over the use of a linear pBR322 molecule for the strand-
breakage assays, but other members of the audience were able to provide assurance that,
despite the fact that the linear molecule was very different to the state of the DNA *in
vivo*, there was no significant difference in the sensitivity of the system. It was thus
appreciated that the sensitivity of the system was not compromised by the convenience
of the use of end-labelled linear molecules.

   The general concensus of opinion was that, in the absence of any proper
understanding of the way in which the 4-quinolones actually work *in vivo*, we must
make the best use possilble of the available assay systems as a means of screening
4-quinolone compounds for potential antibacterial activity. Furthermore, it was
generally agreed that the use of the strand-breakage assay, with labelled DNA substrate,
probably represented the most sensitive system at present available.

Chapter 8

Comparisons of DNA Gyrases from Different Species

N. J. Robillard, B. L. Masecar, B. Brescia and R. Celesk

Exposure of susceptible bacteria to 4-quinolones results in the rapid cessation of DNA replication and eventual cell lysis (Bourguignon *et al.* 1973; Crumplin and Smith 1986; Deitz *et al.* 1966; Rella and Haas 1982). The intrinsic susceptibility of different genera to quinolones, however, varies greatly (Barry and Jones 1984; Barry *et al.* 1984). In most cases inherent characteristics which influence susceptibility to quinolones have not been delineated. In assessing these differences one must first identify each component of the drug-cell interaction which contributes to quinolone activity. The analysis of quinolone resistance mutations in various bacteria have identified two important factors which affect susceptibility.

The isolation of DNA gyrase from *E. coli* (Gellert *et al.* 1976) and its association with nalidixic acid resistance (Gellert *et al.* 1977; Sugino *et al.* 1977) revealed the importance of this enzyme in quinolone action. DNA gyrase, a type II topoisomerase unique to bacteria, is composed of two dimeric subunits. The A subunit, which is sensitive to quinolone inhibition, is responsible for the breakage and reunion of double-stranded DNA, while the B subunit is the site of ATP hydrolysis (Cozzarelli 1980; Sugino and Cozzarelli 1980). DNA gyrase, using energy supplied by ATP, converts relaxed DNA to the supercoiled form. This enzyme, which is essential for DNA replication, also is involved in recombination, repair and transcription (Gellert 1981).

In *E. coli* and other organisms, quinolone-resistance mutations in the gene for DNA gyrase reduce the enzymes' *in vitro* sensitivity to quinolones and result in a proportionate reduction in the quinolone susceptibility of the organism (Aoyama *et al.* 1988; Aoyama *et al.* 1988; Hooper *et al.* 1987; Inoue *et al.* 1987; Robillard and Scarpa 1988; Sato *et al.* 1986). It was recently shown that in both quinolone-resistant derivatives of *E. coli* KL-16 (Yoshida *et al.* 1988) and a uropathogenic *E. coli* clinical isolate (Cullen *et al.* 1989), high-level quinolone resistance was due to either a serine to tryptophan or serine to leucine mutation at amino acid residue 83 (nucleotide 248) of the gyrase A subunit. In addition, two other *E. coli* KL-16 mutants exhibiting low level quinolone-resistance harboured distinct *gyrA* mutations at amino acid residues 67 and 106 (nucleotides 199 and 318, respectively). All of these mutations are clustered in the highly conserved N-terminus region of gyrase A, and near the tyrosine residue at amino acid 122, which has been identified as the binding/cleavage site of DNA (Horowitz and Wang 1987).

Table 8.1. Bacterial strains and plasmids

Strains	Genotype/Phenotype	Reference
E.coli		
H560	*polA endA*	Bachmann collection[a]
DH2	*recA1 relA1 endA1 thi-1 hsdR17 supE44*	"
DH1	*gyrA96 recA1 relA1 endA1 thi-1 hsdR17 supE44*	"
S17-1	*Pro⁻ Res⁻ Mod⁺ Tp' Sm'*	Simon *et al.* (1983)
P. aeruginosa		
PA02	*ser-3*	Holloway collection[b]
PA04701	*cfxA2 ser-3*	This laboratory
PA04702	*cfxA3 ser-3*	" "
PA04703	*cfxA4 ser-3*	" "
PA04704	*cfxA5 ser-3*	" "
S. marcescens		
MP050	Clinical isolate	This laboratory
K. pneumoniae		
MP100	Clinical isolate	" "
Plasmids		
pSLS447	gyrase A clone (pBR322)	Swanberg and Wang (1987)
PLA2917	cloning vector	Allen and Hanson (1985)
pNJR3-2	gyrase A clone (pLA2917)	This study

a. *E. coli* Genetic Stock Center, Yale University School of Medicine.
b. Bruce Holloway, Department of Genetics, Monash University, Clayton, Victoria, Australia.

Although the antimicrobial activity of quinolones is correlated with their ability to inhibit DNA gyrase, they must first penetrate the cell envelope to access their intracellular target. Attesting to this fact is the isolation and characterisation of several quinolone-resistance mutations which affect permeability (Celesk and Robillard 1989; Chamberland *et al.* 1989; Hirai *et al.* 1986; Hirai *et al.* 1987; Hooper *et al.* 1989; Rella and Haas 1982; Robillard and Scarpa 1988).

Recognising that both DNA gyrase inhibition and permeation are important considerations in understanding quinolone potency, we sought to evaluate the role of DNA gyrase and, to a lesser extent, permeability in the quinolone susceptibility of certain Gram-negative bacteria. As representative Gram-negative organisms we chose *P. aeruginosa*, *E. coli*, *S. marcescens* and *K. pneumoniae* (Table 8.1). These genera are clinically relevant and represent a wide range of intrinsic quinolone susceptibilities.

Quinolone Susceptibility

The 4-quinolone susceptibilities of the *E. coli*, *P. aeruginosa*, *S. marcescens*, and *K. pneumoniae* strains from which DNA gyrase was obtained are presented in Table 8.2. Comparison of the MICs for these strains shows *E. coli* H560 to be the most susceptible to the 4-quinolones tested. *K. pneumoniae* MP100 was two- to four-fold less susceptible than *E. coli* H560 to 4-quinolones, while *P. aeruginosa* PA02 and *S. marcescens* MP050 were four- to 16-fold less susceptible than *K. pneumoniae* to 4-quinolones.

Table 8.2. Susceptibility of bacterial strains to quinolones

| Strain | MIC (µg/ml) | | | |
	Ciprofloxacin	Norfloxacin	Enoxacin	Ofloxacin
E. coli				
H560	0.015	0.06	0.125	0.06
P. aeruginosa				
PAO2	0.25	1.0	1.0	2.0
S. marcescens				
MP050	0.5	4.0	4.0	2.0
K. pneumoniae				
MP100	0.06	0.25	0.25	0.125

Sensitivity of DNA Gyrase Enzymes to Quinolones

To determine if the quinolone susceptibility of these strains was paralleled by the quinolone sensitivity of their respective DNA gyrases, we isolated gyrase from each organism (except E. coli enzyme which was kindly provided by Linus Shen) and assayed for inhibition of supercoiling by ciprofloxacin and nalidixic acid. As shown in Table 8.3, the quinolone IC_{50}s were similar for gyrase isolated from each of the organisms (between 0.5 and 1.0 µg/ml for ciprofloxacin). Thus, gyrase sensitivity does not account for the intrinsic differences in the quinolone MICs of these organisms.

Table 8.3. Sensitivity of DNA gyrase enzymes to ciprofloxacin and nalidixic acid

| Source of DNA gyrase | IC_{50} (µg/ml)[a] | |
	ciprofloxacin	nalidixic acid
E. coli		
H560	0.5	50
P. aeruginosa		
PAO2	1.0	50
S. marcescens		
MP050	0.5	25
K. pneumoniae		
MP100	0.5	ND[b]

a. Concentration of drug which inhibits supercoiling by 50%.
b. Not done.

Intergeneric Heterologous DNA Gyrase Assays

To extend the comparative analysis of wild type DNA gyrase from E. coli and P. aeruginosa, we combined the gyrase A or B subunit from E. coli H560 with the A or B subunit of P. aeruginosa PAO2. As shown in Fig. 8.1, gyrase subunits from E. coli and P.aeruginosa complement one another in supercoiling assays. The heterologous combinations were also assayed for inhibition of supercoiling by ciprofloxacin and nalidixic acid. As shown in Table 8.4, the A subunit of E. coli complemented by the B subunit of P. aeruginosa was inhibited by 0.5 µg of ciprofloxacin per ml, while the A subunit of P. aeruginosa complemented by the B subunit of E. coli was inhibited by 1.0 µg of ciprofloxacin per ml. The IC_{50} of nalidixic acid was the same for both subunit combinations. These results show that the DNA gyrase A subunits of E. coli and P. aeruginosa complement the heterologous B subunits without an alteration in quinolone sensitivity.

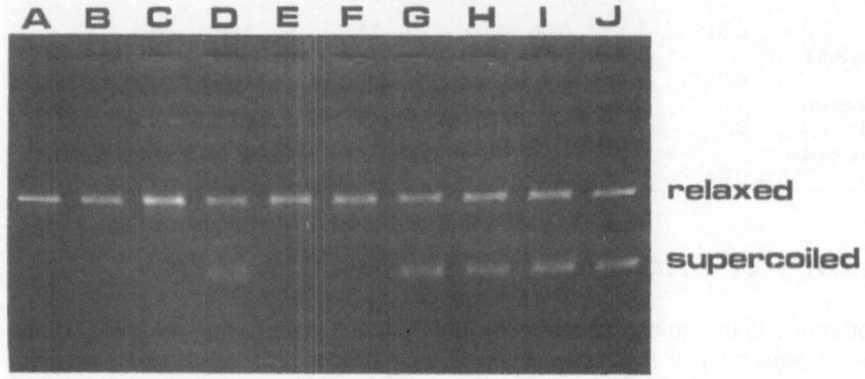

Fig. 8.1. DNA supercoiling of relaxed pBR322 plasmid DNA by reconstituted *E. coli* (E.c.) and *P. aeruginosa* (P.a.) DNA gyrase subunits A and B. (A) No enzyme control; (B) P.a. gyrase A; (C) P.a. gyrase B; (D) P.a. gyrase A and B; (E) E.c. gyrase A; (F) E.c. gyrase B; (G) E.c. gyrase A and B; (H) P.a. gyrase A and E.c. gyrase B; (I) E.c. gyrase A and P.a. gyrase B; (J) *Micrococcus luteus* gyrase control.

Table 8.4. Concentration of ciprofloxacin and nalidixic acid which inhibit reconstituted DNA gyrase from *P. aeruginosa* PAO2 and *E. coli* H560

Source of A subunit	Source of B subunit	IC_{50} $(\mu g/ml)^a$	
		ciprofloxacin	nalidixic acid
H560	PAO2	0.5	50
PAO2	H560	1.0	50

a. Concentration of drug which inhibits 50% of supercoiling.

Analysis of heterologous *E. coli* gyrase A gene expression

Recognising that there is variability in the *in vitro* assay of DNA gyrase and that different IC_{50} values have been reported for both the *P. aeruginosa* and *E. coli* enzymes (Fisher *et al.* 1989; Inoue *et al.* 1987; Robillard and Scarpa 1988; Sato *et al.* 1986), we sought to test our findings *in vivo*. In *E. coli* gyrase A merodiploids, the wild type gyrase A gene is dominant over quinolone-resistant *gyrA* alleles (Hane and Wood 1969; Nakamura *et al.* 1989). This implies that when both the resistant and sensitive gyrase A proteins are present in the cell, gyrase function will be quinolone sensitive. The results of our *in vitro* gyrase assays demonstrate that the A subunit of *E. coli* is capable of replacing the A subunit of *P. aeruginosa*. Furthermore, this hybrid enzyme exhibits a similar level of quinolone sensitivity as the *P. aeruginosa* holoenzyme. Based on these results one might predict that the *E. coli* gyrase A subunit is capable of replacing the *P. aeruginosa* gyrase A subunit *in vivo* (provided the gene is expressed) and that this would not significantly alter the quinolone susceptibility of wild type *P. aeruginosa*. One might also predict that the wild type *E. coli* gyrase A subunit would confer quinolone sensitivity upon a quinolone-resistant *P. aeruginosa* gyrase *in vivo*, but the quinolone susceptibility of this heterodiploid would only decrease to the *P. aeruginosa* wild type level and not to that of *E. coli*.

To determine if these predictions are correct, we cloned the gyrase A gene of *E. coli* into the broad host range plasmid vector pLA2917 for expression studies in *P. aeruginosa*. The recombinant broad host range gyrase A plasmid was designated pNJR3-2. The presence of the gyrase A gene in pNJR3-2 was confirmed by matching the internal *Bgl*II fragments of pNJR3-2 with those of pSLS447 (source of the gyrase A gene). We utilised the dominant sensitivity trait of the wild type gyrase A gene to monitor its expression. To confirm gyrase A expression in *E. coli*, we introduced pLA2917 or pNJR3-2 into the wild type gyrase A strain DH2 as well as the *gyrA* mutant DH1. The MICs of these strains for selected 4-quinolones are presented in Table 8.5 The vector pLA2917 had no effect on the quinolone susceptibility of either strain DH2 or the *gyrA* mutant DH1. pNJR3-2 also had no effect on DH2 but conferred quinolone susceptibility upon the *gyrA* mutant DH1.

Table 8.5. Effect of *E. coli* gyrase A gene on *gyrA*+ and *gyrA E. coli* strains

Strain/plasmid	MIC (ug/ml)			
	Ciprofloxacin	Norfloxacin	Enoxacin	Ofloxacin
DH2 (no plasmid)	≤0.008	≤0.015	0.03	≤0.015
DH2/pLA2917[a]	≤0.008	≤0.015	0.03	≤0.015
DH2/pNJR3-2[b]	≤0.008	≤0.015	0.03	≤0.015
DH1 (no plasmid)	0.03	0.125	0.5	0.125
DH1/pLA2917	0.03	0.25	1.0	0.125
DH1/pNJR3-2	≤0.008	0.03	0.006	≤0.015

a. Vector control plasmid.
b. Gyrase A gene probe.

Table 8.6. Expression of the *E. coli* gyrase A gene in *gyrA*+ and *gyrA P. aeruginosa*

Strain/plasmid	MIC (µg/ml)			
	Ciprofloxacin	Norfloxacin	Enoxacin	Ofloxacin
PAO2 (no plasmid)	0.25	1.0	1.0	2.0
PAO2/pLA2917	0.25	1.0	2.0	2.0
PAO2/pNJR3-2	0.25	1.0	2.0	2.0
PAO4701 (no plasmid)	2.0	8.0	8.0	8.0
PAO4701/pLA2917	2.0	8.0	8.0	8.0
PAO4701/pNJR3-2	0.25	1.0	2.0	1.0
PAO4702 (no plasmid)	4.0	8.0	16.0	16.0
PAO4702/pLA2917	4.0	8.0	16.0	16.0
PAO4702/pNJR3-2	0.25	2.0	2.0	2.0
PAO47033 (no plasmid)	4.0	8.0	16.0	16.0
PAO4703/pLA2917	2.0	8.0	16.0	16.0
PAO4703/pNJR3-2	0.25	2.0	2.0	2.0
PAO4704 (no plasmid)	4.0	8.0	16.0	32.0
PAO4704/pLA2917	4.0	8.0	16.0	16.0
PAO4704/pNJR3-2	0.25	2.0	2.0	2.0

For analysis in *P. aeruginosa*, pLA2917 and pNJR3-2 were introduced via conjugal mating into the wild type gyrase A[+] strain PAO2 as well as several well characterised *gyrA* mutants of this strain (Robillard and Scarpa 1988). Since these mutants were isolated by selection with ciprofloxacin, they have *cfxA* designations. The 4-quinolone MICs for these strains are shown in Table 8.6. The vector pLA2917 had no effect on PAO2 or the *gyrA* mutants. As predicted, pNJR3-2 also had no effect on PAO2. However, this result would also be expected if the *E. coli* gyrase A gene was not expressed in *P. aeruginosa*. In each of four *P. aeruginosa gyrA* mutants, pNJR3-2

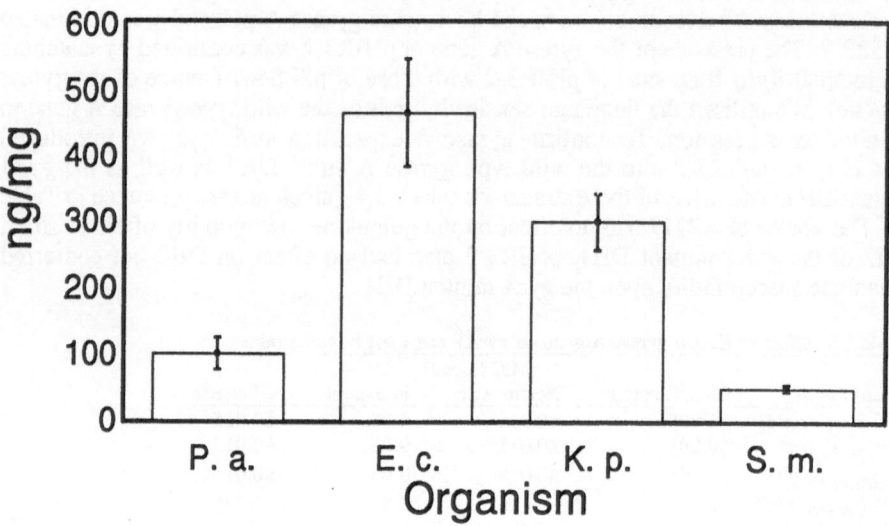

Fig. 8.2. Accumulation of ciprofloxacin. The amount of ciprofloxacin accumulated by *P. aeruginosa* PAO2 (P.a.), *E. coli* H560 (E.c.), *K. pneumoniae* MP100 (K.p.), and *S. marcescens* MP050 (S.m.) cells exposed to 20 μg of ciprofloxacin per ml. ng/mg refers to ng of drug per mg dry weight cells. Values represent the means of four to six experiments.

conferred wild type susceptibility to the 4-quinolones tested, thus demonstrating expression of the *E. coli* gyrase A gene in *P. aeruginosa*. As with *E. coli*, quinolone sensitivity is dominant in these heterodiploids. That pNJR3-2 induced a level of 4-quinolone susceptibility in these *gyrA* mutants which is characteristic of wild type *P. aeruginosa* and had no effect on PAO2, strongly supports our contention that DNA gyrase from *P. aeruginosa* is similar to that of *E. coli* with respect to quinolone sensitivity.

We postulate therefore, that other factors such as permeability play a more substantial role than DNA gyrase in influencing the relative susceptibilities of these organisms to quinolones. Presently, the permeation of quinolones into whole bacteria is difficult to measure because there is no available way to distinguish between binding of drug to the surface of intact cells and drug which has reached the cytoplasm. Recognising this, we still thought it worthwhile to compare the amount of ciprofloxacin accumulated by each wild type organism. We exposed cells to 20μg of ciprofloxacin per ml for 5 min and measured the amount of drug associated with the cells. The results are shown in Fig. 8.2.

E. coli accumulated the most drug (467 ng/mg dry wt) with *K. pneumoniae* accumulating 36% less drug than *E. coli* (300 ng/mg dry wt). *P. aeruginosa* accumulated approximately four-fold less drug than *E. coli* (101 ng/mg dry wt), while *S. marcescens* accumulated nearly ten-fold less drug than *E. coli* (49 ng/mg dry wt).

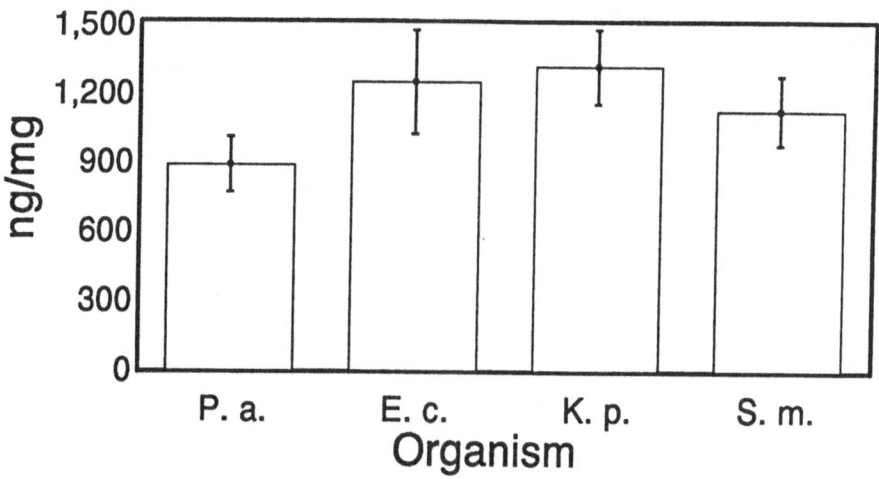

Fig. 8.3. Binding of ciprofloxacin by heat-killed cells. The levels of ciprofloxacin binding to heat-killed *P. aeruginosa* PAO2 (P.a.), *E. coli* H560 (E.c.), *K. pneumoniae* MP100 (K.p.), and *S. marcescens* MP050 (S.m.) exposed to 20 µg of ciprofloxacin per ml. ng/ml refers to ng of drug per mg dry weight cells.Values represent the means of four to six experiments.

Thus the amount of drug accumulated by these organisms followed the same trend as the MIC values, although not necessarily reflecting the magnitude of the difference. In an attempt to quantitate drug binding, and to differentiate this from permeation, we replaced viable cells with heat killed (boiled) cells in the above accumulation experiment. The results are presented in Fig. 8.3. The amount of ciprofloxacin associated with cellular material increased dramatically and was similar for each organism (*P. aeruginosa* and *S. marcescens* still bound less drug than *E. coli* and *K. pneumoniae*). Thus, at least under conditions which render the cells nonviable at 100°C, there is significant ciprofloxacin binding. These results are in contrast to those of Bedard *et al*. (1987) with enoxacin. They found less enoxacin accumulation with heat killed cells. The degree of ciprofloxacin binding which occurs with viable cells is unknown, but these results raise the possibility that quinolone binding to the outer membrane of bacteria may play an important role in permeation. A specific quinolone permeation assay for intact cells would be helpful in addressing this question.

In summary, the intrinsic differences in quinolone susceptibility exhibited by *E. coli*, *P. aeruginosa*, *K. pneumoniae* and *S. marcescens* are not reflected by *in vitro* DNA gyrase inhibition assays. *E. coli* gyrase A and B subunits mixed with gyrase A and B subunits of *P. aeruginosa* also are inhibited similarly by quinolones. Furthermore, these results are corroborated by phenotypic examination of *P. aeruginosa* heterodiploids containing the wild type *E. coli* gyrase A gene. The wild type *E. coli* gyrase A gene, which confers dominant sensitivity in *E. coli* gyrA mutants had no effect on the quinolone susceptibility of wild type *P. aeruginosa*. The same gene conferred quinolone susceptibility on gyrA mutants of *P. aeruginosa*, but only to the *P. aeruginosa* wild type level and not to that of *E. coli*. The implication of permeability as the basis of quinolone susceptibility differences in these organisms is consistent with our ciprofloxacin accumulation data.We would like to expand our studies of quinolone permeability in these organisms to address the role of drug binding in permeation. We also are interested in gaining a deeper understanding of the phenomenon of dominant sensitivity of DNA gyrase genes and the expression of DNA gyrase genes in the heterologous host.

Acknowledgements

We are grateful to Barbara Thurberg for performing the susceptibility determinations presented here. We thank Barbara Painter for her critical review of this work and also Janice Walsh for her assistance with the preparation of this manuscript.

Materials and Methods

Media. ML broth has been described previously (Morrison and Malamy 1970). For ML agar plates, 1.5% agar (Difco Laboratories, Detroit, MI) was added. Mueller-Hinton broth (Difco) was supplemented with $MgCl_2.6H_2O$ and $CaCl_2.2H_2O$ to a final concentration of 25 mg/ml and 50 mg/ml, respectively. Antibiotics, when added for plasmid selection, were used at the following final concentrations: 50 µg/ml, kanamycin; 10 µg/ml, tetracycline for *E. coli*; and 200 µg/ml, tetracycline for *P. aeruginosa*.

Bacterial strains and plasmids. The bacterial strains and plasmids used or constructed in this study are listed in Table 8.1.

Antibiotics. Antibiotics were obtained from the following sources: ciprofloxacin (Miles Inc., Pharmaceutical Div., West Haven CT); nalidixic acid, ampicillin, tetracycline, kanamycin, chloramphenicol (Sigma Chemical Co., St. Louis, MO); norfloxacin (Merck Sharp and Dohme, Rahway, NJ); ofloxacin (Ortho Diagnostics, Raritan, NJ); enoxacin (Warner-Lambert Co., Ann Arbor, MI).

Antimicrobial susceptibilities. Cation-supplemented Mueller-Hinton broth (Difco Laboratories, Detroit, MI) was used to determine minimum inhibitory concentrations (MICs), which were performed according to standard methods (National Committee for Clinical Laboratory Standards 1985).

Isolation of DNA gyrase subunits. *E. coli* DNA gyrase A and B subunits were kindly provided by Linus Shen (Abbott Laboratories, Abbott Park, IL). DNA gyrase A and B subunits from *P. aeruginosa* and *S. marcescens* were isolated as described previously (Robillard and Scarpa 1988; Staudenbauer and Orr 1981) with the following modifications. Gyrase A was eluted from a novobiocin-sepharose column with a 0.05 M to 1.0 M KCl gradient in buffer B (0.05 M KCl, 50 mM N-(2-hydroxyethyl) piperazine-N'-(2-ethanesulfonic acid), 2 mM EDTA, 20% w/v glycerol). Bound gyrase

B was washed extensively with 3 M KCl in buffer B and then eluted with 5 M urea in buffer B. DNA gyrase from *K. pneumoniae* was isolated as described above except the cell extract was passed over the novobiocin-sepharose column directly, omitting the heparin-sepharose column step.

DNA supercoiling assay and IC_{50} determination. Relaxed plasmid pBR322 substrate DNA was prepared by treatment with calf thymus topoisomerase I (Bethesda Research Laboratories, Inc., Gaithersburg, MD) according to the manufacturer's recommended protocol. DNA supercoiling assays were performed as previously described (Mizuuchi *et al.* 1984). One unit of gyrase is defined as that amount of enzyme which catalyses one half maximal supercoiling in 30 min at 37°C in the standard gyrase reaction containing 0.4 µg of DNA. Approximately six units of gyrase A and B were used per reaction. The inhibitory concentration (IC_{50}) is defined as the concentration of antibiotic that inhibits 50% of the supercoiling activity of gyrase in a standard gyrase reaction. Antibiotic was added to each reaction before the addition of gyrase. A control reaction without drug was included in each assay. The IC_{50} was determined by visual inspection of the most supercoiled band in ethidium bromide stained 1.0% agarose gels comparing each reaction to the drug free control.

Construction of a broad host range gyrase A gene probe. pSLS447 contains the wild type gyrase A gene of *E. coli* on a 7 kb *Bam*H1 fragment inserted in the *Bam*H1 site of pBR322 (Swanberg and Wang 1987). pSLS447 was digested with *Bam*H1 and ligated to *Bgl*II-digested pLA2917 (Allen and Hanson 1985) which had been treated with calf intestinal phosphatase. The ligation mixture was transformed into *E. coli* DH1 with selection for tetracycline resistance. Kanamycin sensitive, ampicillin sensitive clones were tested for nalidixic acid sensitising activity (the wild type gyrase A gene confers nalidixic acid-sensitivity on DH1). Recombinant plasmids conferring nalidixic acid-sensitivity were analysed by restriction analysis and transformed into *E. coli* S17-1.

Analysis of gyrase expression in E. coli. *E. coli* DH2 (wild type gyrase A allele) and *E. coli* DH1 (*gyrA* mutant) were transformed with pLA2917 (vector control) or pNJR3-2 (gyrase A probe). The MIC of several quinolones was determined for each strain in the presence or absence of each plasmid. The MICs for strains containing plasmid were performed in the presence of 2 µg of tetracycline per ml.

Analysis of gyrase expression in P. aeruginosa. pLA2917 and pNJR3-2 were introduced into *P. aeruginosa* strains by conjugation from *E. coli* S17-1 (Simon *et al.* 1983) which contains chromosomal *tra* genes capable of mobilising pLA2917 or PNJR3-2. For matings, *E. coli* S17-1 donor strains containing pLA2917 or pNJR3-2 and *P. aeruginosa* recipient strains were grown in ML broth overnight at 32°C with aeration. Donor strains were grown in the presence of 5 µg of tetracycline per ml. One-half ml of donor culture was added to 0.5 ml of recipient culture in a microcentrifuge tube and centrifuged. The mating mixture was suspended in 50 µl of 0.15 M NaCl and plated on ML agar. Mating and expression was allowed to occur during incubation for 5 h at 35°C. The mating mixture was harvested in 3 ml of 0.15 M NaCl and plated on ML agar containing 200 µg of tetracycline per ml. Plasmid-containing *E. coli* donor strains cannot grow at this tetracycline concentration while *P. aeruginosa* recipient strains inheriting pLA2917 or pNJR3-2 can grow. Transconjugants were purified on selection media. The MICs of several quinolones were determined for each strain in the presence or absence of each plasmid. The MICs for strains containing plasmid were performed in the presence of 200 μg of tetracycline per ml.

Ciprofloxacin accumulation. Ciprofloxacin accumulation by bacteria was measured as described previously (Celesk and Robillard 1989). Cells harvested in the logarithmic growth phase were suspended to a density of 260 Klett units. Five ml of cell suspension were exposed to 20 μg of ciprofloxacin per ml for 5 min at 37°C with agitation. One ml samples were centrifuged through silicon oil to remove unassociated drug. Cell pellets were resuspended in PBS (pH 7.2) and ciprofloxacin was extracted by a 7 min incubation at 100°C. The amount of drug accumulated per mg of cell dry weight was determined by bioassay. Drug binding also was measured in heat-killed cells (cells which had been incubated at 100°C for 7 min).

References

Allen LN, Hanson RS (1985) Construction of broad-host-range cosmid cloning vectors: identification of genes necessary for growth of *Methylobacterium organophilum* on methanol. J Bacteriol 161:955-962

Aoyama H, Fujimaki K, Sato K, Fujii T, Inoue M, Hirai K, Mitsuhashi S (1988) Clinical isolate of *Citrobacter freundii* highly resistant to new quinolones. Antimicrob Agents Chemother 32:922-924

Aoyama H, Sato K, Fujii T, Fujimaki K, Inoue M, Mitsuhashi S (1988) Purification of *Citrobacter freundii* DNA gyrase and inhibition by quinolones. Antimicrob Agents Chemother 32:104-109

Barry AL, Jones RN (1984) Cross-resistance among cinoxacin, ciprofloxacin, DJ-6783, enoxacin, nalidixic acid, norfloxacin, and oxolinic acid after *in vitro* selection of resistant populations. Antimicrob Agents Chemother 25:775-777

Barry AL, Jones RN, Thornsberry C, Ayers LW, Gerlach ER, Sommers HM (1984) Antibacterial activities of ciprofloxacin, norfloxacin, oxolinic acid, cinoxacin, and nalidixic acid. Antimicrob Agents Chemother 25:633-637

Bedard J, Wong S, Bryan LE (1987) Accumulation of enoxacin by *Escherichia coli* and *Bacillus subtilis.* Antimicrob Agents Chemother 31:1348-1354

Bourguignon GJ, Levitt M, Sternglanz R (1973) Studies on the mechanism of action of nalidixic acid. Antimicrob Agents Chemother 4:479-486

Celesk RA, Robillard NJ (1989) Factors influencing the accumulation of ciprofloxacin in *Pseudomonas aeruginosa.* Antimicrob Agents Chemother. In Press

Chamberland S, Bayer AS, Schollaardt T, Wong SA, Bryan LE (1989) Characterization of mechanisms of quinolone resistance in *Pseudomonas aeruginosa* strains isolated *in vitro* and *in vivo* during experimental endocarditis. Antimicrob Agents Chemother 33:624-634

Cozzarelli NR (1980) DNA gyrase and the supercoiling of DNA. Science 207:953-960

Crumplin GC, Smith JT (1986) Nalidixic acid and bacterial chromosome replication. Nature 260:643-645

Cullen ME, Wyke AW, Kuroda R, Fisher LM (1989) Cloning and characterization of a DNA gyrase A gene from *Escherichia coli* that confers clinical resistance to 4-quinolones. Antimicrob Agents Chemother 33:886-894

Deitz WH, Cook TM, Goss WA (1966) Mechanism of action of nalidixic acid on *Escherichia coli.* III. Conditions required for lethality. J Bacteriol 91:768-773

Fisher LM, Lawrence J, Josty I, Hopewell R, Margerrison EEC, Cullen ME (1989) Ciprofloxacin and the fluoroquinolones: new concepts on the mechanism of action and resistance. Amer J Med In Press

Gellert M, Mizuuchi K, O'Dea MH, Nash HA (1976) DNA gyrase: An enzyme that introduces superhelical turns into DNA. Proc Natl Acad Sci USA 73:3872-3876

Gellert M, Mizuuchi K, O'Dea MH, Itoh T, Tomizawa J (1977) Nalidixic acid resistance: A second genetic character involved in DNA gyrase activity. Proc Natl Acad Sci USA 74:4772-4776

Gellert M (1981) DNA topoisomerases. Ann Rev Biochem 50:879-910

Hane MW, Wood TH (1969) *Escherichia coli* K-12 mutants resistant to nalidixic acid: genetic mapping and dominance studies. J Bacteriol 99:238-241

Hirai K, Aoyama H, Suzue S, Irikura T, Iyobe S, Mitsuhashi S (1986) Isolation and characterization of norfloxacin-resistant mutants of *Escherichia coli* K-12. Antimicrob Agents Chemother 30:248-253

Hirai K, Suzue S, Irikura T, Iyobe S, Mitsuhashi S (1987) Mutations producing resistance to norfloxacin in *Pseudomonas aeruginosa*. Antimicrob Agents Chemother 31:582-586

Hooper DC, Wolfson JS, Ng EY, Swartz MN (1987) Mechanisms of action of and resistance to ciprofloxacin. Am J Med 82(Suppl 4A):12-20

Hooper DC, Wolfson JS, Souza KS, Ng EY, McHugh GL, Swartz MN (1989) Mechanisms of quinolone resistance in *Escherichia coli*: Characterization of *nfxB* and *cfxB*, two mutant resistance loci decreasing norfloxacin accumulation. Antimicrob Agents Chemother 33:283-290

Horowitz DS, Wang JC (1987) Mapping the active site tyrosine of *Escherichia coli* DNA gyrase. J Biol Chem 262:5339-5344

Inoue Y, Sato K, Fujii T, Hirai K, Inoue M, Iyobe S, Mitsuhashi S (1987) Some properties of subunits of DNA gyrase from *Pseudomonas aeruginosa* PAO1 and its nalidixic acid-resistant mutant. J Bacteriol 169:2322-2325

Mizuuchi K, Mizuuchi M, O'Dea H, Gellert M (1984) Cloning and simplified purification of *Escherichia coli* DNA gyrase A and B proteins. J Biol Chem 259:199-201

Morrison TG, Malamy MH (1970) Comparison of F factors and R factors: existence of independent regulation groups in F factors. J Bacteriol 103:81-88

Nagate T, Komatsu T, Izawa A, Ohmura S, Namiki S, Mitsuhashi S (1980) Mode of action of a new nalidixic acid derivative, AB206. Antimicrob Agents Chemother 17:763-769

Nakamura S, Nakamura M, Kojima T, Yoshida H (1989) *gyrA* and *gyrB* mutations in quinolone-resistant strains of *Escherichia coli*. Antimicrob Agents Chemother 33:254-255

National Committee for Clinical Laboratory Standards (1985) Approved standard M7-A, methods of dilution antimicrobial susceptibility tests for bacteria that grow aerobically. National Committee for Clinical Laboratory Standards, Villanova, PA

Rella M, Haas D (1982) Resistance of *Pseudomonas aeruginosa* PAO to nalidixic acid and low levels of ß-lactam antibiotics: Mapping of chromosomal genes. Antimicrob Agents Chemother 22:242-249

Robillard NJ, Scarpa AL (1988) Genetic and physiological characterization of ciprofloxacin resistance in *Pseudomonas aeruginosa* PAO. Antimicrob Agents Chemother 32:535-539

Sato K, Inoue Y, Fujii T, Aoyama H, Inoue M, Mitsuhashi S (1986) Purification and properties of DNA gyrase from a fluoroquinolone-resistant strain of *Escherichia coli*. Antimicrob. Agents Chemother 30:777-780

Simon R, Priefer U, Puhler A (1983) A broad host range mobilization system for *in vivo* genetic engineering: transposon mutagenesis in Gram negative bacteria. Bio/Technology. (USA) 1/9:784-790

Staudenbauer WL, Orr E (1981) DNA gyrase: affinity chromatography on novobiocin-sepharose and catalytic properties. Nucleic Acids Res 9:3589-3603

Sugino A, Cozzarelli NR (1980) The intrinsic ATPase of DNA gyrase. J Biol Chem 255:6299-6306

Sugino A, Peebles CL, Kreuzer KN, Cozzarelli NR (1977) Mechanism of action of nalidixic acid: Purification of *Escherichia coli nalA* gene product and its relationship to DNA gyrase and a novel nicking-closing enzyme. Proc Natl Acad Sci. USA 74:4767-4771

Swanberg SL, Wang JC (1987) Cloning and sequencing of the *Escherichia coli gyrA* gene coding for the A subunit of DNA gyrase. J Mol Biol 197:729-736

Yoshida H, Kojima T, Yamagishi J, Nakamura S (1988) Quinolone-resistant mutations of the *gyrA* gene of *Escherichia coli*. Mol Gen Genet 211:1-7

Discussion Summary

Some surprise was expressed over the observation of the minimal differences in the sensitivities of the isolated DNA gyrases isolated from different species which clearly failed to explain the differences in the susceptibilities of the intact cells. This was clearly in contrast to earlier findings from other workers who had studied *E. coli* and *Ps. aeruginosa* DNA gyrase. However, the use of heterologous enzyme studies had convinced people of the validity of the findings.

In the study of the isolated enzymes, the use of the novobiocin-sepharose method for the isolation of the DNA gyrases was questioned briefly with regard to the use of high concentrations of urea to elute the enzyme from the column. It was generally considered that such harsh treatment of a protein really ought to have some effect upon the molecule, but no alternative could be suggested, especially as the enzyme, thus treated, seemed to work perfectly and did not seem to differ from the same enzymes isolated by less harsh methods. Consequently, we must regard studies using DNA gyrase isolated by this method as being perfectly acceptable in the absence of any scientific basis for instinctive perceptions.

The demonstration that the differences in the susceptibilities of the species studied in this programme could be accounted for by differences in the uptake or absorption of the 4-quinolone molecules was not regarded as surprising. The methodology of using boiled cells generated some wry comments. Dr Robillard countered such comments by establishing that as yet no alternative methodology could be regarded as satisfactory. No-one could argue with the fact, the data presented, clearly supported the proposition that the cell-wall and membranes played a major role in determining the relative susceptibilities of different species of bacteria to the effects of the 4-quinolones.

In conclusion it was universally agreed that there is a need for the development of a satisfactory method for studying the permeation of intact and viable cells by 4-quinolone molecules. It was hoped that someone would be brave enough to take up this challenge.

Chapter 9

Structure-Activity Relationships of Fluoro-4-Quinolones

L. A. Mitscher, P. V. Devasthale and R. M. Zavod

Introduction

The quinolone anti-infective field is now more than twenty years old and several thousand agents have been prepared from which to assemble generalizations connecting structure and activity. This is still a somewhat risky activity as this field all too often throws up an analog which does not fit the classic pattern and yet has excellent activity. Successfully rationalizing such a discovery often teaches us more than the addition of several more typical examples. The establishment of an incorrect orthodoxy by over interpretation of evidence inhibits such liberating discovery and narrows the scope of investigation. Nonetheless, a periodic critical examination of the cumulative experience in the field is essential in planning new work. In this review the primary emphasis will be on the fluoroquinolones of greatest contemporary interest and on the analogs which do not fit comfortably with the rest.

A problem facing the reviewer is finding a generally useful plane of reference. Inhibition of the function of DNA gyrase is a striking characteristic of the quinolones and, since it is a cell free activity, measurements in this system provide simplified data uncomplicated by questions of the role of additional contributing biological effects, cellular uptake, absorption, distribution, metabolism, excretion, toxicity, species differences, and the like. This strength is however also a weakness as these factors are nontrivial determinants of the potential value of a quinolone and we would like to have comparative measures of them. The DNA gyrases differ in sensitivity from species to species and not all have been purified sufficiently for use in any case. Furthermore, DNA gyrase activity is customarily measured using circular plasmid DNA as the substrate. Elsewhere in this volume it is pointed out that more relevant, though much more complex, measurements would require the use of chromosomal DNA (Crumplin 1989). In the following, *in vitro* activity data will be given the most weight and the available evidence from other characteristics will be brought into the discussion where meaningful insights can be gained thereby. For comparative purposes a common Gram-negative pathogen (*Escherichia coli*), a difficult Gram-negative pathogen (*Pseudomonas aeruginosa*) and a Gram-positive pathogen (*Staphylococcus aureus*) have been selected from the many possible species to be the prime comparison organisms. Potency is affected by media choices, strains chosen, etc., so where possible side-by-side comparisons have been made utilizing data from a single investigation. There are important gaps in the activity spectrum of most quinolones. These include important Gram positive pathogens (*Streptococci*, for example), anaerobes, fungi, and viruses. Data for these organisms is often unavailable so these are reluctantly omitted from this work. The DNA gyrase data involves the enzyme from *E. coli*.

Table 9.1. Classic Quinolones and their *in vitro* Biological Activities (Domagala *et al.*, 1986a)

Name	Ec	Pa	Sa	DNA Gyrase Cleavage	IC50
NAL	6.3	>100	>100	50	>100
PIR	25	>100	12.5	38	>100
PIP	1.6	6.3	50	50	>100
OXO	0.2	6.3	1.6	10	25
CIN	3.1	>100	>100	43	>100
MIL	0.4	12.5	6.3	10	50
ROS	0.2	6.3	0.4	2.5	>10
NOR	0.025	0.2	0.8	1	5.5
PEF 0	.025	0.4	0.2	1	5.5

The breakthrough discovery ushering in the modern quinolone era was norfloxacin with its combination of a 6-fluoro and a 7-piperazinyl moiety (Koga *et al.* 1980). This substance is the benchmark for comparison of subsequent discoveries.

The structures of the most important of the classic quinolones and their comparative potencies are given in Table 9.1. The *in vitro* superiority of oxolinic acid (OXO), miloxacin (MIL) and rosoxacin (ROS) over nalidixic acid (NAL) is readily apparent. The further incremental advancement represented by norfloxacin (NOR) and pefloxacin (PEF) is also obvious (Domagala *et al.* 1986).

The classic (pre-norfloxacin) review of the structure-activity relationships of quinolone anti-infective agents is that of Albrecht (1977). The generalizations abstracted from Albrecht are as follows:

Ring Systems. An aromatic 5,6-annelated-4-pyridone-3-carboxylic acid or a relatively obvious variant on this was essential.
Position 1. Small unbranched aliphatic moieties or their bioisosteric equivalents (OMe, for example) were best with activity often optimizing with ethyl. Vinyl was nearly as good as ethyl. Polar substituents usually decreased activity. Bridging between positions 1 and 8 was satisfactory.
Position 2. Very little work was available but substitutions were usually deleterious except for bioisosteric replacement of CH by N.
Position 3. A carboxyl group or a group converted *in vivo* to a carboxyl group was essential.
Position 4. A carbonyl group was essential.
Position 5. Relatively little work was available, but substituents generally decreased potency.
Position 6. Substitution generally gave potency decreases. Small groups were tolerable such as CH, OR, N and CF.
Position 7. Considerable tolerance for substitution was seen at this center and the substituent could be relatively large. Bridging between positions 6 and 7 was satisfactory.
Position 8. Substitution generally gave potency decreases. Bioisosteric replacement of CH by N was satisfactory.

These rules summarize the position after the first 1000 or so analogs had been investigated. The data in Table 9.1 are consistent with these generalizations. The striking potency of norfloxacin (NOR) and pefloxacin (PEF) against both DNA gyrase and against intact organisms compared with their predecessors explain the intense interest and explosion of activity which followed their preparation. It is interesting in this context to note that they both fit the classic understanding of structure-activity relationships enunciated by Albrecht. Their anti-infective properties are equivalent to those of many useful fermentation-derived agents and the thousands of agents which have been prepared as a consequence are generally faithful to the pattern of a fluoro substituent at C-6 and a basic substituent at C-7.

Now let us see in what ways the classic rules must be modified by the examination of the additional 6000 or so analogs prepared since Albrecht published his review.

Table 9.2. N-1 Substituents; Aliphatic - Non Polar

R	Ec	Pa	Sa	Ref
CH$_3$	0.39	1.56	6.25	Koga et al., 1980
C$_2$H$_5$ *	0.2	0.78	0.78	Koga et al., 1980
CH=CH$_2$	0.1	0.2	1.56	Matsumoto et al., 1984
n-C$_3$H$_7$	0.20	3.13	1.56	Koga et al., 1980
CH$_2$CH$_2$F	0.10	0.78	1.56	Koga et al., 1980
CH$_2$CH=CH$_2$	0.20	1.56	3.13	Koga et al., 1980
CH$_2$Ph	0.78	1.56	1.56	Koga et al., 1980
CH(CH$_2$)$_2$ **	0.05	0.40	3.10	Domagala et al., 1988
CH(CH$_3$)$_2$	0.5	1	1	Bouzard et al., 1989
C(CH$_3$)$_3$ ***	0.06	0.5	0.06	Bouzard et al., 1989
CH$_2$CH(CH$_2$)$_2$	0.5	4	1	Bouzard et al., 1989

```
*      Norfloxacin
* *    Ciprofloxacin
* * *  BMY - 40062
```

Table 9.3. N-1 Substituents; Aliphatic - Polar

R	X	Ec	Pa	Sa	Ref
C$_2$H$_5$ *	CH	0.05	0.39	0.39	Koga et al.,1980
CH$_2$CH$_2$OH	CH	0.39	3.13	1.56	Koga et al.,1980
NHCH$_3$	CH	1.0	1.0	1.95	Wentland et al.,1984
C$_2$H$_5$ **	N	0.1	0.8	3.1	Domagala et al.,1986
CH$_2$CH$_2$OH	N	1.56	50	6.25	Nishimura et al.,1988
(CH$_2$)$_3$OH	N	1.56	50	25	Nishimura et al.,1988

```
*     Norfloxacin
* *   Enoxacin
```

Variations at N-1

Table 9.2 compares a number of quinolones with various non-polar N-aliphatic substituents. From these selected examples, and those in the following tables, three important exceptions emerge. Whereas N-ethyl and small groups of comparable size and nonpolarity are clearly still very useful (Koga *et al.* 1980), an N-cyclopropyl moiety is clearly superior against DNA gyrase function and in intact bacterial cells (Domagala *et al.* 1988a). This would not have been predicted by the Albrecht correlations. This group is bigger than an N-ethyl and, significantly, N-isopropyl is deleterious. The reason for enhanced activity for ciprofloxacin (CIP) and related molecules is still being debated. The addition of an N-cyclopropyl moiety has a favourable influence on bioactivity in almost every series in which it has been included. It is possible that the enhancement is due to the new possibilities for self-association or for hyperconjugation introduced by the cyclopropyl ring. In this sense the group would be analogous, but superior, to N-vinyl substitution. A second exceptional finding is that analogs containing an N-tertiary-butyl group have good activity (Bouzard *et al.* 1989). True, these compounds have less activity against Gram negatives than N-ethyl or N-cyclopropyl, but they have relatively good anti Gram positive activity despite the fact that the classic rules would have predicted confidently that a tertiary-butyl moiety attached to N-1 would give entirely worthless analogs. It is also interesting to note that ciprofloxacin is very effective against Gram positive microorganisms as compared to norfloxacin and that this inclusion broadens the practical activity spectrum of quinolones.

Table 9.3 compares a selected group of N-1 substituents containing polar moieties. It is comforting to note that these decrease potency (Koga *et al.* 1980; Wentland *et al.* 1984). An important question to be answered is whether generalizations arrived at in one ring system can be carried over to other ring systems. It will subsequently be seen that one sometimes can and sometimes cannot. There are enough exceptions in both directions to make it prudent to find out each time a different ring system is employed. In this particular case it is found that polar groups are deleterious at N-1 in both the parent ring system and in the 1,8-naphthyridine series (Domagala *et al.* 1986; Nishimura *et al.* 1988). The N-1 NHMe bioisosteric replacement group for N-1 ethyl, such as is seen with amifloxacin and its analogs, looks to be at variance with this idea. Analogs containing this feature have interesting activity levels and it is the most polar function in this table.

A third iconoclastic series is that employing N-aryl groups (Table 9.4). This substitution pattern was not considered in the classic era but the size of the benzene ring would have made it likely that they would not be active. This would be wrong. While the simple N-phenyl analog is less active than N-ethyl or N-cyclopropyl (Domagala *et al.* 1988a; Chu *et al.* 1985), those with relatively small para groups (with or without small ortho substituents) are nicely active. Fluorine and hydroxyl are particularly good. From a long series, the para-fluoro and -hydroxy and the ortho, para-difluorophenyl analogs are the best (Chu *et al.* 1987; Nishimura *et al.* 1988). *In vitro* difloxacin is more active than norfloxacin against Gram positives but is less active against Gram negatives. These differences are much less significant in *in vivo* studies. Inserting spacer methylenes between N-1 and the benzene ring is deleterious as is enhancing the size of the para substituent. The reason for the attractive biological properties of selected examples of this series is not known. There is no doubt that the aromatic ring is bigger in its long dimension than either an ethyl or a cyclopropyl group. It is also known, however, that the N-1-aryl rings lie about 30 degrees away from orthogonality in quinolones such that π stacking can occur between them (Mitscher, Schlemper and Gollapudi, unpublished X-ray results). Thus additional self association can take place analogously to that seen from X-ray studies of nalidixic acid in which the N-ethyl groups of two adjacent molecules in the same plane approach each other rather closely (Achari and Neidle 1976). Substituent changes at C-8 give three parallel series of analogs in which the same structure-activity considerations hold.

Table 9.4. N-1 Substituents; Aromatic

R	X	Y	Ec	Pa	Sa	DNA gyrase Cleavage	Ref
Et *	CH	H	0.025	0.2	0.8	5.5	Domagala et al., 1988
Et **	CH	CH3	0.025	0.4	0.2	-	Chu et al., 1985
Ph	CH	CH3	0.78	6.2	0.78	-	Chu et al., 1985
4-F-Ph***	CH	CH3	0.2	1.56	0.2	-	Chu et al., 1985
3-F-Ph	CH	CH3	6.2	50	12.5	-	Chu et al., 1985
2-F-Ph	CH	CH3	0.78	6.2	1.56	-	Chu et al., 1985
2,4-DiF-Ph	CH	CH3	0.2	1.56	0.1	-	Chu et al., 1985
4-Cl-Ph	CH	CH3	1.56	12.5	1.56	-	Chu et al., 1985
4-Br-Ph	CH	CH3	6.2	50	3.1	-	Chu et al., 1985
4-OH-Ph	CH	CH3	0.1	0.39	0.10	-	Chu et al., 1985
4-OMe-Ph	CH	CH3	50	200	12.5	-	Chu et al., 1985
4-Me-Ph	CH	CH3	1.56	12.5	1.56	-	Chu et al., 1985
2-Me-Ph	CH	CH3	1.56	25	3.1	-	Chu et al., 1985
2,6-Me2-Ph	CH	CH	>100	>100	>100	-	Chu et al., 1985
4-F-Ph	CF	H	0.1	0.78	0.39	1.3	Chu et al., 1987
Et ****	N	H	0.1	1.56	0.78	-	Nishimura et al., 1988
Ph	N	H	0.2	0.78	1.56	-	Nishimura et al., 1988
CH2Ph	N	H	0.39	1.56	3.13	-	Nishimura et al., 1988
(CH2)2Ph	N	H	50	100	12.5	-	Nishimura et al., 1988
(CH2)3Ph	N	H	12.5	>100	1.56	-	Nishimura et al., 1988
(CH2)4Ph	N	H	>100	>100	>100	-	Nishimura et al., 1988

*   Norfloxacin
* *   Pefloxacin
* * *   Difloxacin
* * * * Enoxacin

Table 9.5. The Effect of Varying N-1 Substituents of Rosoxacin and its 1,8-Naphthyridine Analogs
(Nishimura and Matsumoto, 1987)

X	R	Ec	Pa	Sa
CH	Et *	0.39	3.13	0.78
N	Et	0.2	3.13	0.39
N	CH$_2$CH$_2$F	0.2	1.56	0.39
N	CH$_2$CH$_2$OH	1.56	50	50
N	CH=CH$_2$	0.78	6.25	1.56
N	CH(CH$_2$)$_2$	0.1	0.78	0.2

*   Rosoxacin

The data in Table 9.5 show that these structure-activity relationships hold when C-7 carries a 4-pyridyl moiety as in rosoxacin (the first entry) and its 1,8-naphthyridine analogs.

Table 9.6 gives a longer series of side-by-side comparisons accompanied by DNA gyrase data in a series having a 3-ethylaminoethylpyrrolidinyl moiety (Wentland *et al.* 1988; Bouzard *et al.* 1989; Domagala *et al.* 1988a). This moiety was designed using computer graphics techniques to mimic a piperazinyl group. One notes that in this series there is a better, though still not perfect, correlation between the cleavage assay for anti DNA gyrase activity than there is with IC$_\infty$ data. This suggests that quinolone promoted DNA cleavage is a decisive event in cell killing. Against *E. coli*, N-cyclopropyl (PD-117,558) is the best in this series. It has in particular very good anti-Gram positive potency, including activity against *Staphylococcus spp.* The next best analogs are N-ethyl (CI-934) and N-2-fluoroethyl. This is not very surprising but it is interesting to note that whereas gyrase cleavage data would place them approximately correctly, there are other analogs which gyrase would place as equivalent but which are clearly less active in intact cells (N-trifluoroethyl and N-4-fluorophenyl, for example). The generally accepted rationale for this effect is poorer penetration into cells by the less active agents. While this is reasonable and logical, there is very little direct evidence on the point. Just as activity against *E. coli* is not a good predictor of potency against other microorganisms, good inhibition of DNA gyrase from *E. coli* as seen by enhanced DNA cleavage, is also not a good predictor. Substitution of an N-cyclopropyl ring is dramatically better than substitution with larger N-cycloaliphatic rings. Bioisosteric replacements such as O-methyl and N-methylamino are also satisfactory. The N-methylamino moiety is very nearly the same size as N-ethyl (Domagala *et al.* 1988a).

It has been shown that the N-1 atom cannot be replaced by bioisosteric moieties [C-dimethyl (Hogberg *et al.* 1984a), O (Hogberg *et al.* 1984b) and S (S McCombie,

Table 9.6. N-1 Substituents, continued (Domagala et al., 1988)

| | | | | DNA gyrase | |
R	Ec	Pa	Sa	IC_50	Cleavage
CH_3	0.4	1.6	1.6	7.5	5.0
C_2H_5 •	0.1	1.6	0.1	14	2.5
$CH=CH_2$	0.2	1.6	0.4	2.9	0.8
CH_2CH_2F	0.1	1.6	0.4	7.5	2.5
CH_2CF_3	0.8	6.3	0.4	26	2.5
$NHCH_3$	0.2	0.8	0.4	27	2.5
OCH_3	0.4	6.3	0.8	27	2.5
$CH(CH_2)_2$ ••	0.05	0.4	0.05	1.4	0.25
$CH(CH_3)_2$	0.8	12.5	0.4	28	5.0
$CH(CH_2)_3$	0.2	3.1	0.1	26	2.5
$CH(CH_2)_4$	3.1	12.5	0.4	28	5.0
$CH(CH_2)_5$	6.3	>100	6.3	30	10
Ph	1.6	6.3	0.8	29	7.5
4-F-Ph	0.8	1.6	0.1	14	2.5

•   CI-934
• •   PD-117,558

Table 9.7. C-2 Variants (Matsumoto and Miyamoto, 1989)

X	Ec	Pa	Sa
CH	0.1	0.78	0.39
N	3.13	50	12.5

private communication)]. In sum, then, the classic understanding of the relationship between N-1 substitution and biological activity was too restrictive. We now know that certain significantly larger groups are satisfactory. The relationship is not linear and one often cannot at this stage confidently predict in advance of synthesis which groups would be best. Further, there are diversions in structure-activity relationships between Gram-positive and Gram-negative organisms suggesting that larger groups are better for Gram-positives. This implies that a different series of N-1 substituents would be needed for optimal anti Gram-positive and for anti-anaerobe activity.

Table 9.8. Tethered C-2 Variants (Chu et al., 1986a)

Structure	X	Ec	Pa	Sa
a	H	0.2	1.56	0.78
a	CH3	>100	>100	>100
b	S	0.2	1.56	0.39

Substitutions at C-2

Still very little work has been carried out at this position. Between the publication of Albrecht's review and the appearance of norfloxacin, single substitutions at C-2 of OH and Me were shown to be deleterious (Mitscher et al. 1978). As can be seen in Table 9.7, substitution of N for CH at C-2 significantly reduces the bioactivity of norfloxacin (Matsumoto and Miyamoto, 1989) as it did in the case of cinoxacin vs oxolinic acid (Table 9.1). More recently, as seen in Table 9.8, S-bridging between an N-1 aryl substituent and C-2 has been shown to be compatible with anti-Gram negative activity and, in this particular instance, enhanced anti-Gram positive potency (Chu et al. 1986a). A rather interesting example is seen in Table 9.8. X-ray studies on unbridged N-1 aryl analogs indicate that the aryl ring is about 30 degrees out of orthogonal to the main ring system (Mitscher, Schlemper and Gollapudi, unpublished). The S-bridged molecules are assumed to have these rings coplanar. On the other hand an unbridged analog with 2,6-dimethyl substitution has almost no activity. In this case, the pendant aryl ring is assumed to be essentially orthogonal. These results suggest that groups which perforce must project significantly below the plane of the molecule are inconsistent with activity. Emphasis is placed upon downward projection as deleterious rather than upward projection based upon the results obtained with chiral ofloxacin and chiral methyl flumequine discussed later.

Table 9.9. C-3 Variants

A

	Ec	Pa	Sa	DNA gyrase IC_{50}	Ref
Ciprofloxacin	0.004	0.125	0.25	0.3	Wentland *et al.*, 1988; Domagala *et al.*, 1988
A	0.005	0.02	0.02	0.09	Chu *et al.*, 1989

Table 9.10. C-5 Substituents (Domagala *et al.*, 1988b)

		Ec	Pa	Sa	DNA gyrase IC_{50}	S.c. Cleavage	E.c
R	X						
H	H	0.1	0.2	0.4	2.8	0.5	0.3
H	NH_2	0.013	0.025	0.05	2.8	0.5	0.2
H	$NHCH_3$	0.1	1.6	0.8	13.8	2.5	0.8
H	NHAc	>25	>25	>25	>100	>100	-
CH_3	NH_2	0.013	0.1	0.05	2.8	0.5	1.1
CH_3	$N(CH_3)_2$	>25	>25	>25	>200	>200	-

Table 9.11. C-5 Substituents, continued (Domagala *et al.*, 1986)

X	Ec	Pa	Sa	DNA Gyrase	
				IC$_{50}$	Cleavage
H *	1.6	6.3	50	>100	50
NH$_2$	6.3	25	50	-	100

* Pipemidic acid

Substitutions at C-3

Quite recently it has been shown that the C-3 carboxyl moiety can be exchanged for a bioisosteric isothiazolone moiety bridged from C-2. The proton attached to N in this ring system is quite acidic due to aromatic resonance and the resulting analogs are up to ten-fold more potent *in vitro* across the board as compared to the corresponding analogs with a carboxyl group (Table 9.9) (Chu *et al.* 1989). All other previous carboxyl group substitutions had given poorly active analogs (Yanagisawa *et al.* 1973). Here is yet another instance where the classic rules must be modified in the light of subsequent experience.

Substitutions at C-4

There has been no relevant recent work at this position.

Substitutions at C-5

Data in Table 9.10 show the novel recent finding that small basic substitutions at C-5 very significantly enhance broad spectrum potency in some quinolone series. This effect is more pronounced in the C-7 piperazinyl substituted series than where C-7 has an aminopyrrolidyl moiety. A comparison with the DNA gyrase inhibitory activity suggests that this effect is the result of better penetration into cells rather than better activity at the enzyme level (Domagala *et al.* 1988b). This effect, however, can be seen from the data in Table 9.11 not to hold in the fused pyrimidyl-4-quinolones (Domagala *et al.* 1986). The classic belief that non-basic C-5 substituents are deleterious is reinforced by the sampling in Table 9.12 selected from some benzothiazine analogs related to rufloxacin (Cecchetti *et al.* 1987).

Table 9.12. C-5 Substituents; Benzothiazine Series (Cecchetti et al., 1987)

R	R'	Ec	Pa	Sa
H	H	0.78	50	6.25
Cl	H	>50	>50	>50
H *	N-Me-Pipz.	0.78	12.5	0.78

  *   Rufloxacin

Table 9.13. C-6 Substituents, Norfloxacin Series (Koga et al., 1980)

R	Ec	Pa	Sa
H	0.78	3.13	12.5
F *	0.05	0.39	0.39
Cl	0.20	3.13	1.56
Br	0.39	12.5	3.13
CH_3	0.39	6.25	3.13
SCH_3	0.78	12.5	25
$COCH_3$	100	>100	100
CN	0.39	6.25	12.5
NO_2	0.78	12.5	25

  *   Norfloxacin

Table 9.14. C-6 Substituents, C-7 4-Pyridyl Series (Nishimura *et al.*, 1987)

X	Ec	Pa	Sa
H	0.78	12.5	1.56
F	0.2	3.13	0.39

Table 9.15. C-6 Substituents, Naphthyridine Series

X	Y	Ec	Pa	Sa	DNA gyrase IC$_{50}$	Cleavage	Ref
H	CH	6.3	25	>50	-	18	Domagala *et al.*,1986a
F	CH	0.025	0.2	0.8	1	5.5	Domagala *et al.*,1986a
H	N	6.3	12.5	50	-	75	Domagala *et al.*,1986a
F	N	0.1	0.8	3.1	28	5	Domagala *et al.*,1986a
Cl	N	0.78	6.25	3.1	-	-	Matsumoto *et al.*,1984
CN	N	1.56	6.25	6.25	-	-	Matsumoto *et al.*,1984
NO$_2$	N	6.25	25	6.25	-	-	Matsumoto *et al.*,1984
NH$_2$	N	3.13	6.25	>100	-	-	Matsumoto *et al.*,1984

Substitutions at C-6

A fluoro substituent at C-6 is superior to all tried to date when C-7 has a basic substituent (see the selection in Table 9.13)(Koga *et al.* 1980). As CH is smaller than CF, the effect must be more than steric. The comparison in Table 9.14 shows that fluoridation enhances the activity of rosoxacin (X=H) as well (Nishimura and Matsumoto 1987). The data in Table 9.15 indicates that this effect is found in a large number of analogs containing the 1,8-naphthyridine ring system with a pendant C-7 piperazinyl moiety and is mediated at the DNA gyrase level (Domagala *et al.* 1986; Matsumoto *et al.* 1984). Interestingly with nalidixic acid itself (Table 9.16), a C-7 F atom gives a significant enhancement of activity. The enzyme inhibitory activity of this analog is, however, unimpressive indicating that penetration factors are most likely to be important factors in this case. It is sad that this discovery was not made much earlier. Strategically, the desirability of having a C-7 fluoro substituent has had the corollary effect of decreasing contemporary interest in analogs of pipemidic acid wherein such analogs are structurally impossible.

Table 9.16. C-6 Substituents, Nalidixic Acid Series (Domagala *et al.*, 1986)

X	Ec	Pa	Sa	DNA gyrase IC$_{50}$	Cleavage
H *	6.3	>100	>100	>100	50
F	0.2	12.5	1.6	-	25

* Nalidixic Acid

Substitutions at C-7

Apparently minor substituent changes at C-7 can have profound effects on potency, spectrum, pharmacokinetics and solubility. This part of the molecule is very forgiving so a large number of analogs have been prepared in every ring system with variants here. Small, non-basic substituents are inferior to a piperazinyl moiety across the board (Table 9.17). This effect is particularly pronounced against *Pseudomonas aeruginosa*. It is also consistent with the findings of the classical era (Table 9.1). It is interesting to note that placing a basic moiety at the end of a spacer arm attached to C-7 does little to bridge the activity gap. Aside from this, *E. coli* is sensitive to a wide variety of structural variants at C-7 and *S. aureus* is intermediate. Interestingly, a 4-dimethylaminopiperidinyl moiety conveys activity nearly as well as does methylpiperazyl and this includes activity against *Ps. aeruginosa*. Non basic bioisosteric replacements for piperazinyl rings enhance anti-Gram positive activity but decreases anti-Gram negative action. The substituent on the N-4' atom should be small for optimal *in vitro* activity also. The piperazinyl moiety seems to play a dual role. It not only enhances inhibition of gyrase function but also promotes activity *in vitro*. It is generally seen also that analogs with a methyl group at N-4 of the piperazinyl moiety are less active *in vitro* than those with a hydrogen atom instead. The situation, however, is reversed *in vivo* (Koga *et al.* 1980). It has been found experimentally in

Table 9.17. C-7 Substituents, Norfloxacin Series (Koga *et al.*, 1980)

R	Ec	Pa	Sa
Cl	1.56	100	12.5
CH₃	0.39	50	6.25
* Me—N⟨piperazinyl⟩N—	0.10	1.56	0.39
(CH₃)₂N—	0.39	50	0.78
⟨pyrrolidinyl⟩N—	0.39	12.5	0.20
⟨piperidinyl⟩N—	1.56	50	0.78
O⟨morpholinyl⟩N—	0.20	12.5	0.78
HO—⟨piperidinyl⟩N—	0.20	12.5	0.39
H₂NOO—⟨piperidinyl⟩N—	1.56	100	1.56
Me₂N—⟨piperidinyl⟩N—	0.10	3.13	0.39
HN⟨(C=O)⟩N—	0.39	12.5	3.13
H₂N(CH₂)₂NH(CH₂)₂NH—	6.25	50	>100
Et—N⟨piperazinyl⟩N—	0.10	3.13	0.39
PhCH₂—N⟨piperazinyl⟩N—	0.78	50	0.39

* Pefloxacin

some cases that this methyl group is removed by metabolism. Thus the methyl group seems to enhance oral absorption and then is conveniently removed *in vivo* to give the more active analog. A significant number of analogs variously substituted with metabolically labile groups on the N-4 atom of the piperazinyl moiety have been prepared to exploit this tendency. They are said to be pro-drugs but the classical definition of a pro-drug is an analog which is inactive until metabolized. Thus, strictly speaking, these are not pro-drugs. In Table 9.18 it is seen that the piperazinyl moiety is generally superior also in the N-1 aryl series (Chu *et al.* 1985). The second entry in this table is difloxacin. It can be seen also in the table, however, that an analog containing a 3-aminopyrrolidinyl ring at C-7 is actually more active against all species reported so, clearly, leeway is still available for potency enhancement by molecular manipulation. On the down side, however, these compounds are apparently less water soluble. Increasing the steric bulk of the amino group by alkylation decreases activity in this series. Analogous effects occur in the 1,8-naphthyridine N-1 aryl series as well (Table 9.19)(Chu *et al.* 1986b). In Table 9.19 it can be seen that the 3-aminopyrrolidinyl group is also compatible with potent oral and subcutaneous activity against *E. coli* (Chu *et al.* 1986b). Tosufloxacin is an outstanding N-aryl analog in this class (entry four in the table) and it is nearing commercialization. Table 9.20 recapitulates some of these findings but in the benzopyridine class and DNA gyrase inhibitory data are also available for these analogs. It should be noted that the activity against DNA gyrase does not always parallel the MIC data. The best enzyme inhibitor in the table is norfloxacin and it is also the most potent against *E. coli*. The C-7 methyl analog ranks next

Table 9.18. C-7 Substituents, N-Aryl Series (Chu *et al.*, 1985)

R	Ec	Pa	Sa	Sc	Po
* HN–N piperazinyl	0.05	0.39	0.2	0.6	4.3
** Me–N N	0.2	1.56	0.20	1.6	3.1
Me–N N	0.2	1.56	0.78		
O N (morpholino)	0.39	3.1	0.1		
S N (thiomorpholino)	0.78	3.1	0.05		
N piperidinyl	1.56	6.2	0.2		
HO– N	0.39	6.2	0.1		
H₂N– N	0.02	0.2	0.05		
Me₂N– N	0.39	25	0.39		
HN N (oxo)	0.39	6.2	0.2		

* A-56620
** Difloxacin

Table 9.19. C-7 Substituents, N-Aryl Series, continued (Chu et al., 1986b)

X	Ec	Pa	Sa	S.C.	Oral
HN N— (piperazinyl)	0.05	0.39	0.2	-	-
Me—N N— (4-methylpiperazinyl)	0.39	3.1	0.2	-	-
HN N— Me (3-methylpiperazinyl)	0.1	1.56	0.2	1.4	3.9
* H₂N— (aminopyrrolidinyl)	0.02	0.2	0.05	0.2	1.3

*   Tosufloxacin

against *E. coli* but is rather poor against the enzyme. Three analogs (H, Cl, aminoethylthio) are equivalently active against *E. coli* but differ from each other by an order of magnitude against the enzyme. It is rather hard to reconcile these differences other than to invoke putative penetration differences. It is also seen, perhaps less surprisingly, that no parallelism is seen across the four columns of Table 9.20. The larger cyclic groups with a basic nitrogen are generally more active, especially against the enzyme, but also gyrase cleavage inhibition does not parallel MIC values against *E. coli* (Domagala *et al.* 1986). Tables 9.21 and 9.22 contain another extensive collection of variously C-7 substituted N-aryl benzpyridones and naphthyridinones. Once again it can be seen that many C-7 substituents are compatible with excellent bioactivity, mostly when they contain a basic nitrogen or nitrogens (Narita *et al.* 1986a, 1986b). From this mass of data one concludes that a number of other basic ring systems can be substituted for piperazinyl with useful effect suggesting that substantial leeway for further advances lie in finding such alternatives. Particularly the great leeway provided by flexibility at this center promises to allow for satisfactory manipulation of biopharmaceutical features, including solubility, which presently are a barrier to the development of the ideal fluoroquinolone, "utopiafloxacin".

Table 9.20. C-7 Substituents, Norfloxacin Series, continued (Domagala *et al.*, 1986a)

R	Ec	Pa	Sa	DNA gyrase Cleavage
Piperazinyl *	0.025	0.2	0.8	1
H	0.8	>100	>100	50
Cl	0.8	50	3.1	38
CH₃	0.2	12.5	1.6	25
NHCH₃	3.1	>100	12.5	18
S(CH₂)₂NH₂	0.8	3.1	25.0	5
[pyrrolidinyl structure]	1.6	>100	0.8	5
[pyrrolyl structure]	1.6	12.5	0.4	2.5
[thiazolidinyl structure]	0.2	3.1	0.2	5
[thiomorpholinyl structure]	0.1	0.8	0.006	5

* Norfloxacin

Substitutions at C-8

The data in Table 9.23 show that exchange of N for CH at C-8 in the norfloxacin molecule is attendant with satisfactory results (Bouzard *et al.* 1989). Enoxacin is, indeed, a prominent clinical contender. Other experiments indicate that this may often be accompanied by incremental improvements in oral activity as well. Exchange for a CF moiety gives CI-934, an analog which shows good activity against Gram positives. Exchange of C-8 CH or N for C-8 CF leads to a decrease in anti-gyrase activity but still leads to good *in vitro*, oral and subcutaneous potency (Bouzard *et al.* 1989; Domagala *et al.* 1988b). Smaller groups are generally more effective than larger (Figure 9.24) (Sanchez *et al.* 1988). In this group, the naphthyridine analog holds up rather well. Larger substituents are progressively less satisfactory except against *S. aureus* where one sees again the enhanced bulk tolerance of Gram positives. In the C-7 imidazolyl class, a C-8 OMe substituent enhances anti Gram-positive activity but does not produce good antipseudomonal activity. It is interesting to consider that the OMe case represents a figurative ofloxacin analog cleavage product. Making the substituent at C-8 even larger, as with SMe, gives poorer activity across the board.

Table 9.21. C-7 Substituents, Difloxacin Series (Narita et al., 1986a)

R	Ec	Pa	Sa
HN⏝N—	≤0.05	0.39	0.2
Me—N⏝N—	≤0.05	1.56	0.39
HN⏝N—(d, l), Me	≤0.05	0.78	0.2
Me, HN⏝N—	≤0.05	1.56	0.39
Me, N—(d, l)	0.2	12.5	0.2
HO, H₂N—(d, l)	≤0.05	0.78	0.1
N—(d, l)	≤0.05	1.56	0.2
MeH N, N—(d, l)	≤0.05	6.25	0.78
Me₂N, H₂N—N⏝N—	≤0.05	0.78	0.39

* Difloxacin

Benzoxazines

The benzoxazine class of quinolones is generally less active than the C-8 unsubstituted bicyclic analogs against Gram negative organisms and are about equivalent against Gram positives. Adding a C-Me group enhances activity slightly against Gram negatives but dramatically adds to anti Gram positive activity (Table 9.25) (Hayakawa et al. 1984; Wentland et al. 1988). Adding a second methyl group overdoes it. Interestingly the ring forming effect of converting formally the gem-dimethyl group to

Table 9.22. C-7 Substituents, Naphthyridine Series (Narita et al., 1986b)

R	Ec	Pa	Sa
Enoxacin (N-1 = Et)	≤0.05	0.78	0.78
HN N— (piperazine)	≤0.05	0.78	0.2
HN N— (2-methylpiperazine) +/-	≤0.05	3.13	0.39
Me—N N— (Me)	0.1	3.13	0.39
N— (pyrrolidine) +/-	≤0.05	3.13	≤0.05
HO · pyrrolidine N— +/-	≤0.05	0.2	≤0.05
H₂N pyrrolidine N— +/-	≤0.05	1.56	0.2
MeHN pyrrolidine N— +/-	0.1	12.5	0.39
Me₂N ... H₂N—N N—	≤0.05	0.39	0.2

* Tosufloxacin

a spirocyclopropyl moiety greatly compensates for this depressing effect rather reminiscent of going from N-1 isopropyl to N-1 cyclopropyl. The analogs in Table 9.26 include ofloxacin (entry two), apparently the best of this group. All of the listed analogs with a second basic N are very potent against *E. coli* and mostly have reasonable activity against *S. aureus*. It is against *Ps. aeruginosa* that the superiority of ofloxacin in this group is best expressed (Hayakawa et al. 1984).

Benzothiazines

Replacement of the ring oxygen of the benzoxazines by a sulfur produces the benzothiazines. The best of this group is rufloxacin (the last entry in Table 9.27)

Table 9.23. C-8 Substituents; Comparison of Norfloxacin, Enoxacin, and CI-934
(Bouzard et al., 1989)

Name	X	Ec	Pa	Sa
Norfloxacin	CH	0.13	0.5	0.25
Enoxacin	N	0.13	0.5	0.25
CI-934	CF	0.1	1.6	0.05

Table 9.24. C-8 Substituents; continued

	X	Ec	Pa	Sa	Bf	DNA gyrase IC$_{50}$	S.c. E.c.	Ref
*	CH	0.004	0.125	0.25	2	0.03	0.2	Wentland et al., 1988
* *	N	0.06	0.06	0.25	-	5.3	-	Bouzard et al., 1989; Domagala et al.,1988b
	CF	0.03	0.25	0.13	-	2.8	0.3	Bouzard et al., 1989; Domagala et al.,1988b

*   Ciprofloxacin
* *   Enoxacin

(Cecchetti et al. 1987). While not as potent as ofloxacin (entry one), it has good activity against important veterinary pathogens and good pharmacokinetic characteristics. It is being prepared for introduction for veterinary purposes.

Table 9.25. C-8 Substituents, 3-Aminopyrrolidinyl Series (Sanchez *et al.*, 1988)

X	Ec	Pa	Sa	DNA gyrase Cleavage	S.c E.c.
CH	0.025	0.1	0.1	0.1	0.6
CF	0.013	0.1	0.05	0.1	0.2
CCl *	0.013	0.05	0.05	0.5	0.5
CNO$_2$	0.2	1.6	0.8	-	9
CNH$_2$	0.4	1.6	3.1	18	6
N	0.013	0.05	0.2	1.0	0.6

*   AM-1091

Table 9.26. C-8 Substituents, C-7 Imidazolyl Series (Uno *et al.*, 1987)

X	Ec	Pa	Sa
CH	1.56	12.5	6.25
CF	0.20	6.25	0.39
COCH$_3$	1.56	25	0.39
CSCH$_3$	6.25	25	3.13

Optical Activity

The only presently available molecular theory of the molecular mode of action of the quinolones is based upon transition state inhibitor theory (Shen *et al.* 1989). In this view, the gyrase binds to duplex DNA and creates a localized bubble or melt zone separating the strands. Bubble formation is a necessary step in order to provide access for the drugs to the bases in the interior of the double helix and to prepare the molecule for passage of the rest of the DNA molecule followed by resealing and dissociation to finish the catalytic cycle. Quinolones are postulated to self assemble as a sort of liquid crystal in this enzyme created transition state and to stabilize it by hydrogen bonding

Table 9.27. The Benzoxazine Series

X	R	Ec	Pa	Sa	Ref
HN N—	H	<0.05	3.13	1.56	Hayakawa et al., 1984
HN N— (*)	CH$_3$	<0.05	1.56	0.19	Hayakawa et al., 1984
Me—N N—	CH$_3$	<0.05	0.78	0.78	Hayakawa et al., 1984
N—	CH$_3$	<0.05	0.39	<0.05	Hayakawa et al., 1984
H$_2$N N N—	CH$_3$	<0.05	6.25	0.39	Hayakawa et al., 1984
Me—N N—	-(CH$_2$CH$_2$)-	0.125	4.0	2.0	Wentland et al., 1988
Me—N N—	(CH$_3$)$_2$	0.5	32	2.0	Wentland et al., 1988

* Ofloxacin

and prevent the completion of the reaction. The molecular features required of the quinolones for this interaction are that they line up head to tail in a vertical dimension as a consequence of their amphoteric character (acid on one end and piperazine or substitute on the other). These interactions also dictate their head to tail self association in the second plane. The hydrophilic groups at C-3 and C-4 are exterior and in a position to hydrogen bond to the bases newly exposed by the enzyme. Small non-polar groups or aromatic or hyperconjugative groups on N-1 are necessary to allow close approach and self association on the inside of the multiplex which is proposed to be very lipophilic. These considerations are generally in agreement with the structure-activity relationships discussed above. As DNA is chiral, it seems possible that the putative binding site might also be chiral. In this sense it is interesting to note that recent investigations indicate that in favorable cases introduction of chirality into these totally synthetic molecules can enhance not only potency but give increased water solubility and perhaps also diminish side effects.

Table 9.28. The Benzoxazine Series, continued (Hayakawa et al., 1984)

X	Ec	Pa	Sa
* HN⎯N—	<0.05	1.56	0.19
Me—N⎯N—	<0.05	0.78	0.39
Et—N⎯N—	<0.05	1.56	0.39
n-Pr—N⎯N—	0.19	6.25	0.78
N—	0.78	6.25	0.19
H₂N⎯N—	0.19	6.25	0.39
HO⎯N—	0.19	6.25	0.10
NH₂	0.10	12.5	3.13
N(CH₃)₂	0.10	6.25	0.39

*   Ofloxacin

N-1 Chiral Substitutions

Molecular rigidification should heighten chiral selectivity effects and substituents attached to N-1 should be particularly sensitive to this as the molecules should approach each other quite closely in the Shen hypothesis. In is interesting to note from Table 9.28 that S-ofloxacin is significantly more potent than R-ofloxacin. The effect is seen to be 30-fold against gyrase action, two-fold against *Ps. aeruginosa*, eight-fold against *E. coli* and 20-fold against *S. aureus* (Atarashi *et al.* 1987; Mitscher *et al.* 1987; Wentland *et al.* 1988; Hayakawa *et al.* 1984). The disparity suggests the possibility that uptake may also be a chiral process. Perhaps equally interestingly is the recent reports that S-ofloxacin is up to ten-fold more water soluble than the racemate and that it binds less effectively to GABA receptor preparations and so may be less proconvulsant. These factors suggest that development of chiral ofloxacin may be more

Table 9.29. Benzothiazine Series (Cecchetti *et al.*, 1987)

R	X	Y	Z	Ec	Pa	Sa
CH₃ *	O	F	Me—N⌐⌐N— (N-methylpiperazinyl)	0.19	1.56	0.19
H	S	F	(pyrrolidinyl-N—)	1.56	3.12	<0.39
H	S	F	HN⌐⌐N— (homopiperazinyl)	0.39	12.5	0.78
H **	S	F	Me—N⌐⌐N— (N-methylhomopiperazinyl)	0.78	12.5	0.78

* Ofloxacin
** Rufloxacin

Table 9.30. Optical Activity; N-1; Benzoxazine Series (Atarashi *et al.*, 1987; Mitscher *et al.*, 1987; Wentland *et al.*, 1988; Hayakawa *et al.*, 1984)

	Ec	Pa	Sa	DNA gyrase IC₅₀
d,l	0.5	0.78	0.39	-
R	0.39	12.5	5.0	27
S	0.05	0.39	0.02	0.9

profitable than development of the racemate. In the methyl flumequine case (Table 9.29) the eudismic ratio is still in favor of the S-enantiomer (Gerster *et al.* 1987). The effect is relatively small against *Ps. aeruginosa*. Disappointingly, it is reported that testing of congeners indicates that the more potent series is also the more pro CNS toxic (Gerster *et al.* 1989). Table 9.30 shows that artificial introduction of chirality into ciprofloxacin at N-1 results in significant chiral preference for potency but at the cost of significant loss of overall potency. Interestingly one of the phenylcyclopropyl analogs is exceptionally effective at inhibiting the action of Gram-positive *Micrococcus luteus* (Mitscher *et al.* 1986 and unpublished). Part of the reason for lower activity here could be that added steric bulk between molecules could inhibit self association. A lower eudismic ratio would, in this view, result from lesser intrinsic "fit" and conformational mobility lowering its disymmetric impact.

Table 9.31. Optical Activity; N-1; Flumequine Series (Gerster *et al.*, 1987)

		Ec	Pa	Sa
d,l	*	0.2	6.2	0.05
R		6.2	12.5	3.1
S		0.1	3.1	0.025

*   S-25930

One series which does not fit this otherwise neat pattern is illustrated in Table 9.31. It is reported in meeting abstracts that the most active compounds in this ring system are R. The effect is small (Georgopapadakou *et al.* 1985).

C-7 Chiral Substitutions

Table 9.32 indicates that only a small chiral preference is observed upon introduction of a chiral center in the pyrrolidine series when N-1 is substituted by an ethyl group. The effect is more pronounced in intact cells and in animal protection tests than against the enzyme assay once again raising the possibility of chiral influences on uptake. One further notes that the model of the active conformation would have C-7 significantly away from close contact with either another molecule, the bases in the DNA bubble,

Table 9.32. Optical Activity; N-1; Ciprofloxacin Analogs
(Mitscher et al., 1986; Mitscher et al., unpublished)

	Ec	Pa	Sa	DNA Gyrase IC$_{50}$	Ref
	50	>100	3.10	14.2	a
Ph	12.5	>50	0.78	12.1	a
	>100	>100	>100	34.0	b
Me	25	0.78	25	19.0	b
	0.1	3.1	1.56	2.6	b
	0.05	1.56	1.56	2.7	b

Table 9.33. Optical Activity; C-7; Alkylamino Alkylpyrrolidinyl Series (Culbertson et al., 1987)

X	Ec	Pa	Sa	DNA gyrase IC$_{50}$	Cleavage	S.c. E.c.
-CH$_2$NHEt	0.1	1.6	0.1	13.8	2.5	2
-CH$_2$NHEt	0.1	3.1	0.1	13.8	2.5	3
--CH$_2$NHEt	0.2	3.1	0.2	13.8	2.5	7

Table 9.34. Optical Activity; C-7; Aminopyrrolidinyl Series (Rosen *et al.*, 1988a,b)

X	Ec	Pa	Sa	DNA gyrase IC$_{50}$	S.c. E.c.
-NH$_2$	0.02	0.1	0.02	0.4	0.3
-NH$_2$	0.05	0.2	0.05	1.2	-
--NH$_2$	0.02	0.05	0.02	0.16	0.2

Table 9.35. Optical Activity; C-7; Diastereomeric Pyrrolidinyl Series (Rosen *et al.*, 1988a,b)

R	X	Y	Z	Ec	Pa	Sa	DNA gyrase IC$_{50}$	S.c. E.c.
H	--CH$_2$OH	H	CH	0.78	3.1	0.1	-	-
H	-CH$_2$OH	H	CH	50	>100	3.1	-	-
F	--CH$_3$	NH$_2$	CH	0.05	3.1	0.02	0.9	0.9
F	-CH$_3$	NH$_2$	CH	0.78	5.0	0.39	18	-
F *	--CH$_3$	NH$_2$	N	0.02	0.78	0.02	0.6	0.9
F	-CH$_3$	NH$_2$	N	0.39	12.5	0.2	7.5	-

*   Tosufloxacin

and the bifurcating fork (Culbertson *et al.* 1987). In the N-1 phenyl system, Table 9.33 indicates a larger, but still relatively small, eudismic ratio (Rosen *et al.* 1988a, 1988b). When diastereoisomers are involved (Table 9.34), a significantly larger eudismic ratio is seen as it was at N-1. At this position the chiral influence seems to be associated more with the enzyme than with penetration. Where diastereoisomers are possible (Table 9.35) (notably with tosufloxacin, the penultimate entry) greatly enhanced solubility is seen in comparison with the desmethyl analogs. This is paralleled by increased oral activity (Rosen *et al.* 1988a, 1988b). Rigid analogs in which C-7 substituents are tethered to C-8 are just now appearing in the literature and no data is presently available about their chiral preferences.

Summary

It is now possible to bring the Albrecht generalizations up to date:

N-1 should contain compact lipophilic substituents. Groups which can enhance self association are superior. Chiral groups can lead to further enhancements of activity if the correct enantiomer is selected.

C-2 may contain substituents provided that they bridge to adjacent centers (N-1 or C-3). There is too little evidence to make this a firm recommendation.

C-3 can contain certain carboxyl surrogates but the group should still be acidic.

C-4 has not been further investigated so the original suggestion of a necessary carbonyl has not been challenged.

C-5 may contain a small basic substituent in some series.

C-6 should possess a fluorine substituent, especially when C-7 has a basic group attached.

C-7 should possess a basic group and substantial leeway is available for substitution. Chiral substituents may be useful.

C-8 may be CH, N or CF and enhancements are more likely to be seen *in vivo* than *in vitro*.

Ring annelations between N1-C2, N1-C8, C2-C3, and C6-C7 give good results in some cases.

It will be interesting to see how well these considerations hold up twelve years from today. Doubtless, as has happened in the recent past, unanticipated results will appear from time to time allowing for revision of present theories and enhancing progressive improvements toward improved analogs. One welcome development will be the pursuit of novel ring systems possessing DNA gyrase inhibitory activity. A clear start has been made in this direction. (Sutcliffe 1988). Further, the molecular relationship between DNA gyrase and DNA topoisomerases I and II will also surely clarify as will the action of DNA gyrase on bacterial chromosomal DNA. Finally, the structure-activity relationships of side effects (proconvulsant, joint erosion, chelation effects) will progressively emerge.

Acknowledgement

Preparation of this paper was supported in part by a grant from the NIH, USA, Allergy and Infectious Diseases Institute (AI-13155).

References

Achari A, Neidle S (1976) Nalidixic acid Acta Cryst B32: 600-602

Albrecht R (1977) Development of antibacterial agents of the nalidixic acid type. Prog Drug Research 21:11-104

Atarashi S, Yokohama S, Yamazaki K, Sakano K, Imamura M, Hayakawa I (1987) Synthesis and antibacterial activities of optically active ofloxacin and its fluoromethyl derivative. Chem Pharm Bull (Japan) 35:1896-1902

Bouzard D, DiCesare P, Essiz M, Jacquet JP, Remuzon P, Weber A, Oki T, Masuyoshi M (1989) Fluoronaphthyridines and quinolones as antibacterial agents. 1 Synthesis and structure-activity relationships of new 1-substituted derivatives. J Med Chem 32:537-542

Cecchetti V, Fravolini A, Fringuelli R, Mascellani G, Pagella P, Palmioli M, Segre G, Temi P (1987) Quinolonecarboxylic acids. 2 Synthesis and antibacterial evaluation of 7-oxo-2,3-dihydro-7H-pyrido[1,2,3-de][1,4]benzothiazine-6-carboxylic acids. J Med Chem 30:465-473

Chu DTW, Fernandes PB, Claiborne AK, Pihuleac E, Nordeen CW, Maleczka RE Jr, Pernet AG (1985) Synthesis and structure-activity relationships of novel arylfluoroquinolone antibacterial agents. J Med Chem 28:1558-1564

Chu DTW, Fernandes PB, Pernet AG (1986a) Synthesis and biological activity of benzothiazolo [3,2-a]quinolone antibacterial agents. J Med Chem 29:1531-1534

Chu DTW, Fernandes PB, Claiborne AK, Gracey EH, Pemet AG (1986b) Synthesis and structure-activity relationships of new arylfluoronaphthyridine antibacterial agents. J Med Chem 29:2363-2369

Chu DTW, Fernandes PB, Maleczka RE, Jr, Nordeen CW, Pemet AG (1987) Synthesis and structure-activity relationship of 1-aryl-6,8-difluoroquinolone antibacterial agents. J Med Chem 30:504-509

Chu DTW, Fernandes PB, Claiborne AK, Shen LL, Pemet AG (1989) Structure-activity relationships in quinolone antibacterials: replacement of the 3-carboxylic acid group In: Fernandes PB (ed) Quinolones. Prous, Barcelona, Spain pp 37-46

Culbertson TP, Domagala JM, Nichols JB, Priebe S, Skeean RW (1987) Enantiomers of 1-ethyl-7-[3-[(ethylamino)methyl]-1-pyrrolidinyl]-6,8-difluoro-1,4-dihydro-4-oxo-3-quinoline-carboxylic acid: preparation and biological activity. J Med Chem 30:1711-1715

Domagala JM, Hanna LD, Heifetz, CL, Hutt MP, Mich TF, Sanchez JP, Solomon P (1986) New Structure-activity relationships of the quinolone antibacterials using the target enzyme. The development and application of a DNA gyrase assay. J Med Chem 29:394-403

Domagala JM, Heifetz CL, Hutt MP, Mich TF, Nichols JB, Solomon M, Worth DF (1988a) 1-Substituted 7-[3[(ethylamino)methyl]-1-pyrrolidinyl]-6,8-difluoro-1,4-dihydro-4-oxo-3-quinolinecarboxylic acids. New quantitative structure-activity relationships at N-1 for the quinolone antibacterials. J Med Chem 31:991-1001

Domagala JM, Hagen SE, Heifetz CL, Hutt MP, Mich TF, Sanchez JP, Trehan AK (1988b) 7-Substituted-5-amino-1-cyclopropyl-6,8-difluoro-1,4-dihydro-4-oxo-3-quinolinecarboxylic acids: Synthesis and biological activity of a new class of quinolone antibacterials. J Med Chem 31:503-506

Georgopapadakou NH, Dix BA, Angehern P, Wick A, Olson G (1985) Monocyclic and tricyclic analogs of quinolones: Biological properties. Abstr 25th Intersci Conf Antimicrob Agents Chemother (Minneapolis, MN, 29 Sept-2 Oct) Abstr 129

Gerster JF, Rohlfing SR, Pecore SE, Winandy RM, Stern RM, Landmesser JE, Olsen RA, Gleason WB (1987) Synthesis, absolute configuration, and antibacterial activity of 6,7-dihydro-5,8-dimethyl-9-fluoro-1-oxo-1H,5H-benzo[ij]quinolizine-2-carboxylic acid. J Med Chem 30:839-843

Gerster JF, Rohlfing SR, Rustad NJ, Reiter MJ, Pecore SE, Winandy RM, Landmesser JE (1989) The synthesis and pharmacological profile of the stereoisomers of a tricyclic quinolone antibacterial In: Fernandes PB (ed) Quinolones. Prous Science Publishers, Barcelona, Spain pp 85-98

Hayakawa I, Hiramitsu T, Tanaka Y (1984) Synthesis and antibacterial activities of substituted 7-oxo-2,3-dihydro-7H-pyrido[1,2,3de][1,4]benzoxazine-6-carboxylic acids. Chem Pharm Bull (Japan) 32:4907-4913

Hogberg T, Vora M, Drake SD, Mitscher LA, Chu DTW (1984a) Structure-activity relationships among DNA-gyrase inhibitors. Synthesis and antimicrobial evaluation of chromones and coumarins related to oxolinic acid. Acta Chem Scand A 38:359-366

Hogberg T, Khanna I, Drake SD, Mitscher LA, Shen LL (1984b) Structure-activity relationships among DNA gyrase inhibitors. Synthesis and biological evaluation of 1,2-dihydro-4,4-dimethyl-1-oxo-2-naphthalenecarboxylic acids as 1-carba bioisosteres of oxolinic acid. J Med Chem 27:306-310

Koga H, Itoh A, Murayama S, Suzue S, Irikura T (1980) Structure-activity relationships of antibacterial 6,7- and 7,8-disubstituted 1-alkyl-1,4-dihydro-4-oxoquinoline-3-carboxylic acids. J Med Chem 23:1358-1363

Matsumoto J, Miyamoto T, Minamida A, Nishimura Y, Egawa H, Nishimura H (1984) Pyridonecarboxylic acids as antibacterial agents. 2 Synthesis and structure-activity relationships of 1,6,7-trisubstituted 1,4-dihydro-4-oxo-1,8-naphthyridine-3-carboxylic acid, including enoxacin, a new antibacterial agent. J Med Chem 27:292-301

Matsumoto J, Miyamoto T (1989) Cinoxacin analogues as potential antibacterial agents: synthesis and antibacterial activity In: Fernandes PB (ed) Quinolones, Prous Science Publishers, Barcelona, Spain p115

Mitscher LA, Gracey HE, Clark GW, Suzuki T (1978) Quinolone antimicrobial agents. 1. Versatile new synthesis of 1-alkyl-1,4-dihydro-4-oxo-3-quinolinecarboxylic acids. J Med Chem 21:485-489

Mitscher LA, Sharma PN, Chu DTW, Shen LL, Pernet AG. Chiral DNA gyrase inhibitors. 1 Synthesis and antimicrobial activity of the enantiomers of 6-fluoro-7-(1-piperazinyl)-1-(2-trans-phenyl-1'-cyclopropyl)-1,4-dihydro-4-oxoquinoline-3-carboxylic acid. J Med Chem 29:2044-2047

Mitscher LA, Sharma PN, Chu DTW, Shen LL, Pernet AG (1987) Chiral DNA gyrase inhibitors. 2 Asymmetric synthesis and biological activity of the enantiomers of 9-fluoro-3-methyl-10-(4-methyl-1-piperazinyl)-7H-pyrido[1,2,3-de]-1,4-benzoxazine-6-carboxylic acid (Ofloxacin). J Med Chem 30:2283-2286

Narita H, Konishi Y, Nitta J, Nagaki H, Kobayashi Y, Watanabe Y, Minami S, Saikawa I (1986a) Pyridonecarboxylic acids as antibacterial agents. IV. Synthesis and structure-activity relationships of 7-amino-1-aryl-6-fluoro-4-quinolone-3-carboxylic acids. Yakugaku Zasshi 106:795-801

Narita H, Konishi Y, Nitta J, Kitayama I, Watanabe Y, Miyazime M, Minami S, Yotsuji A, Saikawa I (1986b) Pyridone carboxylic acids as antibacterial agents. V. Synthesis and structure-activity relationship of 7-amino-6-fluoro-1-(fluorophenyl)-4-oxo-1,8-naphthyridine-3-carboxylic acids. Yakugaku Zasshi 106:802-807

Nishimura Y, Matsumoto J (1987) Pyridonecarboxylic acids as antibacterial agents. 9 Synthesis and antibacterial activity of 1-substituted 6-fluoro-1,4-dihydro-4-oxo-7(4-pyridyl)-1,8-naphthyridine-3-carboxylic acids. J Med Chem 30:1622-1625

Nishimura Y, Minamida A, Matsumoto J (1988) Pyridonecarboxylic acids as antibacterial agents. XII. Synthesis and antibacterial activity of enoxacin analogs with a variant at position 1. Chem Pharm Bull (Japan) 36:1223-1228

Rosen T, Chu DTW, Lico IM, Fernandes PB, Shen L, Borodkin S, Pernet AG (1988a) Asymmetric synthesis and properties of the enantiomers of the antibacterial agent 7-(3-aminopyrrolidin-1-yl) -1-(2,4-difluorophenyl)-1,4-dihydro-6-fluoro-4-oxo-1,8-naphthyridine-3-carboxylic acid hydrochloride. J Med Chem 31:1586-1589

Rosen T, Chu DTW, Lico IM, Fernandes PB, Marsh K, Shen L, Cepa VG, Pernet AG (1988b) Design, synthesis and properties of (4S)-7-(4-amino-2-substituted-pyrrolidin-1-yl)quinolone-3-carboxylic acids. J Med Chem 31:1598-1611

Sanchez JP, Domagala JM, Hagen SE, Heifetz CL, Hutt MP, Nichols JB, Trehan AK (1988) Quinolone antibacterial agents. Synthesis and structure-activity relationships of 8-substituted quinoline-3-carboxylic acids and 1,8-naphthyridine-3-carboxylic acids. J Med Chem 31:983-991

Shen LL, Mitscher LA, Sharma PN, O'Donnell TJ, Chu DWT, Cooper CS, Rosen T, Pernet AG (1989) Mechanism of inhibition of DNA gyrase by quinolone antibacterials: A cooperative drug-DNA binding model. Biochemistry 28:3886-3894

Sutcliffe JA (1988) Novel approaches toward discovery of antibacterial agents. Annu Reports Med Chem 23:141-150

Uno T, Takamatsu M, Inoue Y, Kawahata Y, Iuchi K, Tsukamoto G (1987) Synthesis of antimicrobial agents. 1. Syntheses and antibacterial activities of 7-(azole substituted) quinolones. J Med Chem 30:2163-2168

Wentland MP, Bailey DM, Cornett JB, Cornett JB, Dobson RA, Powles RG, Wagner RB (1984) Novel amino-substituted 3-quinolinecarboxylic acid antibacterial agents: synthesis and structure-activity relationships. J Med Chem 27:1103-1108

Wentland MP, Perni RB, Dorff PH, Rake JB (1988) Synthesis and bacterial DNA gyrase inhibitory properties of a spirocyclopropylquinolone derivative. J Med Chem 31:1694-1697

Yanagisawa H, Nakao H, Ando A (1973) Studies on chemotherapeutic agents. I Syntheses of quinoline
    and naphthyridine sulfonamide or phosphonic acid derivatives. Chem Pharm Bull (Tokyo) 21:1080-
    1089

Discussion Summary

This extremely comprehensive overview of the structural development of the
4-quinolones was welcomed by everyone. Although it did not raise any overtly
controversial aspect of the 4-quinolones, the expression of the fact that we still do not
have any clear pattern of structure-activity modelling, despite a massive synthetic effort
over the years, served to confirm to all that this represented the lack of knowledge of
the biological and biochemical components of the equation, rather than lack of
chemical data. The sheer diversity of structures which showed significant activity either
against intact bacterial cells, or in *in vitro* assays with DNA gyrase, was a surprise to
everyone. The fact that new synthetic entities had recently managed to remove even the
last bastion of structure-activity security, namely the absolute need for the carboxylic
acid residue, indicated that the chemists were doing their best to catalyse greater efforts
to elucidate the mechanism(s) of action of these compounds.

The extent of the effects of isomeric considerations in the structure was complicated
by the fact that the results of some studies indicated that the two isomers could be
shown to be additive in some bacterial species, whilst in other species the two
components seemed antagonistic. Such data was considered to be important in that it
stressed the potential benefits of the synthesis of single isomers for development and
pharmacological study. Clearly much interpretive work is still required and it was
considered that the importance of co-operation and collaboration between the medicinal
chemists and the biologists studying the mechanism of action of the 4-quinolones
could not be over-emphasized.

Chapter 10

Aspects of Quinolone-DNA Interactions

L. L. Shen, J. Baranowski, M. Nuss, J. Tadanier, C. Lee, D. T. W. Chu and J. J. Plattner

Introduction

It has been well documented that the common target of quinolone antibacterials is the bacteria-specific type II DNA topoisomerase, i.e. DNA gyrase (for reviews, see Cozzarelli 1980; Gellert 1981; Wang 1985; Drlica and Franco 1988). These drugs share a common mode of action with anti-cancer drugs by forming a ternary complex with the enzyme and the DNA substrate (Gellert *et al.* 1977; Sugino *et al.* 1977; Chen and Liu 1986; Glisson and Ross 1987). The biological consequence of the formation of such a "cleavable complex" is at least two-fold:

i)   the enzyme is inactivated, leading to the arrest of DNA synthesis in bacterial cells

ii)   the cleavable complex formation causes unrepairable DNA damage which is believed to trigger *recA*-dependent SOS repair response and alternatively leading to cell death, presumably by the expression of certain lethal proteins in the cell (Drlica 1984).

Although the important role of the cleavable complex in cell killing was well demonstrated, little is known about the mechanistic events involved in the complex formation or the nature of the drug-receptor interactions. We began the study on the mechanism of action of quinolone antibacterial agents by having ^{3}H-norfloxacin synthesized, and anticipated that the drug would bind selectively and tightly to DNA gyrase. Surprisingly, we found that it did not bind to the enzyme but instead bound to pure DNA (Shen and Pernet 1985). At this preliminary stage of investigation, we observed a specific form of binding, highlighted with a partial saturation phase, near the supercoiling inhibition constant. The binding is relevant to the drug's inhibitory potency, since the binding constants of a number of quinolone analogues correlate well with their K_i values. However, we were still puzzled by the fact that the specific, saturable form of binding can only be observed with the supercoiled DNA (the product of the supercoiling process), but not with the relaxed DNA substrate or the enzyme-substrate complex (without ATP). In more recent publications, we offered an interpretation of these puzzling observations. We proposed, after extensive investigations on the binding specificity and co-operativity to various structural forms of DNA and DNA-enzyme complex, that the drug binds to a DNA site created by the enzyme in the presence of ATP. More specifically, the model suggests that the drug binds to a partially denatured, 4-base-pair staggered cut DNA site created by the enzyme during the intermediate DNA gate-opening step of supercoiling which requires ATP as energy source. A concept of self-association among drug molecules while occupying such a site was proposed; it explains the high binding co-operativity and how the high binding affinity of these small and simple organic molecules is derived (Shen *et al.* 1989a,b,c).

The phenomenon of quinolone binding to the oppositely charged DNA molecule was somewhat surprising. A few controversial observations were reported showing that quinolones do not bind to DNA (LeGoffic 1985; Palu' *et al.* 1988). Our finding, however, was supported by a paper published by Tornaletti and Pedrini (1988) where it was demonstrated that quinolones in the presence of magnesium ions unwind DNA. Our results were also consistent with an early report that interaction between nalidixic acid and pure DNA did occur (Crumplin *et al.* 1980). In this paper, aspects of quinolone-DNA interaction are reviewed. Data will be presented that explain some negative results recently appearing in the literature.

Two Levels of Drug Binding Specificity

The binding specificity of a quinolone to DNA has been extensively investigated in our laboratory. The specificity may be viewed at two levels. The first level of binding specificity is largely governed by the structure of DNA. Quinolones bind preferentially to single-stranded DNA rather than the double-stranded (Shen and Pernet 1985). Base binding preference to double-stranded DNA was studied utilizing three native DNAs of varying GC content, and the results are listed in Table 10.1. It is clearly shown that there is no correlation between the amount of drug binding and the GC content of DNA. The results suggest that when DNA strands are intact, no base preference can be demonstrated.

When DNA strands are separated, the amount of drug binding increases and the second level of binding specificity, i.e. the base binding preference, is revealed. The binding to single-stranded DNA model homopolymers shows a sequence of binding preference to poly(dG), poly(dA), poly(dT) and poly(dC) in decreasing order. The binding to poly(dG) is distinctively greater than that to the other three polydeoxyribonucleotides. One unique structural feature of the guanine base is that it has two common hydrogen-bond donors while the other bases have only one; the results thus suggest the involvement of hydrogen bonds, possibly between the base and the 4-keto and/or the 3-carboxyl group on the quinolone ring. We have also shown that the drug binding to poly(dI) is roughly equal to 20% of that to poly(dG). The structural difference between guanosine and inosine is such that the former possesses an extra amino group on the pyrimidine ring and this amino functional group is an important hydrogen-bond donor in base pairing with the complementary strand. The result once more suggests that the drug binds to unpaired DNA bases via hydrogen-bonding. The binding preference to synthetic polyribonucleotides was also investigated. The same sequence of binding preference was obtained as that of polydeoxyribonucleotides (Shen *et al.* 1989b).

Table 10.1. Binding of ^{3}H-norfloxacin to native DNA with different GC contents[a] (reprinted from Shen *et al.* 1989b)

DNA	GC Content(%)[b]	Tm(oC)[b]	molar binding ratio (x 10^{-4} per NT) at drug concentration of	
			2 mM	6 mM
M. lysodeikticus	72	99.5	3.0 ± 0.52	6.7 ± 0.5
E. coli	50	90.5	2.5 ± 0.28	6.7 ± 0.7
C. perfringens	26.5	80.5	2.4 ± 0.26	6.4 ± 0.6
Linearized ColE1			1.8	6.9

[a] Abbreviations: G, guanine; C, cytidine; Tm, melting temperature; NT, nucleotide.
[b] From Marmur & Doty (1962)

Results described above are consistent with those reported in an earlier publication by Crumplin *et al.* (1980) that nalidixic acid could reversibly bind to single-stranded DNA by a process catalyzed by divalent metal ions. Their studies with synthetic deoxyribose polymers also displayed a preferential binding to guanine residues. Hydrogen-bondings between the two keto groups and the two NH groups of guanine were thus suggested. The binding affinity of nalidixic acid to DNA was about 100-fold weaker than that of norfloxacin (Shen and Pernet 1985). The high binding affinity and/or different charge property (as zwitterions) of norfloxacin may eliminate the requirement of divalent metal ions (Shen and Pernet 1985) which are needed to remove the electrostatic repulsion between nalidixic acid and DNA as suggested by these authors.

Two Types of Binding:Specific vs Non-specific

The various forms of drug binding to DNA, monitored by the two levels of binding specificity described above, may be classified into the following two types according to their binding saturation pattern, binding affinity relative to the biological activity and the degree of cooperactivity:

The Non-specific Binding

The binding of ^{3}H-norfloxacin to single-stranded DNA, and to relaxed forms of DNA, described above are typical examples of non-specific binding, since they lack binding co-operativity and demonstrate no trace of binding saturation. The binding of ^{3}H-norfloxacin to supercoiled ColE1 DNA at high concentrations (up to the drug's solubility of 1 mM) also fits to this type of binding (Shen *et al.* 1989b).

The Specific Binding and the DNA Gyrase Inhibition Model

We have reported that the binding of ^{3}H-norfloxacin observed with supercoiled ColE1 DNA (as illustrated in Fig. 10.1) represents a specific type of drug binding based on the following three characteristics:

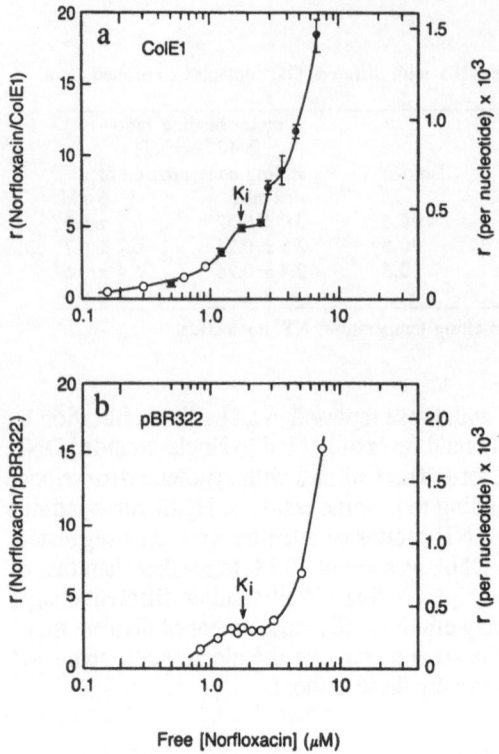

Fig. 10.1a,b. Binding of norfloxacin to plasmid DNA. Binding mixtures contained 4.7 pmol of ColE1 (a) or pBR322 (b) DNA and the indicated amounts of ^{3}H-norfloxacin. Vertical bars represent S.D.s. Different symbols indicate results obtained from different experiments. Supercoiling K_i (against *E. coli* DNA gyrase) are shown for comparison. Reprinted from Shen and Pernet (1985).

i)   the binding is saturable and takes place at the drug's supercoiling inhibitory concentration,

ii)   the binding affinities of some selected quinolones are directly proportional to their supercoiling inhibition constants (Shen and Pernet 1985) and

(iii)   the binding to this saturable site is highly co-operative (Shen *et al.* 1989b).

The characteristic binding pattern described above was observed only with the supercoiled form of covalently closed circular (ccc) DNA (such as ColE1 DNA), but not with the relaxed or linear form. It is evident that there exists in the supercoiled DNA a unique structural feature which serves as the drug's co-operative binding site. We know that the drug binds preferentially to single-stranded DNA. Therefore, the question is: does supercoiled DNA exhibit such a preferred structural feature? The answer is yes. It is known that supercoiled DNA is underwound with reduced linking numbers. It may be speculated that a remnant single-stranded bubble may be retained in the supercoil and it is at this small denatured DNA pocket that the drug binds securely. Another possible candidate is the cruciform (or hairpin) structure that forms with palindromic sequences; this structure is known to be promoted by negative supercoiling (Gellert *et al.* 1978; Lilly 1980; Mizuuchi *et al.* 1982). The cruciform is characterized

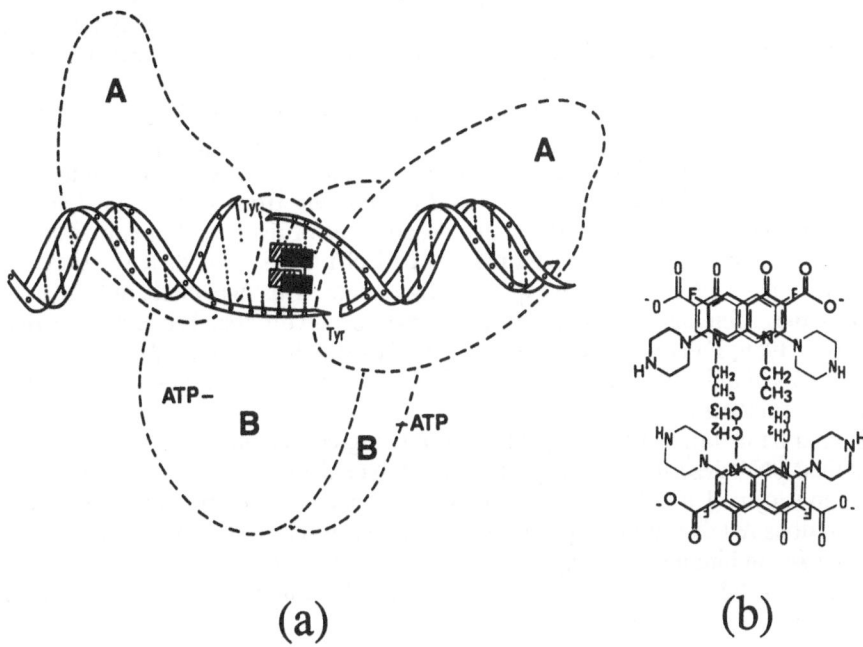

(a)   (b)

Fig. 10.2a,b a Model of the proposed quinolone-DNA co-operative binding in the inhibition of DNA gyrase. Four drug molecules are shown in the diagram, the number of drug molecules involved may be greater depending on the size and the configuration of the binding site. b Detailed drawing of the drug self-assembly in the single-stranded DNA pocket illustrated at the center of the diagram in panel A. Norfloxacin molecules (drawn separately by thick and thin lines) are used for illustrating the two types of proposed interactions: the π-π ring stacking of the quinolone rings and the tail-to-tail hydrophobic interaction between the N-ethyl groups. Such interactions result in a cluster of molecules mimics a phospholipid micelle structure with a hydrophobic core and hydrophilic binding groups exposed on the surface. The angle between the two molecules interacted through N1 hydrophobic tails is not necessarily fixed at 180° so as to provide flexibility in bond pairing. (Information retrieved from Shen *et al.* 1989c)

by single-stranded DNA bubbles that have been demonstrated by the cleavage with S1 nuclease which has specificity to single-stranded DNA (Panayotatos and Wells 1981; Singleton and Wells 1982). This possibility needs to be verified.

Based on the above observations and the crystal structure of nalidixic acid, a quinolone-DNA co-operative binding model for the inhibition of DNA gyrase was proposed (Shen 1986; Shen *et al.* 1989c). This working model has two essential features:

(i)   the unique three dimensional configuration of a drug binding site created on DNA, and

(ii)   the unique ability of the drug molecules to fit into such a site.

Both aspects are equally important to acquire drug binding affinity and specificity. We propose that the bound enzyme induces a binding site for the drug in the relaxed DNA substrate. The binding site is presumably formed during the gate-opening step which requires the binding of ATP. The separated DNA strands between the 4-basepair staggered cuts form a simulated denatured DNA bubble ideal for the drug to bind (Fig 10.2a). On the other hand, drug molecules acquire high binding affinity through a co-operative binding mechanism achieved through interactions between the drug molecules. Two types of interactions, supported by observations in the structure of nalidixic acid crystal, are feasible: the π-π stacking between the quinolone rings and the tail-to-tail hydrophobic interactions between the N1 substitution groups (Fig 10.2b). Such interactions render the drug molecules with multiple sets of hydrogen bond acceptors in a consolidated unit and extend the binding from a unidimensional to multidimensional domain. From the above descriptions of the model, it is reasonable to speculate that the drug binding sites induced by DNA gyrase from different bacterial sources may have different drug receptor configurations. This hypothesis may be used to explain the differential sensitivities of quinolones toward different bacteria (Zweerink and Edison 1986; Fu *et al.* 1986; Shen *et al.* 1989c) and the puzzling observations on the mutations of quinolone-resistant gyrase (for more detailed discussions, see Shen *et al.* 1989c; Fernandes and Shen 1989).

It should be mentioned here that the high binding co-operativity of quinolones to supercoiled DNA described above has been used to explain the absence of norfloxacin binding to supercoiled DNA when the binding experiment was performed at low drug concentrations relative to the apparent binding constant (Palu' *et al.* 1988). The effort to maximize the binding signal by lowering the ligand concentration (thus increasing the receptor/ligand ratio in the reaction mixture) will have an adverse effect due to the existence of high binding co-operativity (Shen 1989).

Direct Effects of Quinolone-DNA Interactions

Unwinding of DNA in a Non-intercalative Way

We have demonstrated that quinolones at high concentration bind to native DNA non-specifically and reversibly (Shen *et al.* 1989d). The preference of the drug binding to single-stranded DNA rather than to double-stranded DNA implies that the quinolones are not DNA intercalators. A recent publication indicates that this class of compounds at high concentration causes DNA unwinding in the presence of magnesium ions (Tornaletti and Pedrini 1988), even though the result also indicates that quinolones are not DNA intercalators. We performed the DNA unwinding experiments using a different approach with supercoiled ColE1 DNA and rat liver DNA topoisomerase I (Pommier *et al.* 1987). With this technique, ethidium bromide (used as positive control) shows unwinding effect at concentration as low as 0.3 mg/ml. Most major quinolones at 50 µg/ml or higher do not show a DNA unwinding effect, supporting the conclusion that quinolones do not intercalate into DNA (Shen *et al.* 1989d).

In the presence of magnesium, the results are different. Quinolones at high concentrations do unwind DNA to a varying degree. This confirms the results obtained by Tornaletti and Pedrini (1988). However, we recently showed that such magnesium-dependent DNA unwinding by quinolones may be antagonized by polyamines at physiological intracellular concentrations (Shen *et al.* 1989, the 29th ICAAC meetings abstract #92). This implies that in intact cells, where numerous DNA binding components are present, such unwinding may not exist after all.

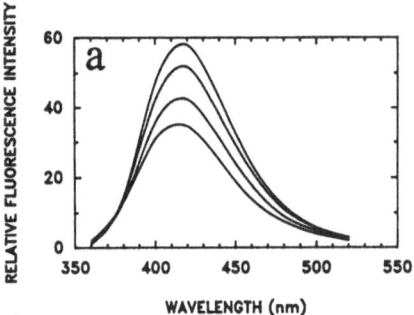
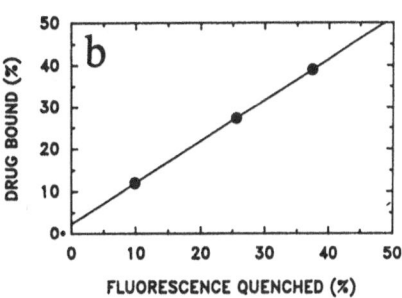

Fig. 10.3a,b. Quenching effect of DNA on the intrinsic fluorescence of norfloxacin. **a** Emission spectra of norfloxacin (1.2 μM) excited at 340 nm. Curves from top to bottom correspond to the spectra of norfloxacin (1.5 ml) added with 0, 10, 30, and 60 μl of thermally denatured calf thymus DNA (3.65 mg/ml) to give nucleotide/drug ratios of 0, 6, 18 and 36, respectively. The addition of DNA caused no more than 4% of volume change. **b** Correlation of the amount of drug bound and the percent of fluorescence quenching. The percent of drug bound at same experimental conditions in panel A was determined by an ultrafiltration technique using ^{3}H-norfloxacin. Both parameters shown have been corrected by the dilution factors. For consistency, same buffer used in previous binding studies (Shen and Pernet 1985; Shen *et al.* 1989a,b) was used for these experiments: 50 mM Hepes, pH 7.4; 20 mM KCl; 5 mM MgCl$_2$; 1 mM EDTA; 1 mM dithiothreitol. (Reprinted from Shen 1989)

Spectroscopic Changes

Quenching of Drug's Intrinsic Fluorescence

It is expected that ligand-DNA interaction would cause spectroscopic changes. Quinolones possess strong fluorescence chromophores (Shen *et al.* 1989b) whilst DNA is essentially non-fluorescent. We capitalized on this fluorescence method to investigate the interaction of quinolones with DNA. As shown in Fig. 10.3a, we found that the intrinsic fluorescence intensity of norfloxacin was quenched when single-stranded DNA was added. The extent of quenching is proportional to the amount of drug bound to DNA (Fig. 10.5b), indicating a near complete fluorescence quenching of the DNA-bound drug. The phenomenon explains the negative results obtained by Palu' *et al* (1988) who showed an absence of the norfloxacin binding to DNA utilizing equilibrium dialysis and fluorescence measurement as a method of quantification. In fact, a measurement of fluorescence quenching such as that shown in Fig. 10.3a is sufficient to indicate the existence of quinolone-DNA interactions (Shen 1989).

Fluorine-19 NMR Spectral Changes

The fluorine-19 NMR spectra of difloxacin in TEN buffer (20 mM Tris-HCl, pH 7.4; 2 mM EDTA; 100 mM NaCl) exhibit two sharp resonances corresponding to the two fluorine atoms of difloxacin (Fig. 10.4, curve A). When difloxacin is titrated with increasing amounts of denatured calf thymus DNA both resonances are dramatically broadened (Fig. 10.4, curves B-F). The natural linewidth of the two fluorine atoms of

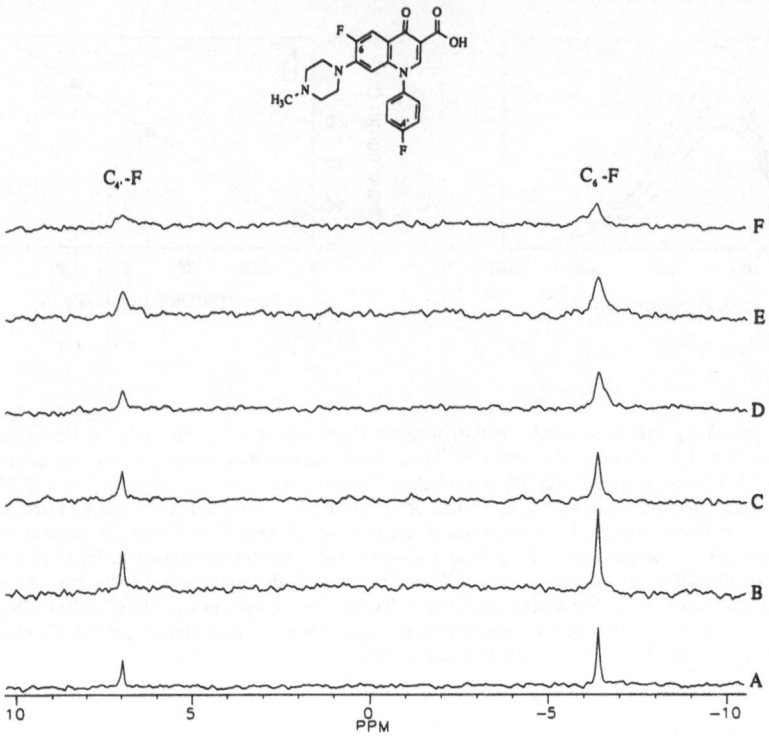

Fig. 10.4. Fluorine-19 NMR spectra of difloxacin and difloxacin-DNA complex. All of the NMR spectra were collected on a GN300 spectrometer using a 5 mm F-19 probe with H-1 decoupling. All of the spectra were collected in double precision using a one pulse experiment with a 50 degree observe pulse and repetition time of approximately two seconds. Each spectrum was processed using a linebroadening of 20 Hz. The calculated natural linewidth for each fluorine resonance (see text) for the solutions A and E was determined taking the linebroadening factor into account. The number of scans for each sample ranged from approximately 15,000 for sample A to 38,000 for sample E. The samples (0.5 ml final volume) were prepared by adding different amounts of denatured calf thymus DNA to a constant amount of difloxacin (final concentration of the drug was 0.42 mM). Denatured DNA was prepared using a thermal denaturation procedure previously described by Shen and Pernet (1985). The DNA stock solution was made in TEN buffer; an aliquot of TEN was added to each sample to keep a constant volume of the solution. 0.05 ml of deuterium oxide was added to each sample for locking the spectrometer during the experiment. The molar ratios of nucleotide to drug are 1.4, 3.6, 7, 10.7 and 14 for curve B to F, respectively. The pH of all solutions in Fig. 10.4 has been checked and shown to be near 7.4. The assignment of peaks for C6-F and C4'-F is based on the spectrum of a compound similar to difloxacin but having fluorine atoms at C6, C4' and C2' (data not shown). Insert shows the chemical structure of difloxacin.

difloxacin in TEN buffer for the 4'-para-fluoro and 6-fluoro resonances increases from 2 and 4 Hz respectively to 44 and 72 Hz for the same resonances in curve E where the nucleotide to drug ratio is 10.7:1. The increased linewidth for the two fluorine resonances in Fig. 10.4 demonstrates that there is an interaction between difloxacin and denatured DNA under these experimental conditions. This observation is contradictory with that reported by LeGoffic (1985) showing the absence of pefloxacin-DNA interaction judged from his fluorine-19 NMR study. Since no experimental condition was specified in that report, investigation on the source of the difference is not possible.

Direct DNA Cleavage

We have observed that the interaction of certain quinolone congeners with DNA causes direct single-stranded DNA nicking and breaking under ambient laboratory lights. These quinolones were synthesized as part of our effort to modify the 3-carboxyl group for improving antibacterial potency (Chu *et al.* 1988). Fig. 10.5a shows the dose-response

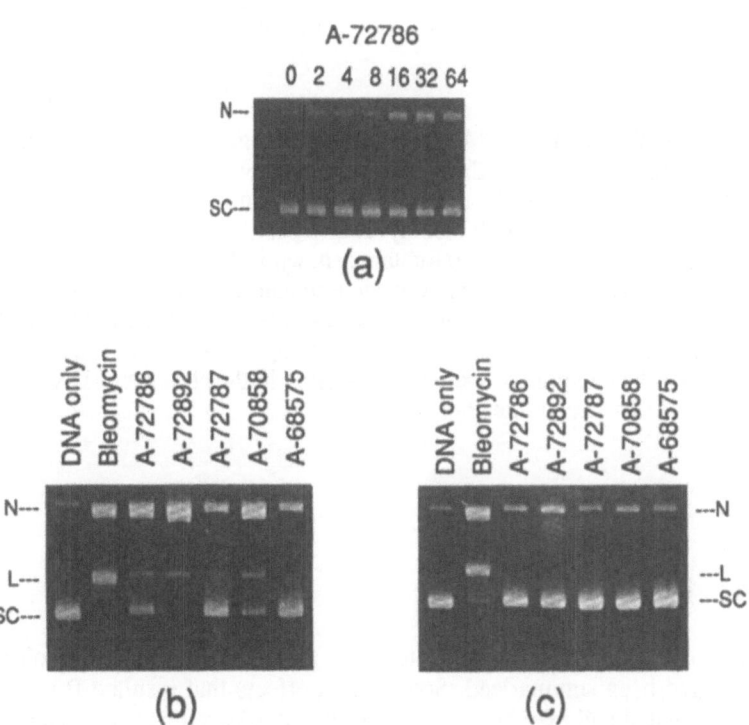

Fig. 10.5. a,b,c The DNA-nicking activity of some selected quinolone derivatives. **a** Dose dependence of DNA nicking by A-72786. Supercoiled ColE1 DNA (0.2 mg) was incubated with increasing amounts of the compound (number on top of each lane indicates the drug concentration in µg/ml) in 20 ml of Tris-HCl (150 mM, pH 7.4) for 60 min under ambient laboratory lighting condition. The nicking activity is visualized by the conversion of supercoiled (SC) band to the relaxed (R) band in a 1% agarose gel stained with ethidium bromide after the run. **b** The DNA nicking activity of a number of derivatives at 64 µg/ml. To amplify the effect, the reaction mixture was placed about 6 inches below a 60-watt fluorescent light tube. The increase in light intensity apparently cause more extensive DNA nicking as evidenced by the appearance of linear (L) DNA band. Bleomycin at same concentration was included for comparison. **c** Same as **b** except that the reaction mixture was shielded from the light.

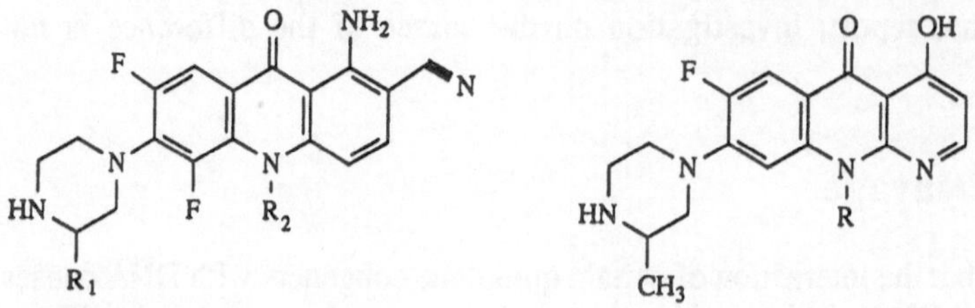

	R_1	R_2		R
A-72787	CH_3	p-FC_6H_4	A-70858	p-FC_6H_4
A-72786	H	c-C_3H_5	A-68575	C_2H_5
A-72892	CH_3	c-C_3H_5		

Fig. 10.6. Chemical structure of quinolone derivatives used in Fig. 10.5.

of the nicking of supercoiled ColE1 DNA after incubating with increasing amounts of A-72786. DNA nicking activities of this compound are evident as shown by the conversion of the supercoiled band to the relaxed band. The nicking activity of a number of other similar derivatives, with bleomycin as a positive control, are shown in Fig. 10.5b. All these tricyclic quinolones (structures shown in Fig. 10.6) cause single-stranded DNA nicking evidently in a light-dependent manner, as the reaction did not take place when the experiments were performed in test tubes shielded from the light (Fig. 10.5c). Pre-irradiation of the compounds before adding DNA did not cause DNA nicking, indicating that the compounds were not converted in the absence of DNA to a new species that nicks DNA strand (data not shown).

Concluding Remarks

In this report, the binding specificity and criteria for the classification of quinolone binding to DNA have been summarized. Some direct effects that resulted from the interaction were presented and may be viewed as further evidence of the existence of quinolone-DNA interaction, in addition to the direct binding results. The interaction, in general, is weak. Its binding to native DNA is limited by the base pairing of the double-helical strands. It causes DNA unwinding only at high drug concentration in the presence of magnesium ions; however, such an effect demonstrated *in vitro* is less likely to be significant *in vivo* since this interaction may be counteracted by other DNA-interacting components, e.g. polyamines. The specific type of binding, which results in the inhibition of DNA gyrase according to our model, takes place at relatively low drug concentrations. The binding site is believed to be a single-stranded DNA pocket induced by the target enzyme. The high binding affinity of quinolones to this site is achieved through a unique drug self-association mechanism, as proposed.

References

Chen GL, Liu LF (1986) DNA topoisomerases as therapeutic targets in cancer chemotherapy. Ann Reports in Med Chem 21:257-262

Chu DTW, Fernandes PB, Claiborne AK, Shen LL, Pernet AG (1988) Structure-activity relationships in quinolone antibacterials: design, synthesis and biological activities of novel isothiazoloquinolones. Drugs Exptl Clin Res 14:379-383

Cozzarelli NR (1980) DNA gyrase and the supercoiling of DNA. Science 207:953-960

Crumplin GC, Midgley JM, Smith JT (1980) Mechanism of action of nalidixic acid and its congeners. Topics in Antibiotic Chem 8:9-38

Drlica K (1984) Biology of bacterial deoxyribonucleic acid topoisomerases. Microbiol Reviews 48:273-289

Fernandes PB, Shen LL (1989) Quinolones: mode of action and mechanism of resistance. In: Actor P (ed) Clinical implications of antimicrobial resistance: mechanism, testing problems and epidemiology. the American Society of Microbiology, Eastern Pennsylvania Branch

Fu KP, Grace ME, McCloud SJ, Gregory FJ, Hung PP (1986) Discrepancy between the antibacterial activities and the inhibitory effects on Micrococcus luteus DNA gyrase of 13 quinolones. Chemotherapy 32:494-498

Gellert M, Mizuuchi K, O'Dea MH, Itoh T, Tomizawa J-I (1977) Proc Natl Acad Sci USA 74:4772-4776

Gellert M, Mizuuchi K, O'Dea MH, Ohmori H, Tomizawa J (1978) DNA gyrase and DNA supercoiling. Cold Spring Harbor Symp Quant Biol 43:35-40

Glisson BS, Ross WE (1987) Pharmac Ther 32:89-106

LeGoffic F (1985) Les quinolones, mecanisme d'action. In: Pocidalo JJ, Vachon F and Regnier B (eds) Les nouvelles quinolones. Editiones Arnette, Paris pp.15-23

Lilley DMJ (1980) The inverted repeat as a recognizable structure feature in supercoiled DNA molecules. Proc Natl Acad Sci USA 77:6468-6472

Marmur J, Doty P (1962) Determination of the base composition of deoxyribonucleic acid from its thermal denaturation temperature. J Mol Biol 5:109-118

Mizuuchi K, Mizuuchi M, Gellert M (1982) Cruciform structure on palindromic DNA are favored by DNA supercoiling. J Mol Biol 156:229-243

Palu' G, Valisena S, Peracchi M, Palumbo M (1988) Do quinolones bind to DNA? Biochem Pharmacol 37:1887-1888

Panayotatos N, Wells RD (1981) Cruciform structures in Supercoiled DNA. Nature 289:466-470

Pommier Y, Covey JM, Kerrigan D, Markovitsm J, Pham R (1987) DNA unwinding and inhibition of mouse leukemia L1210 DNA topoisomerase I by intercalation. Nucleic Acids Res 15:6713-6731

Shen LL, Pernet AG (1985) Mechanism of inhibition of DNA gyrase by analogues of nalidixic acid: the target of the drugs is DNA. Proc Natl Acad Sci USA 82:307-311

Shen LL, Baranowski J, Wai T (1986) Mechanism of inhibition of DNA gyrase by quinolone antibacterials: a cooperative drug-DNA binding model. J. Cellular Biochemistry 1986; Suppl.10B: abstract of UCLA Symposium on DNA replication and recombination, Park City, Utah

Shen LL (1989) A reply: "Do quinolones bind to DNA?" - Yes. Biochem Pharmacol 38:2042-2044

Shen LL, Kohlbrenner WE, Weigl D and Baranowski J (1989a), Mechanism of quinolone inhibition of DNA gyrase. Appearance of unique norfloxacin binding sites in enzyme-DNA complexes. J Biol Chem 264:2973-2978

Shen LL, Baranowski J, Pernet AG (1989b) Mechanism of inhibition of DNA gyrase by quinolone antibacterials. Specificity and cooperativity of drug binding to DNA. Biochemistry 28:2879-2885

Shen LL, Mitscher LA, Sharma PN, O'Donnell TJ, Chu DTW, Cooper CS, Rosen T, Pernet AG (1989c) Mechanism of inhibition of DNA gyrase by quinolone antibacterials. A cooperative drug-DNA binding model. Biochemistry 28:2886-2894

Shen LL, Baranowski J, Wai T, Chu DTW, Pernet AG (1989d) The binding of quinolones to DNA: should we worry about it? In: Fernandes PB (ed) The International Telesymposium on Quinolones, J.R. Prous Science Publishers, Barcelona, Spain, pp 159-170

Singleton CK, Wells RD (1982) Relationship between superhelical density and cruciform formation in plasmid pVH 51. J Biol Chem 257:6292-6295

Sugino A, Peebles CL, Kreuzer KN, Cozzarelli NR (1977) Mechanism of action of nalidixic acid: purification of E. coli nalA gene product and its relationship to DNA gyrase and a novel cicking-closing enzyme. Proc Natl Acad Sci USA 74:4767-4771

Tornaletti S and Pedrini AM (1988) Studies on the interaction of 4-quinolones with DNA by DNA unwinding experiments. Biochim Biophy Acta 949:279-287

Wang JC (1985) DNA topoisomerases. Ann Rev Biochem 54:665-697

Zweerink MM, Edison A (1986) Inhibition of *Micrococcus luteus* DNA gyrase by norfloxacin and 10 other quinolone carboxylic acids. Antimicrob Agents and Chemother 29:598-601

Discussion Summary

The model for the co-operative binding of the 4-quinolones to DNA in the context of a DNA gyrase/DNA complex was generally considered as a working model for the mechanism of action of these compounds. Several members of the audience still retained some doubts about whether such a model could yet be applied specifically to the *in vivo* situation. In particular, the technical problems associated with the establishment of the link between the *in vitro* data and the situation operating within the treated cell were regarded as a significant problem.

There was extensive discussion of the way in which four molecules of a 4-quinolone could be made to fit in an ordered way into the single-strand bubble generated by DNA gyrase. However, Dr Shen was able to show that sufficient space could be available for the packaging of the 4-quinolone molecules. He was also able to show how this type of model for the binding of the drug to DNA helped in the understanding of the physico-chemical correlations with antibacterial activity which had been described by Professor Mitscher.

This was clearly an area of research which necessitated a novel perception of the activities of the 4-quinolones. It now seemed that the long-running confusions, generated by years of conflicting reports of DNA-binding activity, had been resolved by the demonstration of binding and the description of a binding mechanism which correlated with the kinetics of the inhibition of DNA gyrase catalytic activity. Whilst the members of the audience at this meeting were clearly happy with this model as a valid biological explanation of the antibacterial activity of the drugs, there were some reservations over the way this model might be used to explain some mechanisms of resistance associated with changes in the *gyrB* gene product. It was also envisaged that there might still be problems of the perception of such a model by people not intimately involved with the mechanistic study of the 4-quinolones. Clearly the general perception that agents, which bind to DNA, represent a potential hazard would have to be faced. The major barrier to changing this perception was generally thought to be the establishment of a basis for a sequence specificity for the binding and the evidence to suggest that such sequences are not apparent outside the prokaryotes. As with many other presentations at this meeting, the evidence clearly indicated that further detailed study was required.

Chapter 11

Selective Toxicity: The Activities of 4-Quinolones against Eukaryotic DNA Topoisomerases

T. D. Gootz, J. F. Barrett, H. E. Holden, V. A. Ray and P. R. McGuirk

Introduction

A number of new 4-quinolone antibacterials have been recently introduced for clinical use that have greater potency than the older analogs nalidixic acid and oxolinic acid (Andriole 1988; Bergan 1988). Like nalidixic acid, 4-quinolones are believed to inhibit bacterial growth by binding to the A subunit of DNA gyrase, a type II topoisomerase (Gellert *et al.* 1977). Since a homologous enzyme exists in the nucleus of eukaryotic cells, studies have been performed to examine the effects of 4-quinolones on this topoisomerase. Quinolones such as ciprofloxacin, norfloxacin, fleroxacin, and ofloxacin have, in general, been found to be only weak inhibitors of the eukaryotic enzyme in *in vitro* assays that measure the catalytic relaxation, unknotting, and catenation activities of the enzyme (Hussy *et al.* 1986; Riou *et al.* 1986). An alternative approach to assessing the effects of test agents on topoisomerases, involves measuring the extent of drug-enhanced DNA cleavage obtained *in vitro* in the presence of the enzyme. While DNA cleavage assays have been used to measure the effects of anti-tumour agents such as VP-16, VM-26, and ellipticine on eukaryotic topoisomerase II, 4-quinolones have not been extensively examined by this method. We have used four different topoisomerase II assay methods, including a non-radiolabeled DNA cleavage assay, for measuring the effects of some quinolones on purified calf thymus topoisomerase II. An experimental quinolone, CP-67,015, was found to elicit significant levels of DNA cleavage products in the presence of enzyme. The effects of CP-67,015 on mammalian cells in standard *in vitro* and *in vivo* genotoxicity tests are also described and compared to published results obtained with other quinolones.

A preliminary report of this data has appeared elsewhere (Barrett *et al.* 1989).

Materials and Methods

Enzymes, nucleic acids and chemicals. Norfloxacin was kindly provided by Merck Pharmaceutical Co. (Rahway, NJ). Nalidixic acid, oxolinic acid, ellipticine, adriamycin, and tRNA Type XXI were purchased from Sigma Chemical Co (St. Louis, MO). VP-16 (demethylepipodophyllotoxin ethylide-ß-D-glucoside) was kindly provided by the National Cancer Institute. Ciprofloxacin and CP-67,015 were synthesized at Pfizer Central Research by the procedures described in US Patent #4,670,444 (1987) and #4,636,506 (1987), respectively. Knotted P4 DNA was purified from P4 tailless capsids as described elsewhere (Liu *et al.* 1981). Proteinase K was purchased from Boehringer-Mannheim (Indianapolis, IN). pBR322 DNA was purchased from Pharmacia (Piscataway, NJ). All other chemicals were of reagent grade.

Topoisomerase II purification. Topoisomerase II was purified from the thymus of freshly slaughtered calves according to either the procedure of Halligan *et al.* (1985) or Schomburg and Grosse (1986). The Schomburg and Grosse procedure resulted in the isolation of the predominantly unproteolyzed (1.1×10^6 U/mg) and topoisomerase I-free form of the topoisomerase II (Schomburg and Grosse 1986), enabling the recovery of a higher activity enzyme with both catenation activity and sufficient activity to generate drug-free, non-radiolabeled cleavage of pBR322. Topoisomerase II, purified by the Halligan *et al.* procedure, resulted in the isolation of the predominantly proteolyzed form of the enzyme (0.9×10^5 U/mg) with some minor topoisomerase I contamination (which precluded its use in the cold cleavage assay). A unit of eukaryotic topoisomerase II activity is defined as the amount of enzyme required to completely catenate 0.38 µg pBR322 DNA in 60 min at 37°C under the reaction conditions described below.

Catenation assay. This assay was done using a modified version of the procedure of Schomburg and Grosse 1986. Modifications included the following changes in the reaction cocktail: 50 mM Tris-HCl, pH 8.0, 25 mM NaCl, 10 mg/ml bovine serum albumin, 2 mM dithiothreitol (in place of 2-mercaptoethanol), and 0.38 µg of pBR322 DNA per 25 µl reaction mix. The topoisomerase purified by the Schomburg and Grosse procedure, was used in this assay at 4.3 ng/25 µl reaction (4-5 U). Reactions were incubated for 1 h at 37°C, and stopped by the addition of SDS to 0.9%. Samples were deproteinated by incubation with proteinase K (0.8 mg/ml final concentration) for 45 min at 50°C, then electrophoresed in 0.4% TBE agarose gels and stained with ethidium bromide. The concentration of drug to cause 50% inhibition of activity (IC_{50} values) were determined by densitometric analyses of the formation of product (catenated DNA) in the presence of a dilution of compound being tested versus the drug-free control enzyme reaction. Uncatenated DNA is measured as arbitrary units in the reflective (cool white light) mode, versus time, from the photograph of the ethidium bromide-stained gel.

P4 unknotting assay. The assay was a modified version of the unknotting procedure of Liu *et al.* 1981. Modifications included the following changes: the amount of topoisomerase II added to the reaction mix was 1 unit (which was the amount of enzyme required to unknot 50% of the P4 knotted DNA in 30 min at 37°C); all drug dilutions were in dimethyl sulfoxide (final concentration of 1.3%); and after incubating 30 min at 37°C, the reactions were terminated by placing the reaction tubes in an ice water bath, electrophoresed in 0.7% TAE agarose gels and stained with ethidium bromide. Densitometric quantitation of the unknotted DNA is expressed as a percentage of the completely unknotted sample.

Radiolabeled cleavage assay. This assay was performed using [^{32}P]ATP end-labeled pBR322 DNA as substrate (Tewey *et al.* 1984a), with minor modifications. Modifications included the use of 2 mM ATP and between 65 and 780 ng of topoisomerase II (60-70 U) per reaction mix depending on the method of purification [Schomburg and Grosse 1986, and Halligan *et al.* 1985, respectively]. The procedure for end-labeling was done as described elsewhere (Nelson *et al.* 1984). Briefly, *Eco*RI digested pBR322 DNA was labeled with Klenow fragment, alpha-[^{32}P]ATP, and non-radiolabeled dTTP, then cleaved with *Hind*III to generate a single-end labeled fragment of 4332 bp. Two-fold dilutions of test drug were added to the reaction mixture and the radiolabeled cleavage products were separated by electrophoresis. The 50% maximum cleavage concentration (CC$_{50}$) was determined by densitometric tracing of the remaining intact substrate compared to a drug-free control.

Non-radiolabeled cleavage assay. Calf thymus topoisomerase II, purified by the procedure of Schomburg and Grosse, was used to establish a non-radiolabeled topoisomerase II DNA cleavage assay. 0.5 µg supercoiled pBR322 DNA, 22 ng of topoisomerase II (22-24 U), and varying dilutions of the compound being tested were combined in a total reaction volume of 25 µl (comprised of 10 mM Tris, pH 7.5, 10 mM MgCl$_2$, 10 mM KCl; Tewey *et al.* 1984a), and were incubated for 1 h at 37°C. The reaction was stopped by the addition of SDS to 0.1%, followed by deproteination with proteinase K at 0.8 mg/ml for 45 min at 50°C. Samples were electrophoresed in a 1% TBE agarose gel, stained with ethidium bromide, and the amount of cleaved, linear DNA was quantitated by densitometry to determine the CC$_{50}$ for inhibition by a dilution of the compound being tested relative to a control compound.

Densitometric analyses. Densitometric analyses of photographs of ethidium bromide-stained (1.25 mg/ml) DNA agarose gels visualized by transillumination by ultraviolet light, were done in the reflective mode (cool white light) on a Bio-Rad Model 620 Video Densitometer (Richmond, CA).

Genetic toxicology studies. Studies to assess the effects of CP-67,015 in standard genetic toxicology tests were performed as described by Holden *et al.* 1989.

Results

P4 unknotting and catenation assays

The anti-tumour agents (adriamycin, ellipticine, VP-16) and the DNA gyrase inhibitors (nalidixic acid, oxolinic acid, norfloxacin, ciprofloxacin, and CP- 67,015) were studied for their effects on the unknotting and catenation activities of mammalian topoisomerase II purified from calf thymus. Initial time course assays for both enzyme activities were run in the absence of drug in order to define the parameters that provided linear reaction kinetics. Time course studies of these catalytic assays identified linear

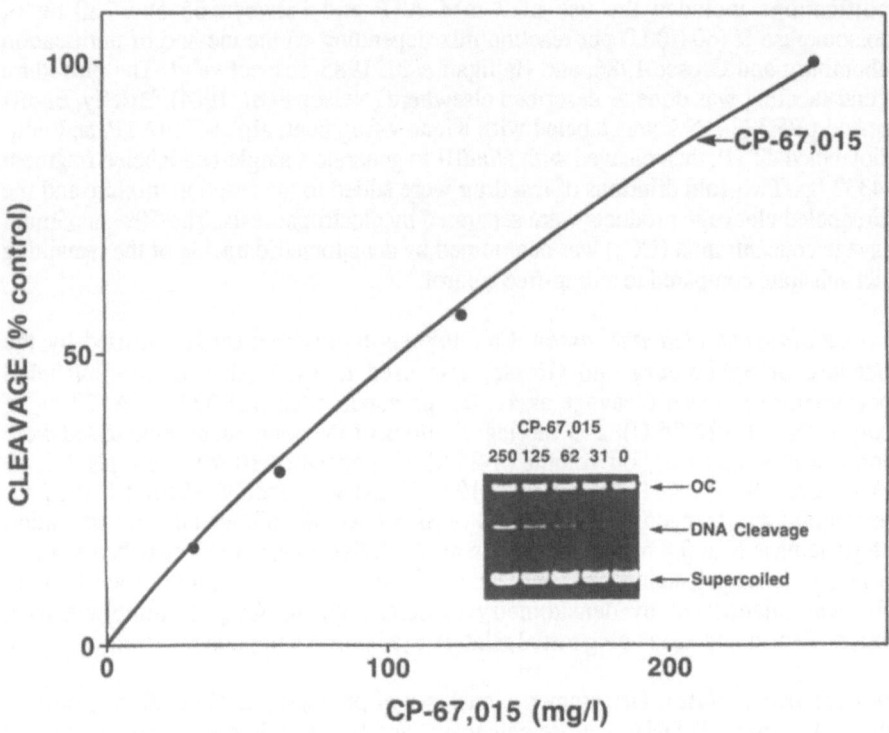

Fig. 11.1. Comparative inhibitory effects of CP-67,015 and ellipticine in the P4 unknotting assay. CP-67,015 is representative of the quinolones tested showing the need for much higher drug levels as compared to ellipticine (representative of the anti-tumour compounds) in testing in the P4 unknotting assay. Inset shows the photograph of the ethidium bromide-stained DNA agarose gel after electrophoresis of reactions inhibited by CP-67,015 and ellipticine, from which the percentage of inhibition of unknotting versus the drug-free control were calculated after densitometric analyses.

regions of the reaction where a 50% completion point could be determined from densitometric plots of photographs of ethidium bromide-stained DNA bands of unknotting and catenation, respectively (Barrett *et al.* 1989). Initial steady-state kinetics were exhibited over at least the first 30 min of the unknotting reaction, with essentially all substrate being converted to product in 60 min. Likewise with catenation, initial catalysis was approximately linear for at least the first 30 min. The parameters of time and enzyme activity that defined a 50% completion point (IC$_{50}$) were used for determining the effect of test drugs on P4 unknotting and catenation of

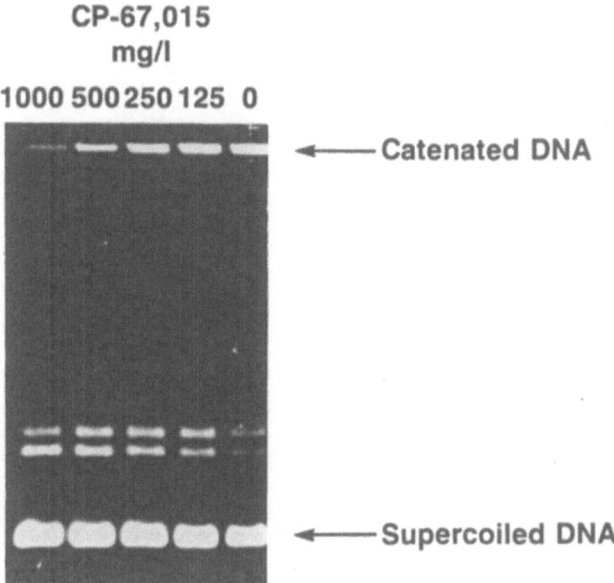

Fig. 11.2. Activity of mammalian topoisomerase II by CP-67,015. Concentration-dependent inhibition of catenation. Shown is the photograph of the ethidium bromide-stained DNA agarose gel afterel ectrophoresis.

Table 11.1. Comparison of Eukaryotic Topoisomerase Assays[a]

Compound	P4 Unknotting[b] IC_{50} µg/ml	Catenation[c] IC_{50} µg/ml	Radiolabeled DNA Cleavage CC_{50}^{d} µg/ml	Non radiolabeled DNA Cleavage CC_{50}^{d} µg/ml
Nalidixic acid	850 ± 90	1850 ± 150	>1000	>2000
Oxolinic Acid	>1000	>1000	>1000	>1000
Ciprofloxacn	140 ± 5	325 ± 5	120	>1000
Norfloxacin	165 ± 15	730 ± 180	>1000	>1000
Adriamycin	0.5	1.4	ND[e]	ND[e]
Ellipticine	4.9 ± 0.7	4.0 ± 0.4	~0.125	ND[e]
VP-16	110 ± 10	70 ± 11	4.5	7.5 ± 4.5
CP-67,015	92.5 ± 5	265 ± 23	33	73 ± 17

[a] Topoisomerase II isolated from calf thymus by the procedure of Schomburg and Grosse 1986

[b] P4 unknotting assay modified from Liu *et al.* 1981

[c] Catenation assay modified from Schomburg and Grosse 1986

[d] CC_{50} is the concentration of drug to induce 50% of the maximal DNA cleavage in this test system ±1 standard deviation

[e] Not determined due to intercalation of compound into substrate.

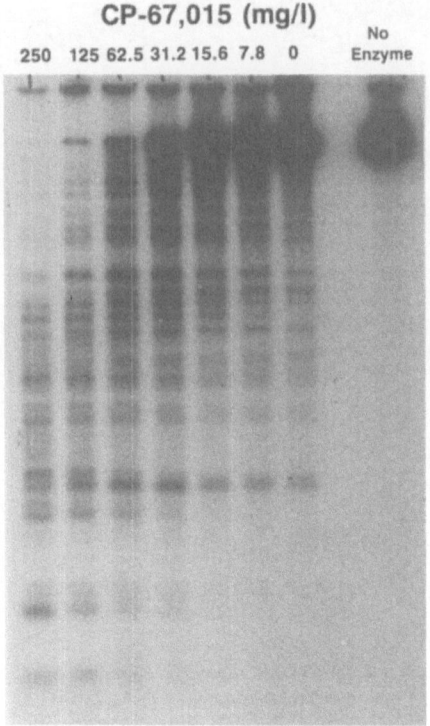

Fig. 11.3. Inhibition of mammalian topoisomerase II by quinolone CP-67,015 in the radiolabeled cleavage assay. Banding pattern of CP-67,015 is shown in the analyses of the autoradiograph (exposed for 72 h) of the drug concentration-dependent cleavage of single-end [^{32}P]ATP-labeled pBR322 by calf thymus topoisomerase II. Shown to the right of the CP-67,015 cleavage data is the enzyme-free control of unhydrolyzed single-end [^{32}P]ATP-labeled pBR322 DNA

supercoiled pBR322. Fig. 11.1 compares the inhibitory activity of ellipticine with that of the quinolone, CP-67,015, tested in the P4 unknotting assay. The degree of unknotting obtained in the presence of each drug was compared to the drug-free control under the conditions of the assay with literature reports (Tewey *et al.* 1984b). The quinolones tested were at least 20-fold less inhibitory (Table 11.1). A similar pattern was observed in the catenation reaction as exemplified by the inhibitory activity of CP-67,015 (Fig. 11.2). Adriamycin and ellipticine were highly inhibitory for topoisomerase II catenation activity with pBR322 as substrate. VP-16 was significantly less inhibitory than the other anti-tumour agents (IC$_{50}$ = 70 µg/ml), while the quinolones tested were not inhibitory at levels less than ~700 µg/ml (Table 11.1), with the exceptions of CP-67,015 (IC$_{50}$ = 265 µg/ml) and ciprofloxacin (IC$_{50}$ = 325 µg/ml).

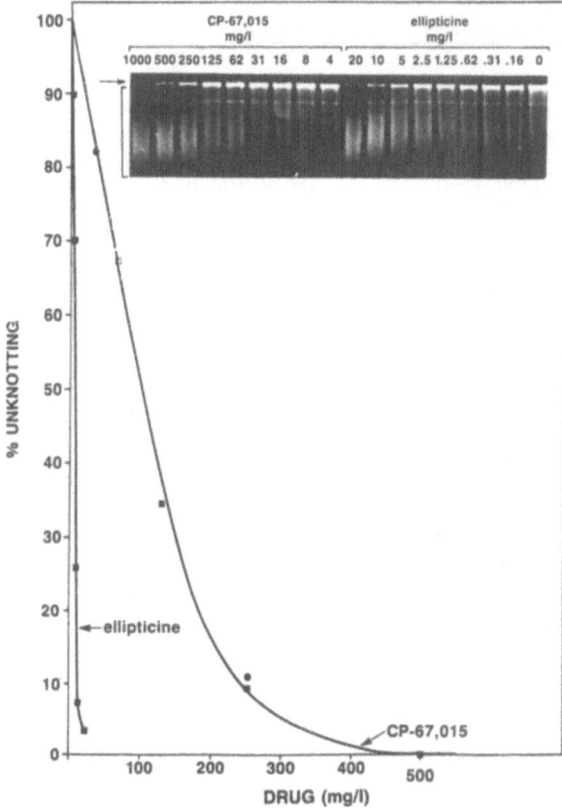

Fig. 11.4. Drug-enhanced DNA cleavage in a non-radiolabeled cleavage assay. Cleavage of supercoiled pBR322 DNA to the linear form by calf thymus topoisomerase II in the presence of CP-67,015 is shown as percentage versus the drug-free control. Shown in the inset is the photograph of the ethidium bromide-stained DNA agarose gel after electrophoresis of the reaction mix containing varying concentrations (mM) of CP-67,015 (with the relative migration of the open circular, cleaved, linear, and supercoiled DNA indicated).

Topoisomerase II-mediated cleavage assays

Since the quinolones tested demonstrated little inhibition of the catalytic activity of topoisomerase II, a radiolabeled cleavage assay was employed to determine whether they enhanced DNA cleavage in the presence of topoisomerase II above levels seen with enzyme alone. $[^{32}P]$ATP-end labeled pBR322 was employed for this purpose. The control compound, the anti-tumour agent VP-16 showed significant concentration-dependent enhancement of topoisomerase II-mediated cleavage of the labeled pBR322 fragment (Table 11.1), consistent with the report from Chen *et al.* 1984. CP-67,015 also induced significant cleavage (Fig. 11.3) with topoisomerase II, although at an 8-fold higher concentration (CC_{50} = 33 µg/ml) than VP-16. It was more potent in this regard than any other quinolone tested, although ciprofloxacin demonstrated radiolabeled DNA cleavage with an $C\dot{C}_{50}$ of 120 µg/ml (Table 11.1), or approximately 4-fold less potent than CP-67,015. The quinolone class of drugs appeared to have a different pattern of "banding" on the autoradiographs in comparison to the anti-tumour drugs,

Table 11.2. Activity of Quinolones In Eukaryotic Test Systems

Test	Quinolone/Active Level	Reference
In vitro Topoisomerase II Drosophila	nalidixic acid (625μg/ml)[a] oxolinic acid (340μg/ml)	Osherhoff *et al.* 1983
Calf thymus	ofloxacin(1300μg/ml) norfloxacin(300μg/ml ciprofloxacin(150μg/ml)	Hussey *et al.* 1986
	CP-67,015 (70μg/ml)[b] ciprofloxacin (>1000μg/ml) norfloxacin (>1000μg/ml)	This manuscript
Unscheduled DNA synthesis	ciprofloxacin (+)[c] norfloxacin (+) ofloxacin (+) pefloxacin (+)	Schluter 1986
Chromosomal breakage		
human lymphocytes	CP-67,015 (50μg/ml)	Holden *et al.* 1989
CHV79	PD-117,579 (10μg/ml) PD-117,962 (45μg/ml)	Theiss *et al.* 1989 (personal comm.)
Alkaline elution	ciprofloxacin (10μg/ml) ofloxacin (80μg/ml)	Bredberg *et al.* 1989
Cell cycle progression inhibition	ciprofloxacin (20μg/ml)	Forsgren *et al.* 1987
Gene Mutation CHV79/HGPRT	PD-117,579 (8μg/ml) PD-117,962 (40μg/ml)	Theiss *et al.* 1989 (personal comm.)
CHO/HGPRT	CP-67,015 (100μg/ml)	Holden *et al.* 1989
Mouse lymphona/ HGPRT	ciprofloxacin (+)[d]	Schluter 1986 and Stahlmann *et al.* 1988
	norfloxacin (+) ofloxacin (+) pefloxacin (+) CP-67,015(38μg/ml)	Holden *et al.* 1989
In Vivo *In Vivo* Cytogenetics Mouse bone marrow cells	CP-67,015 (500 mg/kg)	Holden *et al.* 1989

[a] IC_{50} in relaxation or catenation assays
[b] CC_{50} in non-radiolabeled cleavage assays
[c] Positive response reported
[d] Positive mutagenic response reported

suggesting specific, preferred cleavage sites as has been reported for anti-tumour agents (Chen *et al.* 1984; Nelson *et al.* 1984; Tewey *et al.* 1984a; Yang *et al.* 1985). The enhanced DNA cleavage activity of CP-67,015 was also observed in the topoisomerase II non-radiolabeled DNA cleavage assay using non-radiolabeled pBR322 as substrate (Fig. 11.4). No other quinolones included in this study stimulated DNA cleavage above background levels in the non-radiolabeled DNA cleavage assay when tested at concentrations up to 1000 μg/ml (Table 11.1).

Utilization of the non-radiolabeled DNA cleavage assay was possible due to the high specific activity of the topoisomerase II preparation purified by the Schomburg and Grosse procedure. The general banding pattern of the radiolabeled cleavage assay was the same using topoisomerase II purified by either the Schomburg and Grosse or Halligan *et al.* purification procedures (Farrell and Barrett, unpublished data), suggesting that the specificity of the unproteolyzed and proteolyzed forms of the calf thymus topoisomerase II were the same. Topoisomerase II from either purification procedure was adequate for catalytic assays.

Genetic toxicology tests

CP-67,015 was shown to have strong clastogenic activity in human lymphocytes and CHO cells and was a direct-acting mutagen in L5178Y/TK+/-, CHO/HGPRT, and CHV79/HGPRT cell culture systems at drug levels as low as 50 μg/ml (Table 11.2). CP-67,015 also showed clastogenic activity in bone marrow cells isolated from mice given oral doses of 500 mg/kg for five consecutive days (Table 11.2).

Discussion

In vitro DNA cleavage mediated by mammalian topoisomerase II has been reported to occur (in the absence of drug) at high enzyme to DNA ratios (Liu *et al.* 1983a; Liu *et al.* 1983b) and to be enhanced by anti-tumour compounds such as m-AMSA, ellipticine, VP-16, and VM-26 (Chen *et al.* 1984; Nelson *et al.* 1984; Pommier *et al.* 1985; Minford *et al.* 1986; Auclair 1987). At the same time, these compounds have shown variable inhibitory effects on the catalytic activity of the enzyme (Chen *et al.* 1984; Pommier *et al.* 1985; Auclair, 1987). Since this class of drugs has been associated with DNA breakage observed in cell culture, it has been suggested that enhanced topoisomerase II-mediated DNA cleavage is an important mechanism for the anti-tumour effects of these agents (Chen *et al.* 1984; Nelson *et al.* 1984; Maxwell and Gellert, 1986; Lock and Ross 1987). Quinolone antibacterials are thought to exert their effects by a similar interference with bacterial DNA gyrase (Mizuuchi *et al.* 1978; Lilley 1986; Drlica and Franco 1988), the bacterial homologue of eukaryotic topoisomerase II (Liu *et al.* 1980; Liu *et al.* 1984; Wang 1985; Maxwell and Gellert 1986). Since interaction with topoisomerase II is central to the activity of both quinolones and anti-tumour agents, it is of interest from a safety standpoint (Hussy *et al.* 1986; Phillips, 1987; Christ *et al.* 1988; Hosomi *et al.* 1988) to determine the effects of quinolone antibacterials on mammalian topoisomerase II. This is particularly true in light of the recent controversy over the mechanism of action of 4-quinolone antibacterials as to whether they bind to DNA, gyrase, or the ternary complex of DNA and gyrase (Shen and Pernet 1985; Forsgren *et al.* 1987; Hooper and Wolfson 1988; Tornaletti and Pedrini 1988).

The topoisomerase II inhibitory activity of several anti-tumour compounds and quinolone antibacterials was examined in four *in vitro* assays that measure the catalytic or DNA cleavage activities of a mammalian topoisomerase II enzyme, including a quantitative assay assessing cleavage of unlabeled DNA substrate. Other investigators

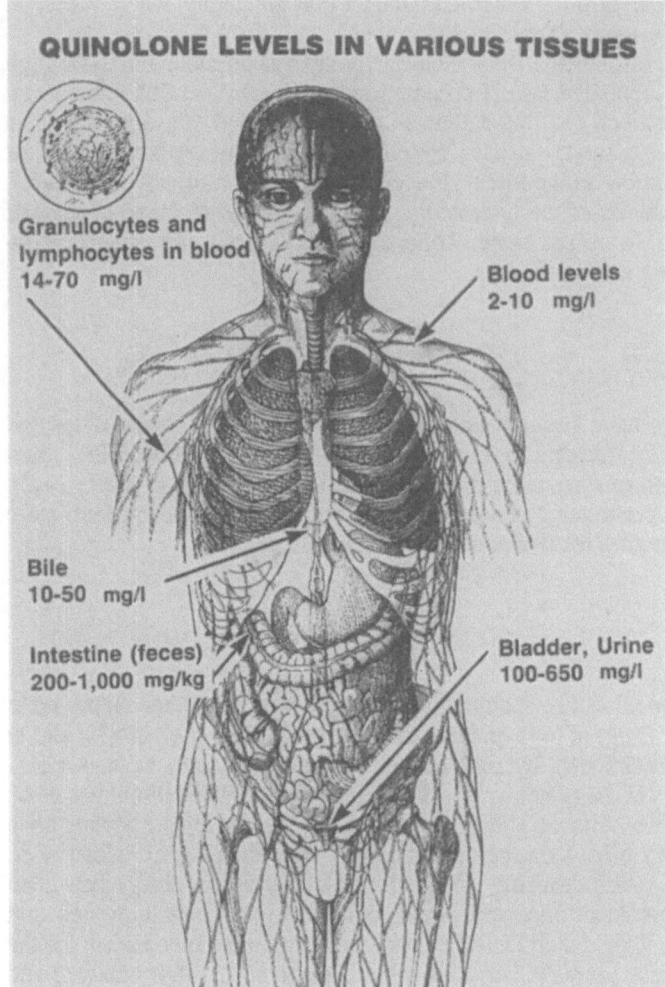

Fig. 11.5. Concentrations of quinolones in some body sites

have shown that anti-tumour compounds such as adriamycin, ellipticine, m-AMSA, and VP-16 are inhibitory to the catalytic activity of topoisomerase II and enhance DNA cleavage by topoisomerase II (Chen *et al.* 1984; Nelson *et al.* 1984; Tewey *et al.* 1984a; Tewey *et al.* 1984b; Riou *et al.* 1986; Auclair 1987). The assays used in the current study have confirmed this. While the effects of quinolone antibacterials on mammalian topoisomerase II have been less well studied, Hussy *et al.* did not find quinolones to be very inhibitory to topoisomerase II catenation activity, with K_i values ranging from 150 to 1000 µg/ml for ciprofloxacin, oxolinic acid, norfloxacin, and nalidixic acid. Likewise, it has been reported that nalidixic acid and pefloxacin are poor inhibitors of catenation/decatenation activities of a trypanosome topoisomerase II (Riou *et al.* 1986).

Our findings are in agreement with these results and indicate that ciprofloxacin, norfloxacin, oxolinic acid, and nalidixic acid are not potent inhibitors of catenation or unknotting activities. They were also not potent enhancers of DNA cleavage mediated by mammalian topoisomerase II, and the significance, if any, of the lower drug concentration of ciprofloxacin that was observed to enhance cleavage in the radiolabeled cleavage assay (Table 11.1) cannot be determined at this time. It is interesting that in one study, a good correlation was shown between the cytotoxicity of quinolones *in vitro* against HeLa cells with their ability to inhibit the relaxation activity of topoisomerase II purified from these cells (Oomori *et al.* 1988). In these studies, ciprofloxacin was found to give an IC_{50} for relaxation inhibition similar to the CC_{50} value obtained in our radiolabeled cleavage assay. Other adverse effects of quinolones on human lymphocytes have also been observed (Forsgren *et al.* 1987).

The activity of CP-67,015 has been found to be in the same order of magnitude as that of norfloxacin and ciprofloxacin with respect to the formation of measurable DNA gyrase cleavable complex and antibacterial potency (Barrett *et al.* 1989). As with other reports (Hussy *et al.* 1986; Riou *et al.* 1986), we found that the enhancement of topoisomerase II-mediated DNA cleavage activity by quinolones against prokaryotic DNA gyrase was approximately 100-fold more potent than against the eukaryotic topoisomerase II counterpart. Although this was also true for CP-67,015, the *in vitro* CC_{50} value for which we observed an increase in eukaryotic topoisomerase II-mediated DNA cleavage was significantly lower than that which has been observed to date with other quinolones. CP-67,015 enhanced DNA cleavage in both the radiolabeled and non-radiolabeled DNA cleavage assays with topoisomerase II at CC_{50} drug levels of 33 and 73 μg/ml, respectively.

CP-67,015 has also shown strong clastogenic activity in human lymphocytes and is a direct acting mutagen in mammalian cells. The *in vitro* activity of CP-67,015 in some standard genetic toxicology tests can be compared to that which has been reported for some other quinolones (Table 11.2).

It is important to note that the positive *in vitro* results seen with other quinolones have been considered as "false" positives, since abnormal effects have not been observed in animal testing, and the quinolones currently in clinical use have been shown to be safe in humans. This is in spite of the relatively high levels (Andriole 1988; Bergan 1988) to which quinolones can reach in some tissues (Fig. 11.5). In this regard, a positive clastogenic effect with CP-67,015 in bone marrow cells was seen in mice only when given at high oral doses of 500 mg/kg (Table 11.2). CP-67,015 is not under clinical development and the relevance of its *in vitro* and *in vivo* genotoxic activity is difficult to assess. While factors other than interaction with topoisomerase II may influence the effects of quinolones on DNA, the marked stimulation of topoisomerase II-mediated DNA cleavage observed with CP-67,015 could be directly linked to the positive genotoxic responses observed with this quinolone.

Acknowledgements

We thank K Drlica, M Gellert, LF Liu, and KM Tewey for assistance in starting-up assays; LF Liu for kindly providing P4 phage, *E. coli* host and titering *E. coli* strains; M Gellert for providing the *E. coli* strains overproducing gyrase A and B subunits; C Farrell, S Haskell, M Frescura and S Sokolowski for technical assistance; A Tait-Kamradt for critical review of this manuscript; J Morris for assistance in preparing the manuscript; and G Welch and E Bellefleur for help in preparing the figures.

References

Andriole VT (1988) Clinical overview of the newer 4-quinolone antibacterial agents. *In:* Andriole VT (ed) The Quinolones, Academic Press, pp 155-200.

Auclair C (1987) Multimodal action of antitumor agents on DNA: The ellipticine series. Arch Biochem Biophy 259:1-14

Barrett JF, Gootz TD, McGuirk PR, Farrell CA, Sokolowski SA (1989) Use of *in vitro* topoisomerase assays for studying quinolone antibacterial agents. Antimicrob Agents Chemo *In Press.*

Bergan T (1988) Pharmacokinetics of fluorinated quinolones. *In:* Andriole VT (ed) The Quinolones, Academic Press, pp 119-154

Bredberg A, Brant M, Riesbeck K, Azou Y, Forsgren A (1989) 4-quinolone antibiotics: positive genotoxic screening tests despite an apparent lack of mutation induction. Mutat. Res. 211:171-180.

Chen GL, Yang L, Rowe TC, Halligan BD, Tewey KM, Liu LF (1984) Nonintercalative antitumor drugs interfere with the breakage-reunion reaction of mammalian topoisomerase II. J Biol Chem 259:13560-13566

Christ W, Lehnert T, Ulbrich U (1988) Specific toxicologic aspects of the quinolones. Rev Inf Dis 10(Suppl. 1):S141-S149

Drlica K, Franco RJ (1988) Inhibitors of DNA topoisomerases. Biochem 27:2253-2259

Forsgren A, Bredberg A, Pardee AB, Schlossman SF, Tedder TF (1987) Effects of ciprofloxacin on eukaryotic pyrimidine nucleotide biosynthesis and cell growth. Antimicrob Agent Chemother 31:774-779

Gellert M, Mizuuchi K, O'Dea MH, Itoh T, Tomizawa J-I. (1977) Nalidixic acid resistance: A second genetic character involved in DNA gyrase activity. Proc Natl Acad Sci USA 74:4772-4776

Halligan BD, Edwards KA, Liu LF (1985) Purification and characterization of a type II DNA topoisomerase from bovine calf thymus. J Biol Chem 260:2475-2482

Holden HE, Barrett JF, Huntington CM, Muehlbauer PA, Wahrenburg MG (1989) Genetic profile of a nalidixic acid analog: A model for the mechanism of sister chromatid exchange induction. Environmental and Molecular Mutagenesis 13:238-252

Hooper DC, Wolfson JS (1988) Mode of action of the quinolone antimicrobial agent. Rev Inf Dis 10(Suppl.1):S14-S21

Hosomi J, Maeda A, Oomori Y, Irikura T, Yokota T (1988) Mutagenicity of norfloxacin and AM-833 in bacteria and mammalian cells. Rev Infect Dis 10(Suppl 1):S148-S149

Hussy P, Maass G, Tummler B, Grosse F, Schomburg U (1986) Effect of 4-quinolones and novobiocin on calf thymus DNA polymerase a primase, topoisomerases I and II, and growth of mammalian lymphoblasts. Antimicrob Agents Chemother 29:1073-1078

Lilley DMJ (1986) DNA topoisomerase inhibitors as anti-microbial agents. Biochem SocTrans 14:489-493

Liu LF(1984) DNA topoisomerases - Enzymes that catalyze the breaking and rejoining of DNA. CRC Crit Rev Biochem 15:1-24

Liu LF, Davis JL, Calendar R (1981) Novel topologically knotted DNA from bacteriophage P4 capsids: studies with DNA topoisomerases. Nuc Acid Res 9:3979-3989

Liu LF, Halligan BD, Nelson EM, Rowe TC, Tewey KM (1983a) Breakage and reunion of DNA helix by mammalian DNA topoisomerase II. In Cozzarelli NR (ed) Mechanisms of DNA Replication and Recombination, AR Liss, Inc, pp 43-53

Liu LF, Liu C-C, Alberts BM (1980) Type II DNA topoisomerases: Enzymes that can unknot a topologically knotted DNA molecule via a reversible double-strand break. Cell 19:697-707

Liu LF, Rowe TC, Yang L, Tewey KM, Chen GL (1983b) Cleavage of DNA by mammalian DNA topoisomerase II. J Biol Chem 258:15365-15370

Lock RB, Ross WE (1987) DNA topoisomerases in cancer therapy. Anti-Cancer Drug Design 2:151-164

Maxwell A, Gellert M (1986) Mechanistic aspects of DNA topoisomerases. Adv Protein Chem 38:69-107

Minford J, Pommier Y, Filipski J, Kohn KW, Kerrigan D, Mattern M, Michaels S, Schwartz R, Zwelling LA (1986) Isolation of intercalator-dependent protein-linked DNA strand cleavage activity from cell nuclei and identification as topoisomerase II. Biochem 25:9-16

Mizuuchi K, O'Dea MH, Gellert M (1978) DNA gyrase: subunit structure and ATPase activity of the purified enzyme. Proc Natl Acad Sci 75:5960-5963

Nelson EM, Tewey KM, Liu LF (1984) Mechansim of antitumor drug action: Poisoning of mammalian DNA topoisomerase II on DNA by 4'-(9-acridinylamino)-methanesulfon-m-anisidide. Proc Natl Acad Sci USA 81:1361-1365

Oomori Y, Tokutaro Y, Aoyama H, Hirai K, Suzue S, Yokota T (1988) Effects of fleroxacin on HeLa cell functions and topoisomerase II. J Antimicrob Chemo 22(Suppl D):91-97

Osherhoff N, Shelton ER, Brutlag D (1983) DNA topoisomerase-II from *Drosophila melanogaster*. J Biol Chem 258:9536-9543

Phillips I (1987) Bacterial mutagenicity and the 4-quinolones. J Antimicrob Chemo 20:771-773

Pommier Y, Minford JK, Schwartz RE, Zwelling LA, Kohn KW (1985) Effects of the DNA intercalators 4'-(9-acridinylamino)methanesulfon-m-anisidide and 2-methyl-9-hydroxyellipticinium on topoisomerase II mediated strand cleavage and strand passage. Biochem 24:6410-6416

Riou G, Douc-Rasy S, Kayser A (1986) Inhibitors of trypanosome topoisomerases. Biochem Soc Trans 14:496-499

Schomburg U, Grosse F (1986) Purification and characterization of DNA topoisomerase II from calf thymus associated with polypeptides of 175 and 150 kDa. Eur J Biochem 160:451-457

Schluter G (1986) Toxicology of ciprofloxacin, In; Neu HC, Wenta H (eds) Proceedings of the First International Ciprofloxacin Workshop. pp 61-70, Excerpta Medica, Amsterdam

Shen LL, Pernet AG (1985) Mechanism of inhibition of DNA gyrase by analogues of nalidixic acid: The target of the drug is DNA. Proc Natl Acad Sci, USA, 82:307-311

Stahlmann R, Lode H (1988) Safety overview: toxicity, adverse effects and drug interactions. In: Andriole VT (ed) The Quinolones, Academic Press, NY, pp 201-233

Tewey KM, Chen GL, Nelson EM, Liu LF (1984a) Intercalative antitumor drugs interfere with the breakage-reunion reaction of mammalian topoisomerase II. J Biol Chem 259:9182-9187

Tewey KM, Rowe TC, Yang L, Halligan BD, Liu LF (1984b) Adriamycin-induced DNA damage mediated by mammalian DNA topoisomerase II. Science 226:466-468

Tornaletti S, Pedrini AM (1988) Studies on the interaction of 4-quinolones with DNA by DNA unwinding experiments. Biochem Biophy Acta 949:279-287

Wang JC (1985) DNA Topoisomerases. Ann Rev Biochem 54:665-697

Yang L, Rowe TC, Nelson EM, Liu LF (1985) *In vivo* mapping of DNA topoisomerase II-specific cleavage sites on SV40 chromatin. Cell 41:127-132

Discussion Summary

It was considered that this work represented an important avenue for study since it was developing two lines of investigation about which we still know relatively little. The development of a potential structure-activity correlation with regard to mammalian cell mutagenicity was clearly very important in the light of the legislative requirements for the future clinical use of this group of agents. The concurrent study of the relative activity of 4-quinolones against mammalian and bacterial type II DNA topoisomerases not only provided further potential correlations with genotoxicity, but also provided significant quantitative evidence in the study of the selective toxicity of the 4-quinolones.

One particular area of interest was whether or not the demonstration of significant mutagenic activity with mammalian cells in culture correlated with demonstrable mutagenic activity in whole animals. This did not seem to be the case when it had been tried in the past, but it was concluded that this might not be surprising in the light of the differences in exposure times and levels of agent used. This was thought to be because of the evident ability of whole animals to both metabolize and excrete 4-quinolones quite efficiently, consequently it is very difficult to achieve similar challenge conditions *in vivo* and *in vitro*. (This argument does of course apply to any compound which is screened for genotoxic activity in an *in vitro* system).

In terms of the methodology used there was some concern expressed over the fact that the DNA substrate for the mammalian topoisomerase was a bacterial plasmid. Even though the use of such a generalized substrate for topoisomerase assays with enzymes from a wide variety of sources has become standard practice, it was considered that it would be useful if some studies could be carried out with a substrate of mammalian origin (like SV40) simply to confirm to the non-DNA topologists that the same results are obtained irrespective of the origin of the DNA. Another area which provoked some interest was the possibility that whilst routine catenation:decatenation assays and supercoiling:relaxation assays have clearly defined end-points, the analysis and quantification of alternative or intermediate structural forms of DNA which can be observed in the agarose gels might provide more information which might distinguish between different 4-quinolones.

It was generally agreed that whilst some techniques of both enzymatic and mutagenic study might not be ideal in a purely scientific sense, these are test systems that have to be used and represent the best available chance of obtaining very valuable information.

Chapter 12

In Vitro Genotoxicity Assessment and the Effects of 4-Quinolones upon Human Cells

G. C. Crumplin

Introduction

The last five years have seen the emergence of several new derivatives of Nalidixic acid which display great antibacterial potency and broad-spectrum antibacterial activity. Recent national and international meetings have clearly demonstrated that this group of compounds, known collectively as the 4-quinolones, are at present the most intensively studied group of antibacterial chemotherapeutic agents. Evidence is rapidly accumulating which suggests that these agents have the potential to become amongst the most exciting and challenging clinical agents to enter clinical use since the advent of the ß-lactams. The 4-quinolones act through a uniquely complex mechanism of action upon DNA topoisomerases. These enzymes control the spatial geometry of the cellular DNA and in molecular biology they are providing new insights into how cells perform the routine mechanical tasks of replicating DNA and transcribing the genetic information. It is becoming increasingly evident that the DNA topoisomerases are fundamentally important in sustaining the normal life-processes of all cellular organisms from the simple bacterium through to *Homo sapiens*.

The 4-quinolone antibacterial agents have become recognized as specific inhibitors of bacterial DNA gyrase (Morita *et al.* 1984; Domagala *et al.* 1986) which is the most complex and sophisticated DNA topoisomerase so far described as well as being an essential enzyme for the growth of bacterial cells (Kreuzer *et al.* 1978). However, whilst DNA topoisomerases and DNA gyrase, as controllers of the DNA topology, are new and challenging subjects for academic study by molecular biologists - to physicians who may have to contemplate the routine use of the 4-quinolone antibacterials, the DNA topoisomerases simply represent a name for the receptor for agents which are potent affectors of DNA systems.

Since both man and his prokaryotic parasites (the bacterial pathogens) depend upon the competent functioning of DNA systems - and 4-quinolones act selectively upon

DNA systems - the 4-quinolones have come under close scrutiny with regard to possible genotoxic effects in man. This is of course an instinctive and correct response, but the very complexity of the activity of these agents makes it difficult to draw objective conclusions from the diversity of published data. This communication is an attempt to present a critical appraisal of the available information relevant to the assessment of the genotoxic potential of the 4-quinolones along with a discussion of the peculiar problems which exist in attempting to interpret such data. Evidence will be presented to indicate that the 4-quinolones, because of their unique mechanisms of action, may present unprecedented problems in the assessment of genotoxic potential through consideration of *in vitro* data.

Outside the confines of the regulatory authorities, clinicians and microbiologists, wishing to make a judgement on the risk/benefit balance of any new group of compounds, must rely upon; (a) trust in the judgement of the regulatory authorities; and (b) personal assessment of available (published) data. Conclusions should be based upon critical analysis and not upon heresay in order to judge fairly. In the case of the 4-quinolones this needs to be done urgently since these compounds represent potentially very useful clinical antibacterial agents. It does however need to be made clear that we should not delude ourselves into thinking that making a judgement will be an easy task. The 4-quinolones are unusual compounds in their activity, they have been subjected to investigation for over 20 years and the history of this research is full of complex surprises. With this group of compounds we also have to contend with the fact that 4-quinolones have grown through serendipity, hence the study of them has not been co-ordinated in any way.

This review is presented from the standpoint of the *in vitro* investigator (and potential patient) who is hopefully free of the prejudices, pressures, and obligations of either the physician, the pharmaceutical company, or the regulatory authority. What follows will I hope be comprehensible to all concerned despite the extensive references made to the mechanics of genetic processes.

The Genotoxic Effects in Question

The 4-quinolones have been developed as antibacterial agents which act selectively upon the DNA replicative apparatus of susceptible bacteria, so it is axiomatic that they affect prokaryotic genetic systems. The specific concern of this communication is the nature of the induced effects in human cell systems, and whether the consequences of these are detrimental, beneficial, or neutral. Since a number of *in vitro* test systems based upon the reponses of prokaryotic cells have been developed as screening systems for genotoxic activity, on the basis of the presumption that prokaryotic DNA systems may be valid indicators of the responses of analagous systems in eukaryotes, it is evident that we should examine the effects of 4-quinolones on prokaryotic DNA systems. Because the 4-quinolones act selectively upon prokaryotic DNA topoisomerases and the functioning of these enzymes affects virtually all genetic processes, the variety of genetic systems which are worthy of examination is very extensive. Table 12.1 lists the range of genetic processes which can be shown to be affected by the action of the 4-quinolones and indicates how they might be implicated in the assessment of the genotoxic potential of the 4-quinolones. All of the mechanisms listed might be directly related to the assessment of the genotoxic potential through DNA-mediated processes using the traditional criteria of carcinogenicity, mutagenicity and teratogenicity. However, evidence of the effects of

Table 12.1. Genetic mechanisms likely to be involved in the induction of genotoxic effects

Genetic mechanism	Direct potential*	Interactive potential**
Point mutation	+	+
Transposition	+	+
Illegitimate recombination	+	+
Generalized recombination	+	+
Error-prone repair induction	+	+
Error-prone repair inhibition	-	+
Error-free repair inhibition	+	+
Direct DNA damaging	+	-
Reverse transposition	+	+
Modification of DNA structure	+	+
Modification of gene-expression	+	+
Modification of immune-response	-(?)	+

*   = indicating the potential for "direct" induction of "genotoxic" sequelae
** = indicating the potential for interaction with other genotoxins

the 4-quinolones upon the regulation of gene-expression in prokaryotes raises the possible consideration of such effects in man. These effects in bacteria are intimately associated with the specific mechanism of action of the compounds and so new criteria need to be incorporated into the consideration of the induced "genetic" effects of the 4-quinolones. These criteria are:

(i)   effects upon development (non-mutagenic)
(ii)   effects upon metabolism
(iii)   meiotic and mitotic effects
(iv)   effects upon immune-responses, and
(v)   CNS-associated effects

All of these processes depend upon competent regulation of gene-expression which, by definition, falls into the domain of potential induced genetic effects to be examined.

The most emotive form of genotoxic effect under review must be the question of carcinogenisis, but even here the links between the carcinogenic event and a specific genetic event are not clear cut. Controversy remains as to whether the most important induced events to be considered are point mutations or spatial rearrangements of the DNA (Cairns 1981; Echols 1981). Consequently we must cater for all opinions and include all potential genetic mechanisms in this critical review.

The interlinking of such a variety of genetic and physiological systems to be jointly considered is clear evidence of the complexity of the problem. Such complexity clearly indicates that we should not expect simplistic answers, and may be taken as a warning of what is yet to come in this communication.

The Basic Assumptions

The examination of the *in vitro* effects of any compound to ascertain the existence of any putative genotoxic potential represents the first step in a complex process of risk-assessment. The nature of the battery of test systems to be used is becoming a matter of established practice developed upon the basis of experience. However, because of the peculiarly complex nature of the activity of the 4-quinolones we must have clear in our minds what basic assumptions are being made routinely. Having established these assumptions we then need to test their validity and applicability to the 4-quinolones.

Because the 4-quinolones act specifically upon the systems involved in the mechanical processing of DNA - and because DNA is presumed to be a uniformly structured molecule throughout the different genera - it is assumed that the enzymes which mechanically process the DNA (the DNA topoisomerases) preserve a functional uniformity. Acceptance of this basic assumption opens up the possibility of using prokaryote and lower eukaryote systems *in vitro* as practicable models of equivalent genetical systems in *H. sapiens*. This means that a second assumption is made simply on the basis of accepting the first assumption. Such "stacking" of assumptions is compensated by the use of mammalian and/or human transformed cell-lines in culture to examine induced effects - but once again assumptions are being made. The new assumptions are that non-human mammalian cell-lines simulate *H. sapiens* cells and that transformed (or tumour) cell-lines are valid models of "normal" human cells.

This second pair of assumptions can be compensated for by the use of non-transformed cells cultured *in vitro* as the basis for a test system. Here again practicality forces the investigator/assessor to make an assumption. In this case the availability of tissues dictates which types of cells can be routinely used and it has to be assumed that any cell-type is a valid genetical model for all of the differentiated cell-types to be found in *H. sapiens*.

Practicality is the causitive agent for the final assumption in a risk assessment. It has to be realized that induced changes consequent upon challenge with the therapeutic level of the agent under examination are likely to be very rare events. The statistically valid quantifying of very rare events demands very large populations for screening, and very large populations (perhaps $>10^9$/sample) are often difficult to attain with eukaryotic systems, even *in vitro*. Consequently investigations are carried out using abnormally high challenge doses to artificially increase the frequency of events and permit statistical validation with smaller target populations. Data derived in this way can be used to undertake a "low dose extrapolation" to provide an estimate of risk at therapeutic exposure levels of the test agent. The whole process of low-dose extrapolation depends upon the assumption that the agent under investigation acts in the same way, upon the same receptor, at both high and low challenge levels.

The listing of these assumptions may appear to be stating the obvious, but we must be aware of them when considering the 4-quinolones. Even though experience has shown us that these assumptions are generally valid, we should not exclude the possibility that in some cases the validity cannot be taken for granted.

A Battery of *In Vitro* Test Systems

The *in vitro* assessment of the potential genotoxicity of chemicals can be based upon the examination of a wide-range of phenomena in prokaryotic and eukaryotic systems. The most generally applied systems include:-

1　　　　Mutagenicity in bacteria (Ames *et al.* 1975; Green and Muriel 1976).
2　　　　Selective toxicity against DNA-repair deficient mutant bacterial strains (Rosenkranz and Liefer 1980; Kada and Hirano 1980).
3　　　　Induction of prophage in bacteria (Moreau and Devoret 1977).
4　　　　Mutagenicity in mammalian cells (Clive *et al.* 1972; Furth *et al.* 1981).
5　　　　Transformation of mammalian cells (Styles 1977).
6　　　　Induction of unscheduled DNA synthesis in mammalian cells (Martin *et al.* 1978).

7   Induction of chromosome changes in mammalian cells (Latt and Schreck 1980).

8   Macromolecular binding of genotoxins to DNA (Garner 1980).

9   Induction of DNA strand breakage in treated cells as estimated by alkaline elution (Kohn 1979).

The use of a battery of such short-term tests is often advocated for the assessment of potential genotoxicity, and detailed examination of the literature on the 4-quinolones indicates that an extensive battery of tests have been undertaken either as specific genotoxicity studies, or in the guise of mechanistic studies in various organisms.

The short-term test systems cited above involve consideration of a wide variety of genetic and pharmacological phenomena, all of which may be implicated in the genesis of genotoxic sequelae either by point mutation or gene rearrangement. However, even with such a variety of phenomena to investigate, examination of the properties of the 4-quinolones has failed to elicit unequivocal results to a bewildering extent.

Do 4-Quinolones get into Human Cells?

Before we attempt to review and examine the evidence for and against 4-quinolones as potential genotoxins we must answer the obvious question - Do 4-quinolones get into human cells exposed to therapeutically attained levels of the agents? The only really valid answer to this question must stem from the examination of cells in their proper context rather than in culture, hence the examination of data from patients is required.

As far as this author is aware there are no published reports of direct measurements of intracellular 4-quinolone levels in people normally dosed with these agents. The only evidence overtly available is indirect circumstantial evidence. For example it is obvious that where a 4-quinolone has been used to successfully treat cases of infection with obligate intracellular pathogens like *Chlamydia trachomatis* that the 4-quinolones must enter the cells to a significant extent in order to exert any effect upon the pathogenic bacteria. Similar evidence can be derived from studies of the effects of 4-quinolones on *Legionella pneumophila* which grows both extracellularly and intracellularly *in vitro*. It has been shown that ofloxacin is very active against this organism in its intracellular niche when tested in both an animal model and in infected human monocytes *in vitro* (Saito *et al.* 1985). The activity of a 4-quinolone against intracellular bacteria has also been used as a measure of the uptake of ciprofloxacin by mouse peritoneal macrophages. Using *Staphylococcus aureus* as the indicator organism it has been shown that macrophages actually concentrate ciprofloxacin to between two and three times the extracellular levels (Easmon and Crane 1985). Using the new 4-quinolone AM833 these workers have also shown that whilst neutrophils concentrate this 4-quinolone 4-fold, macrophages only concentrate it 2-fold (Easmon and Crane 1986) thus indicating that it is not reasonable to extrapolate between cell-types. This evidence of concentration of 4-quinolones in certain types of tissue may be taken as an indication that even where the intracellular mammalian receptor is intrinsically less sensitive than the bacterial equivalent, effective levels for activity can still be obtained in a mammalian cell if the mechanism for concentration exists.

In vivo evidence of cellular uptake of 4-quinolones in *H. sapiens* stems from an examination of the adverse effects of 4-quinolones (where reported). The best documented adverse effect of a 4-quinolone is the nalidixic acid-induced photosensitivty of the skin exposed to ultraviolet irradiation. Phototoxic bullous eruptions associated with nalidixic acid therapy were first reported in 1964 (Zelickson 1964), and subsequently several reports of dermal photoreactions have appeared (Birkett *et al.* 1969; Louis *et al.* 1973; Ramsay and Obreshkova 1974; Burry 1974). This particular reaction is a particularly useful model since its occurence is indicative of intracellular activity, and the fact that the principal induced effect of UV-irradiation is the induction of photoproducts (usually photodimers) in DNA. It seems likely that the photosensitivity reaction induced by nalidixic acid represents either a sensitization of the cellular DNA to UV-damage, or an induced reduction in the capacity of the cells to carry out repair of the DNA damage. In either case this is indicative of an effect of a 4-quinolone upon the DNA systems of cells exposed to therapeutic levels of the agent. This conclusion clearly indicates the necessity for examining the possibility of an irreversible adverse effect upon the genetic capacity of cells in patients treated with a 4-quinolone.

Available "Genotoxicity" Data

Examination of the literature on 4-quinolones yields the following contradictory *melange* of results which I have subdivided simply upon the basis of the test systems involved:-

Data from Bacterial Systems

Mutagenisis studies: Using standard protocols to measure the induction of reverse mutations at the *his* locus in *S. typhimurium* (Ames strains) and at the *trp* locus in *E. coli* WP2, McCoy *et al* (1980) were unable to detect any change in mutant yield using nalidixic acid. These workers concluded that nalidixic acid could be classified as non-mutagenic on the basis of these results. A similar result has been obtained with the use of the Ames strains for ofloxacin (Mayer and Bruch 1985) and for norfloxacin and AM-833 (Hosomi *et al.* 1988). However, earlier work in *E. coli* K12 showed a positive mutagenic effect of the drug when measuring the frequency of reverse mutation to streptomycin-independence (Cook *et al.*1966). Similar results were obtained for nalidixic acid in *E. coli* WP2 *uvrA* for reverse mutation at the *trp* locus (Witkin and Wermundsen 1979) and for mutation at the *gal* locus in *E. coli* K12 with a range of 4-quinolones (Phillips *et al.* 1987). These results from mutant frequency measurements in bacteria are borne out by the demonstration of a positive mutagenic effect in T4 bacteriophage (Vigier 1974), and in various Staphylococcal bacteriophages (German *et al.* 1969).

The disparity in the conclusions drawn from mutant yield and mutation frequency studies is understandable on the basis of the differences of interpretive protocols used, but much more confusing is the finding that nalidixic acid may be antimutagenic. Using the *mutD5* mutator mutation (which increases spontaneous mutation frequencies by a factor of $<10^5$), and screening for mutation frequencies at a variety of loci in *E. coli* K12, it has been shown that treatment of the *mutD5* cells with nalidixic acid significantly reduces the exceptionally high spontaneous mutation frequencies characteristic of this strain (Degnen 1974).

Data from Specific Eukaryotic Genotoxicity Studies

Early studies of the effects of nalidixic acid upon human cells yielded contradictory results. Rosenkranz and Lambek (1965) reported positive changes in chromosome structure of human cells exposed to nalidixic acid, but subsequent studies have reported the absence of detectable effects upon the chromosomes of human cells exposed to Nalidixic acid (Stenchever et al. 1970) or ofloxacin (Shimada et al. 1984).

Examination of the chromosomes from children treated therapeutically with alidixic acid indicated the induction of sister chromatid exchanges (SCEs) by the drug (Kowalczyk 1980). Unfortunately this report contained no control studies and prompted a re-examination of SCE induction by nalidixic acid and its active metabolite hydroxynalidixic acid in vitro. These studies showed no evidence of SCE induction by either compound (Lejeune and Couturier, personal communication). Examination for SCE induction in CHO cells in vitro by ofloxacin has also yielded a negative finding (Shimada et al. 1984) as has the examination of CHO cells and V-79 cells treated with AM-833 or norfloxacin for the induction of chromosome aberrations (Hosomi et al. 1988).

Recent studies with the novel analogue CP-67015 (see Fig. 12.2) have shown that this analogue, which is non-mutagenic in bacterial assays, shows strong clastogenic activity in human lymphocytes and in CHO cells, with induced chromosome damage in vivo in mouse bone-marrow cells. Surprisingly, SCE studies in vitro in human lymphocytes or CHO cells showed only weak SCE inducing activity even at levels of agents which produce severe chromosome breakage (Holden et al. 1989). In vivo there was also no significant elevation of SCE in mouse bone marrow following parenteral treatment with CP-67015. In the same study these workers also showed that CP-67015 was a direct acting mutagen in mammalian cell assays using (a) L5178Y/TK $^{+/-}$ gene mutation, (b) CHO/HGPRT gene mutation and (c) V79/HGPRT forward mutation assay.

Using the dominant lethal mutation assay system in mice no mutagenic effects of nalidixic acid, ofloxacin, miloxacin, or norfloxacin have been detected (Shimada et al 1980; 1985; Tokunaga et al 1979; Irikura et al 1981). These negative results in a well tried and tested assay system are however contradicted by the findings of Czinn et al. (1981) who reported the positive induction of genetic abnormalities in the American sea urchin by nalidixic acid. The issue is further confused by the finding of a negative inductive effect for ofloxacin in the mouse micronucleus assay system (Shimada et al. 1984).

The genotoxic potential of ofloxacin has also been assessed by assaying for the induction of mutation to 6-thioguanine resistance in V-79 cells in vitro (Mayer and Bruch 1985; Miltenberger 1985), and measurement of the induction of unscheduled DNA synthesis in human fibroblasts (Shimada et al. 1984). Both of these studies yielded negative results indicative of no significant genotoxic potential.

As far as this author is aware the only available data on the in vivo genotoxicity of a 4-quinolone is a study of the carcinogenicity of nalidixic acid in mice (Kanisawa et al. 1974). In this study there was no evidence of any induction of tumors or tumorigenic alteration in any organs.

The confusing nature of this data has been further complicated by recent unpublished reports of a positive result with ciprofloxacin obtained in the rat hepatocyte unscheduled DNA synthesis assay system and on some (but not all) mammalian cell in vitro mutagenicity studies. However even this presentation of scientific evidence by rumour is complicated by other reports attributed to independent work which casts some doubts upon the validity of any simplistic interpretation of the rat hepatocyte data. Under normal circumstances a positive result in a UDS assay system is accompanied by a similarly positive induction of DNA strand breakage (as evidence of the alkaline elution method). It is reported that in lymphocyte cells treated *in vitro* with ciprofloxacin a positive response could be demonstrated for UDS induction accompanied by no evidence of the induction of DNA strand breakage at challenge doses >10 times the serum level of the drug attained with normal dosing. The absence of DNA strand breakage has also been observed in mice challenged with AM-833 (Hosomi *et al.* 1988). This disparity could be taken as evidence that the interpretation of the UDS result as a simple indicator of potential genotoxicity may require significant reconsideration. This indication is borne out by the recent studies of Melcion and Cordier (1986) who studied the induction of UDS in primary rat hepatocytes by nine of the 4-quinolones (pefloxacin, ciprofloxacin, enoxacin, pipemidic acid, norfloxacin, ofloxacin, piromidic acid, oxolinic acid and nalidixic acid) and found no evidence of UDS induction, only the induction of dose-related decreases in UDS. However these workers found evidence to indicate that with the use of the autoradiographic technique false-positive results could be obtained if proper account is not taken of changes in the cytoplasmic grain count on the autoradiographs. The conflicting *in vitro* and *in vivo* evidence derived from studies of the effects of the 4-quinolones upon DNA repair in mammalian cells has also been highlighted by the recent work of McQueen and Williams (1987) with studies of rat hepatocytes which showed for ciprofloxacin a positive *in vitro* induction of DNA repair with a negative result *in vivo* after subcutaneous administration of high doses. In fact the existence of such conflicts has been used to support the proposal that the *in vitro* data can be regarded as an example of "false positive" results (Schluter 1986).

A recent study (Maura and Pino 1988) of the DNA-damaging and mutagenic potential of oxolinic and pipemidic acids in an *in vivo* system, the granuloma pouch assay, showed that neither of these agents exhibited any mutagenic activity (assayed as forward mutations at the *hgprt* -locus) when administered either orally or sytemically. However, whilst systemic administration of the test agents resulted in no apparent induction of DNA strand breakage (assayed by alkaline elution) oral administration of high doses resulted in evidence of strand-breakage in the granuloma cells. It is however interesting to note that whilst evidence was also found of DNA strand-breakage induction in the liver and kidney cells of animals sacrificed 12 hours after administration, there was no evidence of breakage in animals sacrificed after 24 hours. With similar results indicative of repair of the induced strand-breakage in the graniloma cells it is evident that the 4-quinolone induced lesions in the rat are repairable. However, the most significant finding of this study is the observation that DNA damaging activity was only evident after oral administration, thus indicating that activity is only apparent after metabolic activation and thus it is either one or more metabolites alone (or in combination with the parent moiety) which demonstrates a genotoxic potential in this assay system.

Bacterial Mutagenicity?

Because of the clear disparity between the negative and positive results published for the estimation of mutagenicity of nalidixic acid in *E. coli* strain WP2 clarification experiments were carried out using the same strain of *E. coli*.

The potent bactericidal activity of nalidixic acid against Gram negative bacteria (Crumplin and Smith 1975) presents a significant problem in obtaining sufficient

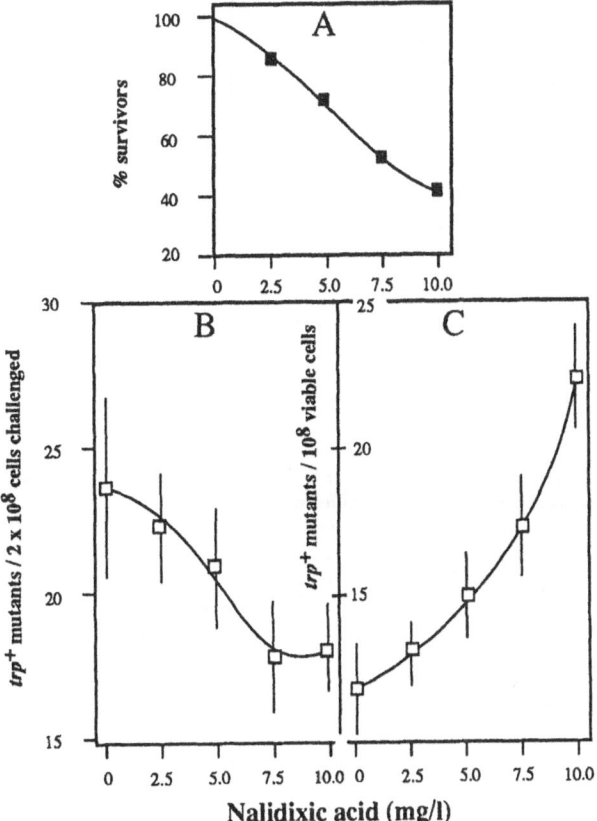

Fig. 12.1. Comparison of results from induced mutant yield (B) and induced mutant frequency (C) in *E. coli* WP2 *wt* with varying doses of nalidixic acid measuring reversion at the *trp* locus.

survivors for the estimation of mutant yield (absolute numbers of mutants). Consequently direct measurements were made of survival and mutant yield to facilitate the estimation of induced mutant frequencies. Furthermore, since nalidixic acid has been shown to exert locus specific effects upon transcription (Smith *et al.* 1978; Sanzey 1979), mutant yield was measured at three independent loci, namely reversion to tryptophan-independence, and forward mutations to ampicillin and streptomycin resistance. An important divergence from the normal practice of using repair-defective indicator strains was the use of a DNA repair proficient strain. This was done because; (a) repair-defective strains are more sensitive to the inhibitory effects of the drug (Winshell and Rosenkranz 1970; McDaniel *et al.* 1978; McCoy *et al.* 1980) and their use would reduce the numbers of surviving cells for screening; and (b) in bacteria nalidixic acid inhibits DNA gyrase, itself is involved in DNA repair (Crumplin 1981).

Using *E. coli* strain WP2, cells were challenged with various doses of nalidixic acid for one hour at 37°C before washing and resuspending in fresh drug-free medium for one hour prior to diluting and plating on selective and non-selective media. Fig. 12.1 shows the results obtained when mutational events were monitored at the *trp* locus.

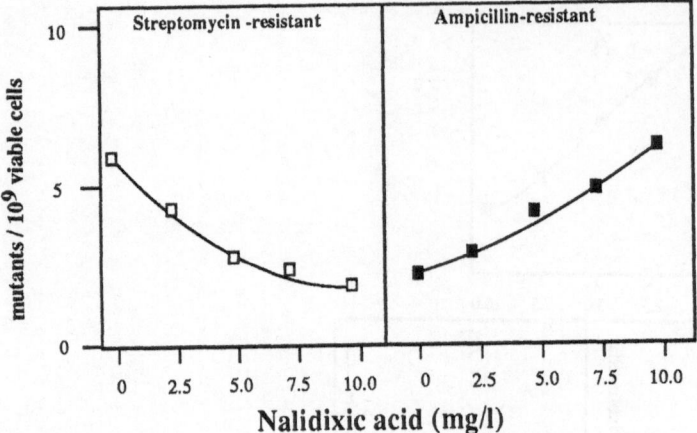

Fig. 12.2. Comparison of induced mutation frequencies in *E. coli* WP2 *wt* with nalidixic acid and scoring for mutation to either streptomycin-resistance or ampicillin resistance.

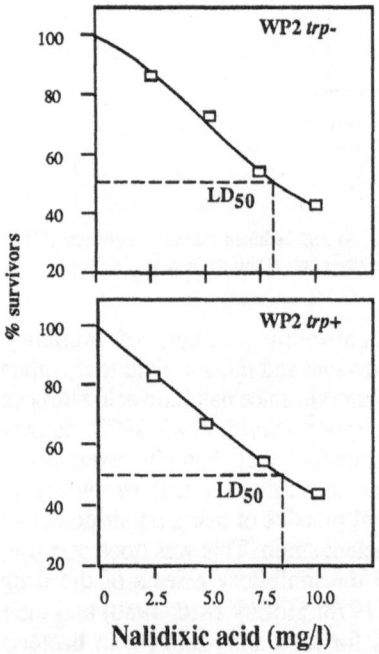

Fig. 12.3. Comparison of the susceptibilities of tryptophan-requiring and tryptophan-independent isolates of *E. coli* WP2 to challenge with nalidixic acid.

Fig. 12.1B shows that there was no absolute increase in the yield of mutants able to grow in the absence of tryptophan, if anything there was a dose-related decline in yield. This result is not in essence different to that obtained at the same locus by McCoy et al (1980). However when consideration is also given to the effects of nalidixic acid upon survival of the this organism (see Fig.12.1A), and mutant frequencies per survivor calculated, it can be seen in Fig. 12.1C that challenge with nalidixic acid elicited a positive response. Statistical examination of the data by t-test shows that nalidixic acid at 5 µg/ml or more resulted in a significant increase in mutant frequence ($p = 0.001$). Estimates of mutant yield and mutant frequency at the two drug-resistance loci showed no increase in mutant yield at either locus (see table 12.2). Conversion of the mutant yield data to mutant frequencies per survivor showed a small increase in mutant frequency at the ampicillin-resistance locus, and a small decrease at the streptomycin-resistance locus (see Fig. 12.2). Estimations at these two loci did not yield sufficient mutants for a critical statistical treatment, but the results do suggest that the mutagenic effects of nalidixic acid in bacteria may be locus-specific. This indication bears out the locus-specific effects of the drug upon transcription and the locus-specific effects of the gyrA mutation (resistance to nalidixic acid by modification of the receptor protein) upon spontaneous mutation to other drug resistances (Chao 1978).

Table 12.2. Data from study of the mutagenic effects of Nalidixic acid at drug-resistance loci

NAL (mg/ml)	% survivors	STR^r mutants yield/10^{10} cells plated	frequency/ 10^9 viable bacteria	AMP^r mutants yield/10^{10} cells plated	frequency/ 10^9 viable bacteria
0	100.00	58	5.8	21	2.10
2.5	86.0	36	4.20	22	2.60
5.0	70.0	19	2.74	28	3.90
7.5	52.0	12	2.24	24	4.62
10.0	41.0	7	1.68	25	6.10

The marginal effect in induced mutant frequency to tryptophan-independence might possibly be explained by a selective effect of the drug in killing tryptophan-dependant cells more readily than the revertant cells. This possibility was examined by carrying out one-hour treatments of five representative clones from each of (i) the parental strain and (ii) mutant clones isolated in the initial experiment. Monitoring of survivors permitted the derivation of the dose-response curves shown in Fig. 12.3. Comparison of the LD$_{50}$ values obtained showed that there was no significant difference between these values for the two genotypes, viz:- for tryptophan dependent parents

$$LD_{50} = 8.125 \pm 0.28 \ \mu g/ml$$

and for tryptophan-independent mutants

$$LD_{50} = 8.340 \pm 0.24 \ \mu g/ml$$

These results clearly demonstrate that in absolute terms nalidixic acid can be shown to be a weak mutagen at certain loci in bacteria if every effort is made to maximize the sensitivity of the test system. The obvious problem in assessing the mutagenicity of this type of compound is the toxicity of the test compound which by definition must seriously reduce the potential yield of mutants. This problem is of course exacerbated by the possibility that the DNA repair processes, which by their error-proneness effect the ultimate mutational events, may themselves be inhibited by a 4-quinolone. Such inhibition would not only reduce survival, it could also artificially reduce the yield of mutational events. It would thus seem preferable to attempt to estimate mutant frequency amongst survivors rather than relying upon estimates of total mutant yield. Only by allowing for the toxic effects can any small mutagenic effect be detected.

The problem of the toxicity of the test compound is clearly demonstrated here with the use of a DNA repair-proficient test organism. The use of this type of genetic background is in contrast to the routinely used screens which use repair-deficient tester strains (eg. *E. coli* WP2 strains, and the Ames stains of *S. typhimurium*). The use of repair-defective strains for the screening of the 4-quinolones in particular is counterproductive for two specific reasons: firstly it is known that some types of repair-defective mutant are not only more sensitive to the inhibitory effects of nalidixic acid upon growth (Winshell and Rosenkranz 1970; McDaniel *et al.* 1978), but also more specifically hypersensitive to the lethal sequelae of inhibition of the cell-growth (McDaniel *et al.* 1978); secondly, many of the tester strains used in routine assays have been constructed to carry the plasmid pKM101 which further increases the mutability of the bacteria. This plasmid is derived from R46 which is a tra$^+$ plasmid of compatibility group N, pKM101 retains the tra$^+$ phenotype which has been shown to increase the sensitivity of host bacteria to the toxic effects of nalidixic acid (Crumplin and Smith 1981). Hence, even though the use of a repair-defective mutant strain (±pKM101) may effect a higher initial incidence of mutation, this property may be nullified by a greater tendency of the cells to die during exposure to a 4-quinolone - and dead mutants are notoriously difficult to monitor!

A further potential problem in obtaining an accurate measure of the mutagenic effects of 4-quinolones is the distinct possibility that the drug may selectively inhibit the transcription of mutant genes (Smith *et al.* 1978). The existence of this possibility indicates that any method used should involve screening at several independent loci and should not use the incorporation of the test drug in the final selective medium. To overcome this latter problem the experiments carried out here utilize a finite exposure period in liquid medium followed by removal of the drug by thorough washing before plating out on selective media. This simple procedure should minimize the risk of any selective transcriptional events.

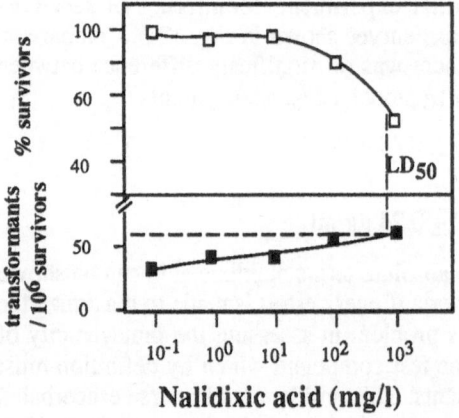

Fig. 12.4. Results of the induction of transformation in BHK21 cells by nalidixic acid.

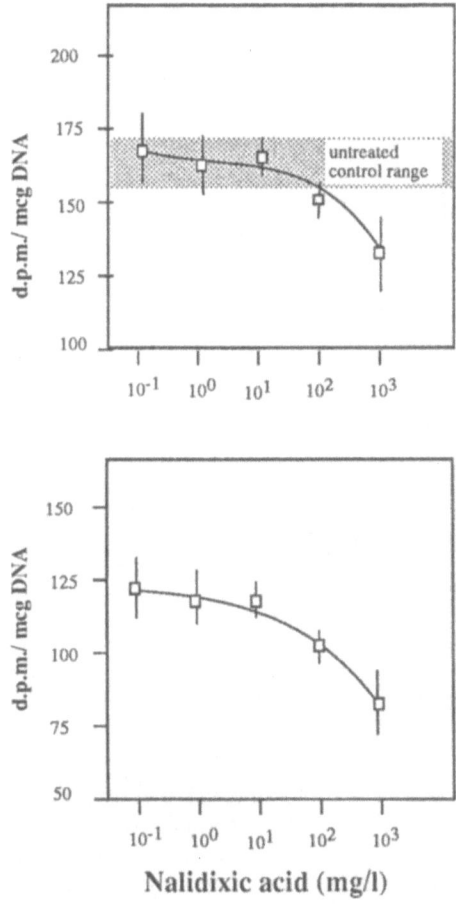

Fig. 12.5. Measurement of the induction of unscheduled DNA synthesis in TK6 cells by nalidixic acid.

Some Gap-Filling Mammalian Genotoxicity Screens

The potential genotoxic activity of nalidixic acid has been assessed in four mammalian *in vitro* test systems with the following results:-

Transformation of BHK21 cells: Using the experimental protocol of Styles (1977) in the absence of an induced rat-liver S9 metabolic system nalidixic acid was tested using challenge doses up to 1 mg/ml. As can be seen in Fig. 12.4 there was no evidence of a positive response using the criterion of a five-fold increase over background frequency at the LD$_{50}$ (Ashby *et al.* 1980). Control studies carried out with 4-nitroquinoline-n-oxide (4NQO) were clearly positive by the same criterion.

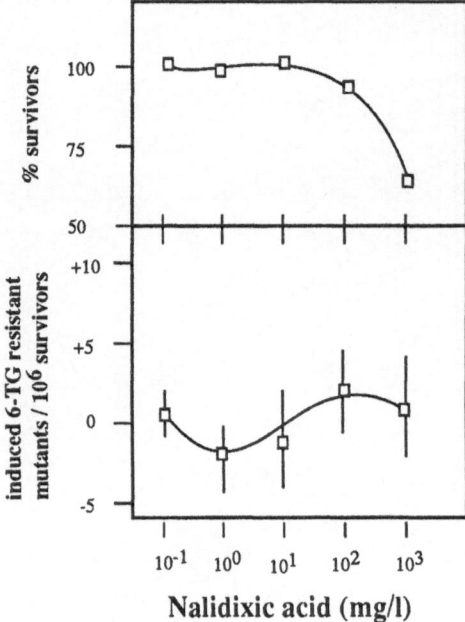

Fig. 12.6. Measurement of the induction of mutation in TK6 cells to 6-Thioguanine resistance by nalidixic acid.

Induction of unscheduled DNA synthesis (UDS): The protocol developed by Martin *et al* (1977; 1978) for the measurement of UDS in HeLa cells was modified to facilitate the use of the human lymphoblast cell-line TK6 which grows in suspension culture. Examination of the effects of nalidixic acid in this test system showed that the drug did not induce any significant levels of UDS in these cells (see Fig. 12.5). Control studies with 4NQO yielded a clear positive result at concentrations above $5 \times 10^{-7}M$. The negative result obtained with nalidixic acid is perhaps not entirely unexpected since this drug has been shown to inhibit DNA repair in mammalian cells (Mattern and Scudiero 1981).

Mutation of TK6 cells to 6-thioguanine resistance: Measurement of mutation in this diploid human lymphoblast cell line using the microtitre plate method (Furth *et al.* 1981) showed that nalidixic acid exerted no detectable mutagenic effect in this test system (see Fig. 12.6). Control studies with 4NQO showed a positive induction of mutation in this sytem. Similar point mutation studies of nalidixic acid using the mouse-lymphoma cell-line L5178Y with assessment of induced mutation frequency at the TK locus and to ouabain resistance have been carried out using standard protocols (Clive and Spector 1975). The results obtained in this study showed no difference from those obtained with the TK6 cell-line assay. However, examination of induced mutation at the *hgprt* locus with varying doses of ciprofloxacin showed that under normal experimental conditions there was no evidence of a mutagenic effect, but under the conditions imposed by a growth medium depleted of Mg^{2+} ions ciprofloxacin exerted a weak, but statistically significant, mutagenic effect. This observation is clearly reminiscent of the modulation of the bactericidal effect of 4-quinolones by Mg^{2+}

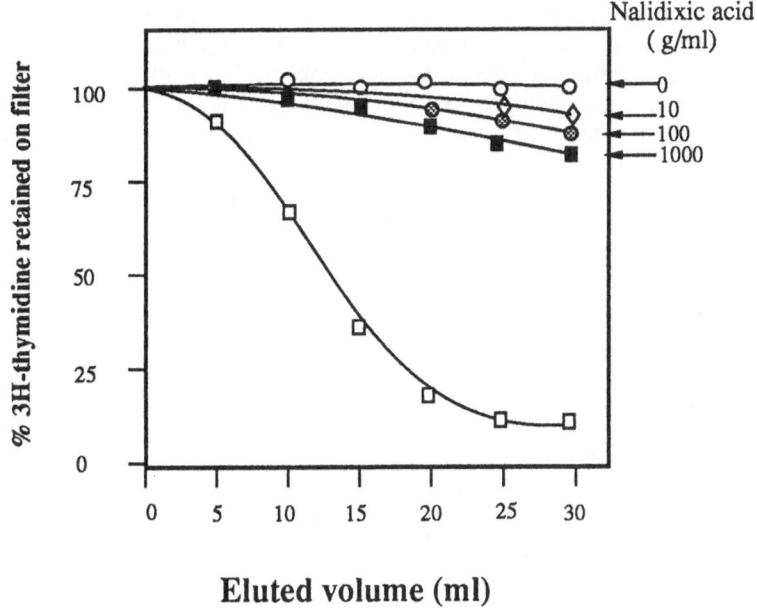

Fig. 12.7. Measuement of the induction of DNA strand breakage in TK6 cells by nalidixic acid.

and Mn^{2+} ions in urine (Ratcliffe and Smith 1984), and may be considered as indicating that the 4-quinolones may be weak mutagens under certain stringent cellular environmental conditions.

Induction of DNA strand-breaks assayed by alkaline-elution: Using the technique described by Kohn (1979) to assay DNA from TK6 cells exposed to nalidixic acid for 24 hours the results shown in Fig. 12.7 were obtained. No evidence of induced DNA strand-breakage is apparent even at doses in excess of the levels attained during therapy. As with the assay systems described above, control studies were carried out with 4NQO, which gave a positive result in the alkaline elution system.

In all of these mammalian cell assay systems the highest challenge doses used are well in excess of the tissue levels attained in man after oral dosage (McChesney *et al.* 1964). When all of the above assays were repeated in the presence of a phenobarbitone-induced rat liver S9 metabolic supplement there was no significant variation from the negative results shown above for the S9-free assays.

This small collection of mammalian cell assays for genotoxicity all show convincingly negative results, and it might be concluded that there is no evidence of significant genotoxic activity.

Extrapolation from Tumour/Transformed Cells?

The idealised *in vitro* genotoxicity screening system for any agent should involve the use of human cells, but questions of practicality, reproducibility and comparability indicate the use of immortal cell-lines. By definition such systems are likely to involve the use of either transformed or tumour cell-lines. The use of such cells makes the very reasonable assumption that the molecular receptor in these cells will be comparable to that available in "normal" target cells. In most cases this seems a sensible assumption, but with the 4-quinolones there may be a significant problem. Experimental evidence from mechanistic studies, and intuition, strongly suggest that the most probable receptor in human cells for these agents is a DNA topoisomerase, and this presumption is crucial.

DNA topoisomerases act to control the spatial conformation of the cellular DNA, and it is the competent regulation and functioning of these enzymes which helps to maintain the DNA conformation. The action of these enzymes is particularly crucial to mitochondrial DNA where histones are not available to contribute to the regulation. If we consider the organization of mtDNA as a convenient indicator of competent DNA

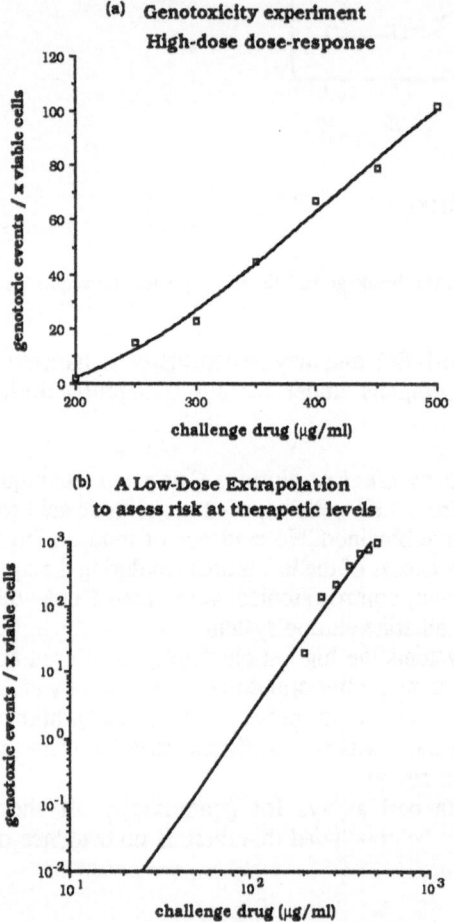

Fig. 12.8. The principle of the carrying out of a high-dose estimation of an induced genotoxic event followed by a low-dose extrapolation to assess the risk of low challenge doses.

topoisomerase functioning, we can assess the comparability of tumor/transformed cell lines with normal cells by examination of the mtDNA from the two types of cell.

Admittedly little work of this nature has been done up to the present, but a comparison of mtDNA conformation between leukaemic and normal cells has shown significant differences between the mtDNA from the two cell types (Firkin and Clark-Walker 1979). These differences are almost certainly due to either the presence of an altered (abnormal) DNA topoisomerase, or significant defects in topoisomerase enzyme regulation in the tumour cells. As yet we cannot differentiate between these two options, but it has recently been suggested that there are differences in the DNA topoisomerases themselves between tumour and non-tumour cells (Priel et al. 1985). It has also been shown that virally transformed cells have altered levels of DNA topoisomerases (Chow and Pearson 1985). Since it is already well known that alterations in the target DNA topoisomerases in bacteria drastically affect 4-quinolone susceptibility, it would seem unwise to try and equate tumour/transformed cell-line susceptibilities with those which might be expected in normal cell-lines.

Low-Dose Extrapolations?

The concept of the low-dose extrapolation involves the use of sophisticated statistical methods to extend a dose-response curve derived from high-dose studies to facilitate the estimation of acceptable levels of carcinogenic risk (Hartley and Sielken 1977) (see figs. 12.8A and 12.8B). As an applied technique it is reliant upon the existence of a single dose-related mechanism of action for the agent under investigation such that increasing the dose increases the biological effect. This technique cannot be applied to

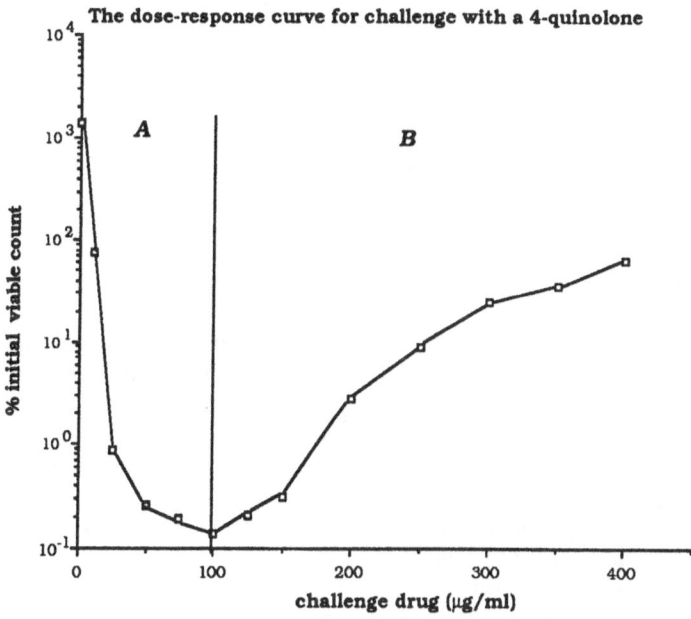

Fig. 12.9. The dose-response curve for survival of susceptible bacteria to a typical 4-quinolone (*E. coli* K12 versus nalidixic acid) showing the biphasic nature of the dose-response curve.

the 4-quinolones as they are characterized by displaying a paradoxical dose-response curve in bacteria whereby increasing doses are increasingly bactericidal up to a maximal-lethal dose above which increasing doses display decreasing toxic activity (Crumplin and Smith 1975; Stevens 1980; Smith 1984; Ratcliffe 1985) (see Fig. 12.9). This property is the result of the existence of two independent, dose-related mechanisms of action upon DNA replication and RNA synthesis (Crumplin and Smith 1975). It is not known whether this paradoxical dose-response is manifested in eukaryotic cells, but the possibility cannot be excluded. Hence, until such time as we know the nature of the dose-response properties for normal human cells any high-dose studies reliant upon a low-dose extrapolation for interpretation must be contra-indicated.

Extrapolations between 4-Quinolones?

The structural homogeneity of the 4-quinolones, and the common property of the inhibition of bacterial DNA gyrase *in vitro* suggests strongly that information derived from the study of the mechanism of action of one 4-quinolone is applicable to them all. The only anticipated differences would relate to the relative potency of each individual agent. Unfortunately recent studies of ciprofloxacin, ofloxacin and amifloxacin have indicated the existence of significant differences in the nature of the effects induced in treated cells (Smith 1984; Crumplin 1985). Furthermore the recent work of Shen and Pernet (1985) on the binding of 4-quinolones to DNA clearly identifies norfloxacin as being significantly different to other 4-quinolones, a factor which is borne out by the observation that whilst nalidixic acid treatment of *E. coli* can induce the relaxation of plasmid DNA molecules *in vivo* even in the absence of DNA topoisomerase I (Mirkin *et al.* 1984) the induction of relaxation by norfloxacin is dependent upon the presence of topoisomerase I (N Cozzarelli, personal communication). In addition the activities of amfonelic acid against human topoisomerase I (Bond 1988) and CP-67015 against calf thymus topoisomerase II in certain assay systems (Barrett *et al.* 1987) suggest that certain 4-quinolones may display what might be termed "anomalous" inhibitory activities.

These results clearly suggest that extrapolating from one 4-quinolone to another is fraught with risks.

Conclusions

The simple question that we have to try and answer is - is there evidence of significant genotoxic sequelae resulting from the exposure of cells to 4-quinolone agents? The overwhelming amount of evidence available refers to the use of Nalidixic acid as the representative 4-quinolone and as is evident from the above sections a large body of data is available to aid us in our assessment. I shall attempt to draw some basic conclusions on the bases of the two types of experimental evidence - namely the general genetic data, and the data drawn from experiments which represent the legislatively acceptable protocols applicable to genotoxic risk-assessment.

Genetic experiments in prokaryotes designed to investigate either the mechanism of action of the 4-quinolones, or the role(s) of DNA gyrase in various genetic processes, clearly indicate that these agents are specific inhibitors of DNA replication. However the principal area of interest for us resides not in inhibitory processes, but in inductive processes. It has been shown that nalidixic acid exerts a weak mutagenic effect in bacteria, but that the results are unpredictable by virtue of some degree of gene-specificity (in some instances the agent can act as an apparent anti-mutagen). Similar contradictory evidence of an inductive effect can be found with regard to transposition with certain transposable elements being inducible whilst others are inhibited. The same familiar pattern is to be found in the examination of the effects of the 4-quinolones upon gene-expression with the range of effects covering induction, unaffected through to inhibited. The capacity for stimulating and inhibiting DNA repair systems merely complements the data from mutation, transposition, and gene-expression, and serves only to leave us thoroughly confused.

The only positive conclusion that can be drawn would seem to be that the 4-quinolones generally demonstrate specific activity upon DNA systems with the inhibition of DNA synthetic step(s) being the principal observable. Whilst the vast body of other data can be taken as indicative of a general specificity of activity upon DNA processing systems, and might be taken by the pessimistic amongst us as grounds for concern, no firm evidence of a specific inductive genotoxic effect is available. We have to be reliant upon the available evidence of a clear margin of selective toxicity in the clinical situation.

Examination of the effects of 4-quinolones, as represented by nalidixic acid, upon genetic processes in eukaryotic organisms shows that the data are much more limited than for prokaryotes, but that there is extreme diversity in the range of organisms and systems studied. It is apparent that the 4-quinolones display less overt toxicity against eukaryotes whilst still having the capacity to significantly perturb the systems at high doses of drug. It is however fairly clear that it is very difficult (maybe even impossible) to predict the consequences of 4-quinolone exposure in one organism on the basis of the observed effects in another unrelated eukaryote. The problems of attempting to extrapolate between eukaryotes is perhaps exemplified by the range of observed effects upon chloroplasts in different organisms. Here the effects range from an irreversible bleaching in euglena, reversible bleaching in lemna, and no evidence of bleaching in nicotiana.

Probably the only valid conclusion that can be drawn is that nalidixic acid is capable of entering into eukaryotic cells (though in some cases almost certainly with very low efficiency) and producing an observable effect in a genetic and/or metabolic system. On the basis of the limited evidence available it might be suggested that extrachromosomal genetic elements (mitochondria and chloroplasts) seem to be more susceptible to the effects of nalidixic acid than do chromosomal genetic systems. However it could equally well be argued that this slight bias in the available evidence stems from the fact that extrachromosomal parameters have proved to be easier to examine experimentally and that many chromosomally mediated parameters just have not been looked at. Whether or not this is true is immaterial, a fair judgement would be that the 4-quinolones can exert significant effects which may stem from effects upon the regulation of gene-expressions in eukaryotes, and that the situation is extremely complex and far from being resolved. Intuition indicates that the most probable target for the 4-quinolones in eukaryotic cells is a DNA topoisomerase and that evidence is accumulating which shows enzymes of this type play a vital role in the management of eukaryotic DNA structures and in the control of genetic systems. There is also evidence to suggest that the DNA topoisomerases may play a significant role in determining the phenotypic differences between tumour cells and non-tumour cells (Priel *et al.* 1985), a feature which further emphasizes the significance of these enzymes to our particular problem in this communication.

Perhaps of more immediate concern is an attempt to draw a reasonable conclusion or set of conclusions from data derived from direct *in vitro* genotoxicity experiments which normally conform to the legislative requirements for establishing the safety of any agent under test. In this communication a list of a battery of *in vitro* test systems was given, the majority of which have been carried out on nalidixic acid (with some of the results presented here). A first step is to summarize the results of such a battery of tests on a single 4-quinolone agent in order to derive a general concensus (see table 12.3):

Table 12.3. Summary of the results of short-term test study data for Nalidixic acid

Test system	Result
1 Mutagenicity in bacteria	+/-
2 Selective toxicity against DNA repair-deficient mutant bacteria	+
3 Induction of prophage in bacteria	+/-
4 Mutagenicity in mammalian cells	-
5 Transformation of mammalian cells	-
6 Induction of unscheduled DNA synthesis	-
7 Induction of chromosome changes in mammalian cells (e.g. SCE)	-
8 Induction of DNA strand breakage in mammalian cells (alkaline elution)	-
9 DNA binding	+/-

The results listed in table 12.3 fall into 3 basic categories, namely bacterial systems (nos 1-3), mammalian cell systems (nos 4-8), and *in vitro* (no 9).

The three bacterial test systems yield one positive result (selective toxicity against DNA-repair-deficient bacteria) plus two sets of contradictory results which are purely dependent upon experimental protocol. The +/- results should really be discarded from the assessment, and even the remaining positive should be viewed with some scepticism because of the extreme potency of the lethal effects of the 4-quinolones upon bacterial cells. The fact that these agents have been developed as selective bactericidal agents logically contra-indicates the use of such organisms with a view to extrapolating to human cell systems.

The battery of five mammalian short-term test systems show conclusive negative results in four/five tests, with the question of SCE induction as the only indeterminate finding. However the fact that the one positive result stems from an uncontrolled study weights the data slightly in favour of a conclusion of no positive inductive effect. Overall these results are indicative of a negative result from mammalian cell studies, which instinctively should be the best indicators of possible consequences in patients treated with the agent.

The remaining study of macromolecular binding to DNA *in vitro* again yields a $\pm$ result depending upon the experimental protocol used. In the case of nalidixic acid the best experimental evidence for a positive binding to DNA is to be found in the studies of binding to single-stranded DNA in the presence of Cu^{2+} ions. These conditions may be regarded as potentially artifactual, whilst other studies under less stringent conditions yield only negative results. The most recent and sensitive studies of Shen and Pernet (1985) showing the binding of norfloxacin to DNA have failed to detect evidence of nalidixic acid binding under similar conditions (L Shen, personal communication). Hence we must at present regard DNA-binding studies with nalidixic acid as inconclusive.

Overall we can probably best conclude that in the case of nalidixic acid as the representative 4-quinolone the main body of evidence, in particular the mammalian cell studies, indicates that there is no clear evidence of a significant genotoxic potential with regard to the classical sequelae of carcinogenisis and/or mutagenisis. However, we must not forget the important inference of possible genotoxic activity generated by a metabolite (alone, or in combination with other factors) of a 4-quinolone as indicated in the recent study by Maura and Pino (1988). This single inference casts significant doubts upon any conclusions drawn upon the basis of studies of the parent molecules of 4-quinolones since *in vivo* most 4-quinolones have one to five metabolites in man (Lode *et al.* 1987). In fact we know little or nothing about the biological effects of the metabolites of the 4-quinolones and since it is unreasonable to attempt extrapolation between 4-quinolones we cannot make the presumption that metabolites act in the same way as the parent molecules.

In the context of genotoxic effects involving developmental features affected by nalidixic acid. The published data on the effects of nalidixic acid upon cartilage development in experimental animals provides the main basis of concern. However, whilst these results may preclude the use of new 4-quinolones as routine agents for the treatment of paediatric and childhood infections, the results are reputedly reversible, and are not allied to any evidence of teratogenic effects.

There appears to be no evidence of any detrimental effect of nalidixic acid upon the competent functioning of the immune system of animals (or man) challenged with the drug. Consequently we are at present forced to conclude that there is no evidence of a detectable risk in this context.

The conclusions put forward thus far have been based upon the evidence from *in vitro* test systems; to put these into some form of realistic perspective we must remember that nalidixic acid has been administered to patients of all ages in relatively high doses for continuous periods ranging from a few days to several years. Since 1963 approximately 10^9 courses of treatment have been administered, and over this period there have been no confirmed reports of genotoxic sequelae. In the context of genotoxic risk-assessment this represents a crude form of "mega-man" experiment and may be interpreted as indicating that the probability of an induced genotoxic effect is thus far $<1 \times 10^{-8}$ for the doses used therapeutically . This probability is in line with levels of acceptable risk for proven genotoxins and carcinogens as determined by present techniques of low-dose extrapolation from animal experiments designed to estimate the "virtually safe exposure level". Hence it seems reasonable to conclude that the clinical findings support the *in vitro* genotoxicity assessment data as indicative of negligible genotoxic risk attributable to nalidixic acid as a representative 4-quinolone antibacterial agent.

Can the 4-Quinolones be Assessed by Normal Criteria?

The overwhelming majority of compounds requiring screening for potential genotoxicity can be subjected to a preliminary screen through the use of *in vitro* short-term tests. The normal practice of using a battery of different tests to assess differing criteria in both prokaryotic and eukaryotic systems seems to provide an acceptable level of efficiency in identifying those compounds worthy of closer scrutiny in animal studies. In any given battery of tests the criteria routinely used to determine compounds worthy of closer scrutiny are usually combinations of positive results in:-

Bacterial mutation induction
Selective inhibition of repair-defective bacteria
Mutation induction in mammalian cell-culture
Unscheduled DNA synthesis induction
Chromosome aberration / SCE induction
DNA strand breakage induction
DNA binding

The question we must attempt to answer is whether these criteria, and the standardized protocols used, are applicable to the valid assessment of the 4-quinolones.

The problem we have to face with the 4-quinolone group of antibacterial agents is that their mechanism(s) of action are radically different to those of unrelated compounds subjected to this type of examination. The DNA topoisomerases, which are the presumptive targets or receptors for these agents, are uniquely important to living cells, with a relationship to DNA systems that is peculiarly emotive in raising a question mark over the possible effects of the 4-quinolones upon normal human cells. Unfortunately for the assessors of potential genotoxicity the 4-quinolones, when subjected to a battery of short-term test, yield a confusing set of results which conform to no obvious pattern. It is becoming increasingly apparent that many of the *in vitro* test systems used routinely contain features which may preclude them from being objectively acceptable as valid screens. Examples of this are to be found in the unpredictability of bacterial test systems, with their artificially induced features designed to 'sensitize' the systems, which are essentially de-sensitized with regard to such potent bactericidal agents. Further examples are to be found in the use of transformed mammalian cell lines, and tumour cell lines, which may provide different targets for the challenge agent which are non-comparable with the normal human enzymes.

The detailed examination of the mechanism of action of the 4-quinolones, and the development of resistance to them, in bacteria has over the years indicated that these drugs do not conform to the expected rules in most respects. The sheer diversity of the observable effects in bacteria, accompanied by the paradoxical dose-response effects, have necessitated the evolution of far more stringent experimental design, with a willingness to accept the unexpected results as being valid. In genetic terms the 4-quinolones break more rules than they conform to by affecting virtually every genetic system we choose to examine. Despite the extensive research already carried out we still have no real concept of how the 4-quinolones act as simple bactericidal agents in mechanistic terms - which is really surprising after more than 20 years of searching. We do perhaps have to remember that in the special case of DNA gyrase, research workers effectively had to resort to the invention of a completely new enzyme in order to describe one target for these drugs in the basic process of selectively inhibiting DNA synthesis. Even with this enzyme being available for consideration we can still identify the same functional enzyme from different bacterial species as being either intrinsically sensitive or resistant to the effects of individual 4-quinolones. The necessity for such drastic action as describing a completely new enzyme should have set the tone for our expecting the unexpected when we attempt to characterize the effects of 4-quinolones upon human cells. The non-comparability of the target between bacterial species suggests that we may face the same problem in mammals, in which case the choice of any system other than a normal human cell system introduces a strong element of luck into the likelihood of our obtaining an indication applicable to man.

The mechanistic criteria applied in the routine short-term test systems have, over recent years, proved to be acceptably reliable indicators of potential genotoxicity. However, significant questions remain with regard to the cell-lines used in many of the protocols when testing the 4-quinolones. To provide a valid assessment of potential genotoxicity for the 4-quinolones *in vitro* it is essential that the mechanistic criteria like mutagenicity and clastogenocity be assessed in a background that can be clearly established as being objectively comparable to the potential human cell target. In fact it has been suggested that for agents purported to act via DNA topoisomerase II any mutations induced are likely to be non-point mutations because of the nature of the double-strand DNA scissions induced, hence any mutagenic response will be acutely locus-dependent (De Marini *et al*. 1986). This inference invalidates the use of the *hgprt* locus which is hemizygous, and point mutations in bacteria (e.g. *his* locus as used in the Ames tester strains) and suggests that the only valid predictive test systems must reflect clastogenicity and not point mutagenicity .

In general terms it would thus seem reasonable to suggest that the use of existing bacterial screening systems is singularly inappropriate in the case of the 4-quinolones and that test systems involving point mutagenisis in any organism or cell should be excluded from consideration. If we also consider the difficulty of justifying any form of linearized low-dose extrapolation of data generated in high-dose studies because of the paradoxical nature of the dose-response curves of the 4-quinolones, it is evident that there are significant grounds for treating these agents as a special case.

Where Do We Go from Here?

The mixture of data, conclusions and opinion presented above, when assessed in the context of the practical consideration of assessing any individual 4-quinolone agent, would seem to permit the credibility of the following:-

It would seem reasonable to suggest that each 4-quinolone needs to be investigated at a mechanistic level in bacteria in order to establish whether or not it is comparable to other chemically related agents since recent studies have indicated that agents like ciprofloxacin and ofloxacin may be very different to nalidixic acid in how they work (and upon what intracellular system they are targetted). Hence comparability must be established to provide at least a guideline of precedents.

However, even the establishment of comparability between 4-quinolones in bacterial systems should not be given too much weight since the real question relates specifically to effects upon mammalian (viz. *Homo sapiens*) genetic systems. Whilst intuition might indicate that the most potent antibacterial 4-quinolones should also be considered as likely candidates for highest potency against analagous mammalian systems - intuition is a poor guide without proper consideration of the relative selective toxicities of old and new generation 4-quinolones. A simple comparison of nalidixic acid and ofloxacin as inhibitors of the activity of DNA topoisomerase II from bacteria and HeLa cells shows that the relative molar potencies (ofloxacin : nalidixic acid) are for *E. coli* topoisomerase II = 0.016 : 1 (data from this laboratory) and HeLa topoisomerase II = >4 : 1 (Ikeda *et al*. 1987). Hence in this instance greater potency against the bacterial target enzyme is inversely related to potency against the mammalian target enzyme, a situation which refutes the intuitive presumption and clearly indicates the need to consider each 4-quinolone on an individual basis.

With regard to the studies already carried out it seems evident to this author that far greater credence should be given to short-term tests not involving bacteria since only these tests are not compromised by the unusually diverse, and unpredictable, "antibacterial" properties of this group of agents. McCoy and his co-workers (1980) expressed concern over the non-correlation between the effects of nalidixic acid in the bacterial DNA repair assay and mutagenicity assays, but still regarded nalidixic acid as a suspect agent. The results reviewed in this communication may be interpreted as suggesting that existing bacterial system methodology is perhaps more suspect than the 4-quinolone test agents. Perhaps the 4-quinolones should be investigated further - not in a concerted search for some form of positive induction, but - as an aid in developing more reliable screening procedures for assessing the genotoxic potential of chemotherapeutic agents.

The data available outside the confines of the confidential files of the pharmaceutical companies and regulatory authorities clearly indicates the existence of an enormous volume of relevant data for consideration. These data indicate that the presumptive intracellular target for the 4-quinolones is involved in virtually all DNA related genetic functions ranging from the necessities of bacteriophage replication and lysogeny (Itoh and Tomizawa 1977; Marians *et al.* 1978; Mizuuchi *et al.* 1978) through to involvement in the structuring of eukaryotic chromatin (Earnshaw and Heck 1985; Earnshaw *et al.* 1985). The problem of demonstrable selective toxicity against prokaryotic DNA topoisomerases does still need to be established with regard to normal human enzymes. The task of distilling the data to yield an assessment of the genotoxicity of the 4-quinolones is complex and only gives rise to the conclusion that there is at present no clear evidence to indicate that these agents are significantly genotoxic to man. However, objective conclusions are elusive as a result of the diversity of the data and of the compounds, and the potential invalidity of many of the test systems used in formal investigations.

It seems that the question posed in the title of this communication can only be given the following qualified answer:- Use of the short-term test systems available at present provide no incontrovertible evidence indicative of there being a significant risk associated with the use of 4-quinolones under the controlled conditions of prescribed dosing with therapeutic levels over defined (usually short) treatment periods. Investigation of these agents does however raise significant questions over the validity of data derived from experiments performed upon transformed and tumour cell-lines.

This answer unfortunately cannot be taken as an absolute conclusion since it only represents a judgement based upon the available data. Because of the extensive interest being taken at present in the development of new 4-quinolones, further investigative data will clearly become available over the following months (and years). Not only will new data emerge from the use of existing methodologies and investigative criteria, but also the continued use of the 4-quinolones in fundamental genetical research is likely to continue to provide new insights into the mechanics of DNA management *in vivo* simply by following the precedent set by nalidixic acid in the identification and characterization of DNA gyrase. The study of DNA topoisomerases and their roles in the fundamental genetic processes of living systems is at present one of the most dynamic areas of biology, we are therefore likely to have to continue to revise our views of therapeutic agents which act specifically upon such important enzymes. This also implies the need to be aware of the likelihood that we should also consider the possibility of revising the ways in which we construct the risk/benefit analysis for potentially invaluable therapeutic agents.

References

Ames BN, McCann J and Yamasaki E (1975) Methods for detecting carcinogens and mutagens with the Salmonella/mammalian microsome mutagenicity test Mutat Res 31:347-364

Ashby J, Styles JA and Paton D (1980) Studies *in vitro* to discern structural requirements for carcinogenicity in analogues of the carcinogen 4-dimethylaminoazobenzene (butter yellow). Carcinogenesis 1:1-7

Barrett JF, Gootz TD, McGuirck PR, Farrell C, Sokolowski S and Frescura M (1987) Use of *in vitro* topoisomerase II assays in studying nalidixic acid derivatives. XXV ICAAC (New York)

Birkett DA, Garrett M and Stevenson CJ (1969) Phototoxic bullous eruptions due to Nalidixic acid. Brit J Dermatol 81:324-344

Bond CM (1988) Studies of DNA topoisomerases of human origin PhD Thesis University of York)

Burry JN (1974) Persistent phototoxicity due to Nalidixic acid. Arch Dermatol 109:263

Cairns J (1981) The origin of human cancers. Nature 289:353-357

Chao L (1978) An unusual interaction between the target of nalidixic acid and novobiocin. Nature 271:385-386

Chow K-C and Pearson GD (1985) Adenovirus infection elevates levels of cellular topoisomerase I Proc Natl Acad Sci 82:2247-2251

Clive D, Flamm WG, Machesko MR and Bernheim NJ (1972) A mutational assay system using the thymidine kinase locus in mouse lymphoma cells. Mutat Res 16:77-87

Clive D and Spector JFS (1975) Laboratory procedure for assessing specific locus mutations at the TK locus in cultured L5178Y mouse lymphoma cells. Mutat Res. 31 17-29

Cook TM, Goss WA and Deitz WH (1966) Mechanism of action of Nalidixic acid on *Escherichia coli*: V Possible mutagenic effect. J Bacteriol 91:780-783

Crumplin GC(1981) The involvement of DNA topoisomerases in DNA repair and mutagenisis. Carcinogenisis 2:157-160

Crumplin GC (1985) The mechanisms of action of 4-quinolone antibacterials. ASM annual meeting (Las Vegas)

Crumplin GC and Smith JT (1975) Nalidixic acid: an antibacterial paradox Antimicrob. Ag. Chemother 8:251-261

Crumplin GC and Smith JT (1981) The effect of R factors on host cell responses to Nalidixic acid: I Increased susceptibility of Nalidixic acid sensitive hosts. Antimicrob Chemother 7:379-388

Czinn SJ, Speck WT and Rosenkranz HS (1981) Abnormalities in the development of the American sea urchin induced by Nalidixic acid. Mutat Res.91:119-121

Degnen GE (1974) A conditional mutator gene,*MutD*, in Escherichia coli. PhD Thesis Princeton University

De Marini DM, BrockKL, Doerr CL and Moore MM (1986) Mutagenicity of topoisomerase II active drugs due to clastogenic mechanisms.First Conference on DNA topoisomerases in cancer chemotherapy (New York Nov 19-20)

Domagala JM, Hanna LD, Heifitz CL, Hutt MP, Mich TF and Solomon P (1986) New structure-activity relationships of the quinolone antibacterials using the target enzyme. The development and application of a DNA gyrase assay. J Med Chem 29:394-404

Earnshaw WC and Heck MMS (1985) Localization of topoisomerase II in mitotic chromosomes. J Cell Biology 100:1716-1725

Earnshaw WC, Halligan B, Cooke CA, Heck MMS and Liu LF (1985) Topoisomerase II is a structural component of mitotic chromosome scaffolds. J Cell Biology 100:1706-1715

Easmon CSF and Crane JP (1985) Uptake of ciprofloxacin by macrophages. J Clin Pathol 38:442-444

Easmon CSF and Crane J (1986) Uptake of Ro 23-6420 by phagocytic cells XXVIth ICAAC (New Orleans) abstr 947

Echols H (1981) SOS functions, cancer and inducible evolution. Cell 25:1-2

Firkin FC and Clark-Walker GD (1979) Abnormal mitochondrial DNA in acute leukaemia and lymphoma,Brit J Haematol 43:201-206

Furth EE,Thilly WG, Penman BW, Liber HL and R and WM (1981) Quantitative assay for mutation in diploid human lymphoblasts using microtitre plates. Analyt Biochem 110:1-8

Garner RC (1980) *In vitro* assays to predict carcinogenicity? In Short-term test systems for detecting carcinogens, (eds. KH Norpoth and RC Garner) pub. Springer-Verlag pp 5-18

German A, Panouse-Perrin J and Ardouin AC (1969) Mutagenic action of Nalidixic acid on Staphylococcus phage.CR Acad Sci Paris 268:1827-1829 Green MHL and Muriel WJ (1976) Mutagen testing using *trp*+ reversion in *Escherichia coli*. Mutat Res 38:3-32

Green MHL and Muriel WJ (1976) Mutagen testing using trp^+ reversion in *Escherichia coli*. Mutat Res 38:3-32

Hartley HO and Sielken RL (1977) Estimation of "Safe Doses" in carcinogenic experiments. Biometrics 33:1-30

Holden HE, Barrett JF, Huntington CM, Muehlbauer PA and Wahrenburg MG (1989) Genetic profile of a naidixic acid analog: A model for the mechanism of sister chromatid exchange induction.Environmental Mutagenesis 13:(in press)

Hosomi J and Irikura T (1981) The mutagenicity of AM-715 *in vitro*.Chemotherapy 29 (S-4):938-945

Hosomi J, Maeda A, Oomori Y, Irikura T and Yokota T (1988) Mutagenicity of norfloxacin and AM-833 in bacteria and mammalian cells Rev Infect Dis. 10 (suppl. 1):S148

Ikeda S, Yazawa M and Nishimura C (1987) Antiviral activity and inhibition of topoisomerase by Ofloxacin, a new quinolone derivative. Antiviral Research 8:103-113

Irikura T, Suzuki H and Sugimoto T (1981) Mutagenicity studies of AM-715 in animals Chemotherapy 29 (S-4);932-937

Itoh T and Tomizawa J-I (1977) Involvement of DNA gyrase in bacteriophage T7 DNA replication.Nature 270:78-80

Kada T and Hirano K (1980) Screening of environmental chemical mutagens by the rec⁻ assay system with *Bacillus subtilis*.in Chemical mutagens Principles and methods for their detection vol.6. (ed FJ de Serres) pub Plenum Press:149-173

Kanisawa M, Katoh H and Aiso K (1974) Carcinogenicity of potassium 1-methyl-7-[2-(5-nitro-2-furyl)-vinyl] -4-oxo-1,4-dihydro-1,8-naphthyridine-3-carboxylate in ICR mice.Gann 65:1-11

Kohn KW (1979) DNA as a target in cancer chemotherapy: Measurement of macromolecular DNA damage produced in mammalian cells by anticancer agents and carcinogens.In Methods in Cancer Research vol.XVI. (Academic Press):291-345

Kowalczyk J (1980) Sister chromatid exchanges in children treated with Nalidixic acid. Mutat Res 77:371-375

Kreuzer KN, McEntee K, Geballe AP and Cozzarelli NR (1978) Lambda-transducing phages for the *nalA* gene of *E. coli* and conditional lethal mutations. Molec Gen Genet:167:129-137

Latt SA and Schreck RR (1980) Sister chromatid exchange analysis. Amer J Hum Genet 32:297-313

Lode H, Hoffken G, Prinzing C, Glatzel P, Wiley R, Olschewski P, Sievers B, Reimnitz D, Borner K and Koeppe P (1987) Comparative pharmacokinetics of new 4-quinolones. Drugs 34:(Suppl.1) 21-25

Louis P, Wiskemann A and Schulz KH (1973) Bullous photodermatitis following Nalidixic acid. Hautartz 24:445-448

Marians KJ, Ikeda J-E, Schlagman S. and Hurwitz J (1978) Role of DNA gyrase in FX replicative form replication *in vitro*. Proc Natl Acad Sci 74: 1965-1968

Martin CN, McDermid AC and Garner RC (1977) Measurement of unscheduled DNA synthesis in HeLa cells by liquid scintillation counting after carcinogen treatment.Cancer Letters 2:355-360

Martin CN, McDermid AC and Garner RC (1978) Testing of known carcinogens and non-carcinogens for their ability to induce unscheduled DNA synthesis in HeLa cells. Cancer Res 38:2621-2627

Mattern MR and Scudiero DA (1981) Dependence of mammalian DNA synthesis on DNA supercoiling. III Characterization of the inhibition of replicative and repair-type DNA synthesis by novobiocin and Nalidixic acid. Biochim Biophys Acta 653: 248-258

Maura A and Pino A (1988) Evaluation of the DNA-damaging and mutagenic activity of oxolinic and Pipemidic acids by the granuloma pouch assay. Mutagenisis (in press)

Mayer D and Bruch K (1985) Kein hinweis fur mutagenitat von ofloxacin. Infection (in press)

McChesney EW, Froelich EJ, Lesher GY,Crain AVR and Rosi D (1964) Absorption, excretion and metabolism of a new antibacterial agent, Nalidixic acid.Toxicol. App Pharmacol 6:292-309

McCoy EC, Petrullo LA and Rosenkranz HS (1980) Non-mutagenic genotoxicants: Novobiocin and Nalidixic acid, two inhibitors of DNA gyrase. Mutat Res 79:33-43

McDaniel LS, Rogers LH and Hill WE (1980) Survival of recombination deficient mutants of *Escherichia coli* during incubation with Nalidixic acid. J Bacteriol 134:1195-1198

McQueen CA. and Williams GA (1987) Effects of quinolone antibiotics in tests for genotoxicity.Amer J Med 82, Suppl 4A:94-96

Melcion C and Cordier A (1986) Absence of genotoxicity of quinolones to mammalian cells.International Symp New Quinolones (Geneva)

Miltenberger HC (1985) cited in: Ofloxacin: Keine Anhaltspunkte fur mutagene Wirkungen.Arztliche Praxis 37 :3482

Mirkin SM, Zaitsev EN, Panyutin IG and Lyamichev VI (1984) Native supercoiling of DNA: The effects of DNA gyrase and ω protein in *E. coli*. Molec Gen Genet 196:508-512

Mizuuchi K, Gellert M and Nash HA (1978) Involvement of supertwisted DNA in integrative recombination of bacteriophage lambda .J Molec Biol 121:375-392

Moreau P and Devoret R (1977) Potential carcinogens tested by induction and mutagens of prophage λ In *Escherichia coli* K12.In Origins of human cancer book C (eds HH Hiatt and JD Watson) Pub Cold Spring Harbor Labs:1451-1472

Morita J, Watanabe K and Komano T (1984) Mechanism of action of new synthetic nalidixic acid-related antibiotics: Inhibition of DNA supercoiling catalyzed by DNA gyrase.Agric Biol Chem 48:663-668

Phillips I, Culebras E, Moreno F and Baquero F (1987) Induction of SOS response by 4-quinolones.J Antimicrob Chemother 20:631-638

Priel E, Aboud M, Feifelman H and Segal S (1985) Topoisomerase II activity in human leukemic and lymphoblastoid cells. Biochem Biophys Res. Comm 130:325-332

Ramsay CA and Obreshkova E (1974) Photosensitivity from Nalidixic acid. Br J Dermatol 91:523-528

Ratcliffe NR (1985) Bacterial responses to antigyral agents. PhD Thesis University of London

Ratcliffe NR and Smith JT (1984) The mechanism of reduced activity of 4-quinolone agents in urine. FAC:563-569

Rosenkranz HS and Lambek C (1965) *In vivo* effects of nalidixic acid on the DNA of human diploid cells in tissue culture. Proc Soc Exp Biol Med 120:549-552

Rosenkranz HS and Leifer Z (1980) Detection of carcinogens and mutagens with repair-deficient bacteria. In Chemical mutagens - Principles and methods for their detection vol.6 (ed FJ.de Serres) Plenum press:109-147

Saito A, Sawatari K, Fukuda Y, Nagasawa M, Koga H, Tomonaga A, Nakazato H, Fujita K, Shigeno Y, Suzuyama Y, Yamguchi K, Izumikawa K and Hara K (1985) Susceptibility of *Legionella pneumophila* to Ofloxacin *in vitro* and in experimental Legionella pneumonia in guinea pigs.Antimicrob Ag Chemother 28:15-20

Sanzey B (1979) Modulation of gene-expression by drugs affecting DNA gyrase. J Bacteriol 138:40-47

Schluter G (1986) Toxicology of Ciprofloxacin, Proc. First Int. Ciprofloxacin Workshop (eds HC Neu and H Weuta) pp 61-67 (pub. Exerpta Medica, Amsterdam)

Shen LL and Pernet AG (1985) Mechanism of inhibiton of DNA gyrase by analogues of Nalidixic acid: The target of the drug is DNA. Proc Natl Acad Sci 82:307-311

Shimada H, Morita H and Akimoto T (1980) Lack of induction of dominant lethal mutations in male mice by Nalidixic acid. Mutat Res 7:165-170

Shimada H, Ebine Y, Kurosawa Y and Aranchi T (1984) Mutagenicity studies of DL-8280, a new antibacterial drug.Chemotherapy 32:1162-1170

Shimada H, Ebine Y, Kurosawa Y and Aranchi T (1985) Dominant lethal study in male mice treated with ofloxacin, a new antibacterial drug. Mutat Res 114:51-55

Smith CL, Kubo M and Imamato F (1978) Promoter specific inhibition of transcription by antibiotics which act on DNA gyrase. Nature 275:420-423

Smith JT (1984) Awakening the slumbering potential of the 4-quinolone antibacterials. Pharm J 233:299-305

Stenchever MA, Powell W and Jarvis JA (1970) Effects of Nalidixic acid on human chromosome integrity. Amer J Obstet Gynaec 107:329-330

Stevens PJE (1980) Bactericidal effect against *E. coli* of Nalidixic acid and four structurally related compounds. J Antimicrob Chemother 6:535-542

Styles JA (1977) A method for detecting carcinogenic organic chemicals using mammalian cells in culture. Brit J Cancer 36:558-563

Tokunaga Y, Asaba M and Kato R (1979) Dominant lethal mutation test of Miloxacin, a new synthetic antibacterial agent. Pharmacometrics 18:37-45

Vigier PRR (1974) Effet mutagene de l'acide nalidixique sur le bacteriophage T4.Mutat Res 25:25-32

Winshell EB and Rosenkranz HS (1970) Nalidixic acid and the metabolism of *Escherichia coli*. J Bacteriol 104:1168-1175

Witkin EM and Wermundsen IE (1979) Targeted and untargetedmutagenesis by various inducers of SOS functions in *Escherichia coli*. Cold Spring Harbor Symp Quant Biol 43:881-886

Zelickson AS (1964) Phototoxic reactions with Nalidixic acid. J Amer Med Ass 190:556-557

Discussion Summary

The whole area of the study of the potential genotoxicity of the 4-quinolones seemed to be an area of distinct uncertainties. There were marked uncertainties expressed over the significance of *in vitro* mutagenicity studies in mammalian cells on the grounds of the very high levels of agent used and, in some cases, over the extended period of exposure apparently used. Since the new 4-quinolones in particular are so potent as antibacterials that serum and/or tissue levels often peak at less than 10μg/ml and sustain such levels for relatively short periods of time, the correlation of very high, and sustained, levels in the *in vitro* studies with the *in vivo* situation was clearly difficult.

Further uncertainties are apparent when the question of the potential effects of magnesium ion variation upon apparent mutagenicity *in vitro* is raised since the levels of this ion are never recorded or reported in the literature. However, since some patient populations, like alcoholics, often present with very low inter-cellular magnesium, the possibility of an elevated potential risk of mutagenicity in such patients clearly meant that we ought to try and elucidate both mechanisms and significance of such *in vitro* observations.

Despite the evident uncertainties which were generally expressed over the relationships between *in vitro* risk assessment and the clinical situation, there was certainty expressed over the obvious need for some form of standardisation. Although serious efforts have already been made, and are still being made, to standardize methodologies and interpretation in genotoxicity for all chemicals, it was accepted that the 4-quinolones represented a group of compounds which might require rather different consideration. In particular the fact that the compounds displayed two independent, dose-related mechanisms of action which, in principle, precludes the use of the low-dose extrapolation procedure was clearly seen as a problem which still required serious consideration.

In general it was thought that the application of well-motivated and meticulous studies, using generalized methodologies and protocols, had contributed more to our expression of the need for such studies than to our understanding of the activities of the 4-quinolones. Only the better elucidation of the intra-cellular effects of 4-quinolones, and the consequences of the effects, could help us attain a proper understanding of the potential genotoxic risk of these compounds. However, the concensus of belief was that those 4-quinolones which had already reached the clinical market and those which were likely to do so in the near future, probably represented a minimal genotoxic risk in the light of the complete absence of any apparent risk associated with the widespread use of nalidixic acid over the last 30 years.

Chapter 13

Mechanisms of Resistance to 4-Quinolones

D. C. Hooper and J. S. Wolfson

Introduction

With expanding clinical use of the 4-quinolone antibacterial agents, resistance to these agents has been identified (Wolfson and Hooper 1989a,b), and its mechanisms have been the subject of increasing study. In this article we review the current information on the mechanisms of bacterial resistance to 4-quinolones. Information related to the frequency of selection of 4-quinolone resistance and the possibilities of plasmid-mediated 4-quinolone resistance are covered in other chapters in this book and will not be addressed here in detail.

Alterations in *Escherichia coli* DNA Gyrase

DNA gyrase, the intracellular target of the 4-quinolones, is a tetrameric enzyme (Gellert *et al.* 1977; Sugino *et al.* 1977). Although the two A and two B subunits of the holoenzyme are required to reconstitute all of its known activities, there is a partitioning of functions (Gellert 1981). DNA supercoiling by DNA gyrase is accomplished through the concerted and processive double-strand cleavage, translocation of another double-strand through the cleavage site, and rejoining of duplex DNA strands at the expense of ATP hydrolysis. The gyrase A subunit directly mediates strand cleavage and rejoining (Sugino *et al.* 1980), and the B subunit mediates energy transduction, with ATP hydrolysis regenerating the enzyme conformation needed to initiate the next cycle of DNA cleavage, strand passing, and rejoining (Sugino *et al.* 1978; Mizuuchi *et al.* 1978).

Gyrase A Subunit

The *gyrA* gene (located at 48 min on the *E. coli* genetic map), encoding the DNA gyrase A subunit, was initially identified as a locus (*nalA*) conferring resistance to nalidixic acid (Hane and Wood 1969). Since these initial studies, a number of *E. coli* *gyrA* mutants selected for resistance to 4-quinolones have been characterized (Gellert *et al.* 1977; Sugino *et al.* 1977; Yamagishi *et al.* 1981; Hirai *et al.* 1986; Hooper *et al.*

1986; Hooper *et al.* 1987; Wolfson *et al.* 1987; Yoshida *et al.* 1988). These mutations confer resistance only to 4-quinolone agents, and, in general, the increments in resistance to nalidixic acid tend to be greater than the increments in resistance to the newer fluoroquinolones.

Recently the *gyrA* genes from wildtype *E. coli* and four mutants selected in the laboratory for resistance to either nalidixic acid or pipemidic acid (Yoshida *et al.* 1988), or occurring in a patient treated with the newer fluoroquinolone derivative enoxacin (Cullen *et al.* 1989), have been cloned and their nucleotide sequence determined (Table 13.1).

Table 13.1. Alterations in the DNA gyrase subunit A affecting 4-quinolone resistance[a,b]

Study	Selecting agent	Amino acid change (position)	Nalidixic acid	MIC (µg/ml) Norfloxacin	Ciprofloxacin
Yoshida *et al.* 1988	None	Wildtype	3 .1	0.05	0.0125
	NAL[c]	(67) Ala -> Ser	25	0.20	0.05
	NAL	(83) Ser -> Leu	400	0.78	0.39
	PPA	(83) Ser -> Trp	400	0.78	0.39
	PPA	(106) Gln->His	12.5	0.10	0.05
Cullen *et al.* 1989	ENX[d]	(83) Ser -> Trp	500	1.0	0.25

[a] Amino acid changes inferred from changes in nucleotide sequence of the cloned genes.
[b] The gyrase A polypeptide has 875 amino acids. Tyrosine at position 122 is within the enzyme active site for 4-quinolone-induced DNA cleavage.
[c] Abbreviations: NAL, nalidixic acid; PPA, pipemidic acid; ENX, enoxacin; Ser, serine; Leu, leucine; Trp, tryptophan; Gln, glutamine; His, histidine; Ala, alanine.
[d] Clinical isolate.

In each of the mutants selected in the laboratory, single nucleotide changes were identified producing inferred amino acid changes at positions 67, 83 (two mutants), and 106, all in the amino terminal portion of the 875-amino acid gyrase A polypeptide and in proximity to position 122 at which the tyrosine involved in gyrase-mediated DNA cleavage is located (Horowitz and Wang 1987). This tyrosine is covalently linked to a 5' phosphate of the DNA when gyrase-mediated DNA strand-passing is interrupted by a quinolone and protein denaturants and thus appears to be in the active site of the enzyme.

There were many nucleotide differences between the *gyrA* gene cloned from the resistant *E. coli* strain isolated from a patient treated with enoxacin and that of wildtype *E. coli* K-12, but of the several nucleotide differences responsible for differences in inferred amino acid sequence, the change from serine to tryptophan at position 83 alone was shown by subcloning to be sufficient to account for the resistance phenotype (Cullen *et al.* 1989). Thus three of the five spontaneous mutant *gyrA* alleles studied thus far have had alterations in serine-83, tryptophan substitutions in two and a leucine substitution in one. The levels of resistance conferred by these two types of substitutions at position 83 were similar and higher than those conferred by the changes at positions 67 and 106.

Gyrase B Subunit

The *gyrB* gene (83 min), encoding the gyrase B subunit, was initially identified as a locus (*cou*) conferring resistance to coumermycin A1 and novobiocin (Gellert *et al.* 1976). These agents are thought to be competitive inhibitors of the ATPase activity of DNA gyrase mediated by the B subunit (Sugino *et al.* 1978), and are structurally distinct from the 4-quinolones. Alterations in the gyrase B subunit, however, have been shown to affect susceptibility to 4-quinolones.

Table 13.2. Alterations in the DNA gyrase subunit B affecting 4-quinolone resistance[a,b]

Study	Selecting agent	Amino acid change (position)	MIC (μg/ml)		
			Nalidixic acid	Norfloxacin	Ciprofloxacin
Yamagishi et al.	None	Wildtype	4	0.05	0.005
1987	NAL[c]	(426) Asp -> Asn[d]	20	0.075	0.015
	NAL	(447) Lys -> Glu[e]	75	0.005	0.001

[a] Amino acid changes inferred from changes in nucleotide sequence of the cloned genes.
[b] The gyrase B polypeptide has 804 amino acids.
[c] Abbreviations: NAL, nalidixic acid; Asp, aspartic acid; Asn, asparagine; Lys, lysine; Glu, glutamic acid.
[d] Results in charge change of +1.
[e] Results in charge change of -1.

Two *gyrB* alleles (*nalC* and *nalD*) conferring resistance to nalidixic acid have been identified (Yamagishi et al. 1981) and more recently cloned and sequenced (Yamagishi et al. 1986). Single nucleotide changes in the midportion of the *gyrB* gene were found in each allele, producing inferred amino acid changes of opposite charge. Although both changes conferred resistance to nalidixic acid, they differed in their effects on susceptibility to fluoroquinolones with a piperazinyl substituent at position 7 of the 4-quinolone molecule (Smith 1984). At position 426, a change from aspartic acid (1-) to asparagine (1+) (*nalD*) produced a small increment in resistance to the piperazinylated 4-quinolone congeners, but at position 447, a change from lysine (1+) to glutamic acid (1-) (*nalC*) conferred hypersusceptibility to these congeners. Such findings suggest the possibility that the positively charged piperazinyl substituent of such 4-quinolones might by electrostatic interactions involving the B subunit directly bind more strongly to a complex of DNA and DNA gyrase holoenzyme constituted with the mutant gyrase B subunit [glu-447 (1-)] and less strongly to a complex containing the mutant gyrase B subunit [asn-426 (1+)]. Alternatively, these changes in the gyrase B polypeptide might act indirectly by altering the conformation of the gyrase A subunit and its putative site of 4-quinolone binding.

Possible Molecular Mechanisms of Resistance by Alterations in DNA Gyrase

Studies of additional *gyrA* and *gyrB* resistance alleles occurring spontaneously, and after site-directed mutagenesis, should define further the domains of the gyrase subunits responsible for 4-quinolone resistance. The means by which these changes in the gyrase molecule are translated into 4-quinolone resistance may be postulated to be several. Simplest to envisage is the alteration of a 4-quinolone binding site on the gyrase protein or its complex with DNA such that binding occurs with lower affinity. In the model for 4-quinolone action proposed by Shen and associates (Shen et al. 1989a; Shen et al. 1989b) in which 4-quinolones bind primarily and cooperatively to a single-strand DNA pocket created by DNA gyrase, alterations in DNA gyrase might also be envisioned to produce resistance by either blocking access of the 4-quinolone to the DNA pocket or by producing an altered pocket, the conformation of which binds 4-quinolones with lesser affinity or cooperativity. Physicochemical studies of complexes of 4-quinolones, mutant and wildtype DNA gyrases, and DNA are likely to help differentiate among these possibilities.

Relative Occurrence of *gyrA* and *gyrB* Mutations Conferring 4-Quinolone Resistance

In the single study which has surveyed the relative occurrence of *gyrA* and *gyrB* mutations in *E. coli* (Nakamura *et al.* 1989), among mutants selected for resistance to nalidixic acid [at 4-fold the minimal inhibitory concentration (MIC), n = 20], there were equal numbers of *gyrA* and *gyrB* mutations identified by loss of resistance upon introduction of a plasmid encoding either a genetically dominant wildtype *gyrA* or *gyrB* gene. Among mutants selected (at about 4-fold the MIC) with enoxacin, a congener which contains a piperazine moiety, three were *gyrA* and two were *gyrB*. The highest levels of resistance with both selections occurred in the *gyrA* mutants, and some of the *gyrB* mutants selected with nalidixic acid were hypersusceptible to enoxacin. Thus the relative occurrence of *gyrA* and *gyrB* mutations is likely to be affected by the level of resistance selected and the agent used. Among eight resistant clinical isolates studied (for which the selection was not specified), five were *gyrA*, one was *gyrB*, and two were indeterminate.

Alterations in DNA Gyrase in Other Species

Among Gram-negative bacteria, mutations in the *gyrA* gene conferring resistance to 4-quinolones have also been identified in *Pseudomonas aeruginosa* (Rella and Haas 1982; Hirai *et al.* 1987; Inoue *et al.* 1987; Robillard and Scarpa 1988) and *Haemophilus influenzae* (Setlow *et al.* 1985). Gyrase A subunits partially purified from resistant strains of *P. aeruginosa* (Inoue *et al.* 1987; Robillard and Scarpa 1988), *Citrobacter freundii* (Aoyama *et al.* 1988), and *Serratia marcescens* (Fujimaki *et al.* 1989) have been shown to confer resistance to DNA gyrase DNA supercoiling activity when mixed with gyrase B subunits purified from their respective susceptible strains. The DNA gyrase DNA supercoiling activity purified from 4-quinolone-resistant derivatives of *Enterobacter cloacae* selected in an animal model of infection was more resistant to 4-quinolones, but the subunit responsible for resistance was not determined (Lucain *et al.* 1989). Alterations in the gyrase B protein from these species have not yet been shown to confer 4-quinolone resistance.

In Gram-positive bacteria, *gyrA* mutations have been shown to confer 4-quinolone resistance in *Bacillus subtilis* (Sugino and Bott 1980). A DNA fragment from a highly 4-quinolone-resistant strain of *Staphylococcus aureus* has been cloned and shown to confer 4-quinolone resistance in *E. coli* (Ubukata *et al.* 1989). This 5.5 kilobase DNA fragment, when used as a probe, hybridized at low stringency with the cloned *gyrA* and *gyrB* genes of *E. coli*, indicating that these two genes are closely linked in *S. aureus* as they are in *B. subtilis* (Lampe and Bott 1985). It has not yet been determined if the resistance in *S. aureus* was conferred by mutation in the *gyrA* gene, the *gyrB* gene, or both.

Other Mutations and Mechanisms of 4-Quinolone Resistance in *E. coli*

Mutations other than those in the *gyrA* and *gyrB* genes have been described with selection with nalidixic acid and the newer fluoroquinolones (Table 13.3). Many of these mutations appear to confer resistance by alterations in drug permeation. Mutants capable of inactivating 4-quinolones have not been described.

Table 13.3. 4-quinolone resistance mutations in *E. coli* other than *gyrA* and *gyrB* mutations

Selecting agent	Mutation	Map location[a]
Nalidixic acid	*nalB*	58
	nalD	85.5
	crp	74
	cya	85
	icd	25
	purB	25
	ctr (ptsHI)	52
Norfloxacin	*nfxB*	19
	norB	34
	norC	8
Ciprofloxacin	*cfxB*	34
Tetracycline or chloramphenicol	*marA*	34

[a] Location on *E. coli* genetic map in min.

Mutations Selected with Nalidixic Acid

Among mutations selected with nalidixic acid, a hydrophobic congener, the *nalB* (Hane and Wood 1969; Bourguignon *et al.* 1973) and *nalD* (distinct from the *nalD* allele of *gyrB* discussed above) (Hrebenda *et al.* 1985) loci are thought to produce resistance by reduced drug permeation. For both of these mutants (and not *gyrA* mutants), treatment with ethylenediaminetetraacetic acid (EDTA), which disrupts the lipopolysaccharide (LPS) layer of the bacterial outer membrane, abolishes the resistance of DNA synthesis to nalidixic acid in these mutants. In the *nalD* mutant, accumulation of nalidixic acid and glycerol were also reduced relative to the parent strain.

Several other mutations conferring low levels of resistance to nalidixic acid have been described -*crp* (Kumar 1976), *cya* (Kumar 1976; Kumar 1980), *icd* (Helling and Kukora 1971; Kumar 1980), *purB* (Helling and Adams 1970; Kumar 1980), and *ctr* (Helling and Adams 1970) - but the mechanisms of resistance are not known. Some nalidixic acid-resistant mutants have been reported to produce a new haemolysin similar to γ-haemolysin (Walton and Smith 1969).

Mutations Selected with Newer Fluoroquinolones

After selection for resistance to the newer hydrophilic fluoroquinolones, norfloxacin and ciprofloxacin, by either serial passage on increasing concentrations of drugs (Hooper *et al.* 1986, 1987) or by plating on agar (Hirai *et al.* 1986), several non-gyrase resistance loci have been identified: *nfxB* (19 min), *cfxB* (34 min), *norB* (34 min), and *norC* (8 min). These mutations have in common a pleiotropic resistance phenotype including low levels of resistance to fluoroquinolones (fourfold) and structurally unrelated agents

such as tetracycline, chloramphenicol, and some ß-lactams. In addition, each of these
mutants exhibited substantial reductions in the *OmpF* porin outer membrane protein,
which serves as a diffusion channel for small molecules from the extracellular fluid to
the periplasmic space. For the *nfxB* and *cfxB* loci, expression of the *ompF* gene
(21 min) is reduced at the post-transcriptional level (Hooper *et al.* 1989). In the *norC*
mutant, alterations in LPS were also found (Hirai *et al.* 1986).

These findings suggested reduced 4-quinolone permeation as a mechanism of
resistance. Although it has not yet been possible to measure 4-quinolone permeation
directly, this hypothesis is supported by the reduced accumulation of norfloxacin by
these mutants (Hirai *et al.* 1986; Hooper *et al.* 1989) and by *ompF* mutants (Bedard *et
al.* 1987; Cohen *et al.* 1988a, 1989). For the *nfxB* and *cfxB* mutants, the rate and
steady-state level of norfloxacin accumulation were both reduced. A reduced rate of
accumulation may be explained by the reduction in the number of *OmpF* diffusion
channels, because fluoroquinolones with piperazinzyl substituents (including
norfloxacin and ciprofloxacin) are sufficiently small and have zwitterionic charge
configurations known to allow diffusion of other antimicrobial agents through the

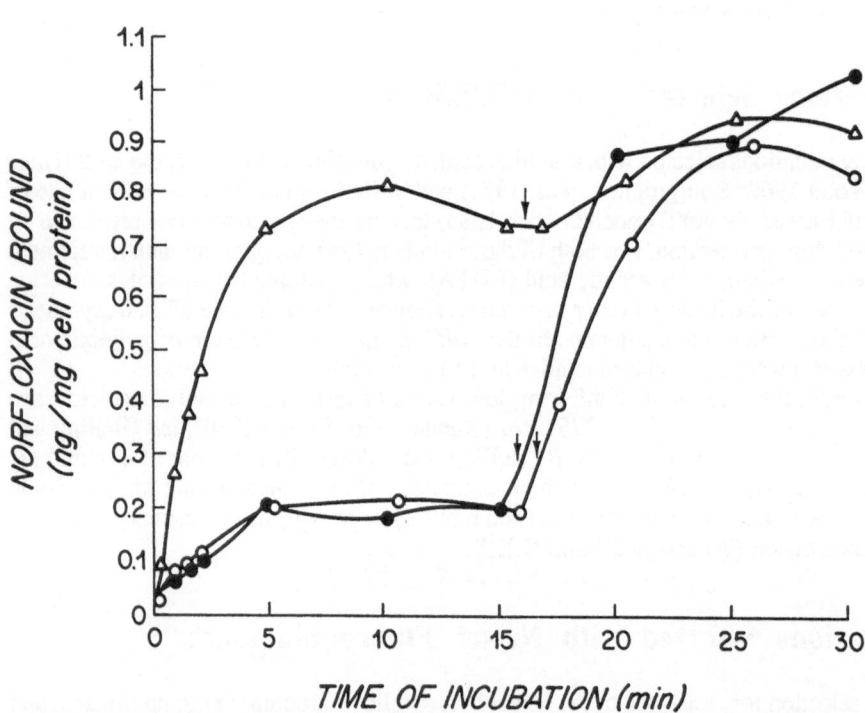

Fig. 13.1. Accumulation of norfloxacin by wildtype (Δ), *nfxB* (●), and *cfxB* (o) mutants.
Carbonyl cyanide *m*-chlorophenylhydrazone (50 μM) added at the arrow. Taken from Hooper *et al.*
1989 with permission.

OmpF porin (Nikaido 1985). Reductions in the steady-state level of 4-quinolone accumulation, however, are not expected from the observed level of reduction in diffusion rates alone and suggested that other factors are also operative. One such additional factor affecting accumulation appears to be an energy-dependent mechanism for reducing 4-quinolone accumulation first recognized in wildtype *E. coli* (Cohen *et al*. 1988a). After treatment with energy inhibitors, norfloxacin accumulation in *nfxB* and *cfxB* mutants increases to wildtype levels (Hooper *et al*. 1989) (Fig. 13.1).

This effect in wildtype cells appears to result from an energy-dependent, carrier-mediated efflux system present at the bacterial inner membrane, because ^{3}H-norfloxacin accumulation by everted (inside-out) inner membrane vesicles requires an energy source and is saturable with increasing concentrations of unlabeled norfloxacin, but not tetracycline (Cohen *et al*. 1988a). Because of the energy-dependence of the reduced norfloxacin accumulation seen in *nfxB* and *cfxB* mutants, such a system may also be involved in the reduced norfloxacin accumulation observed in these mutants. Reduced accumulation in these mutants also requires the presence of an outer membrane, because its removal by formation of spheroplasts by treatment with lysozyme and EDTA results in similar accumulation by mutant and wildtype cells (Hooper *et al*. 1989). Thus at least two components, an energy-dependent process (possibly related to the efflux system located at the inner membrane) and the outer membrane, are needed for expression of the differences in accumulation between *nfxB* and *cfxB* mutants and wildtype cells. No mutants with an altered efflux system have been identified, and thus it is not yet possible to link 4-quinolone resistance, or reduced accumulation, and the efflux system directly.

Mutations Selected with Other Agents

Mutants in the *marA* locus (34 min) were originally selected for resistance to tetracycline and chloramphenicol (George and Levy 1983a,b) and were recently shown to have reduced expression of the *ompF* gene by increased expression of the *micF* antisense RNA (Cohen *et al*. 1988b), which hybridises with *ompF* mRNA and is thought to destabilize its binding to the ribosome and thus antagonize its translation (Mizuno *et al*. 1984; Misra and Reeves 1987). *marA* mutants also have low increments in resistance to nalidixic acid (George and Levy 1983a) and newer fluoroquinolones (Cohen *et al*. 1989) and accumulate less norfloxacin than wildtype strains (Cohen *et al*. 1989). The *cfxB* mutation is closely linked to, and may be an allele of, the *marA* locus (Hooper *et al*. 1987, 1989).

Results with *marA* mutants, however, also indicate the need to invoke additional factors to explain the level of 4-quinolone resistance and the amount of reduction of norfloxacin accumulation. Strains with *ompF*(null) mutations (created by transposon insertion) exhibit levels of 4-quinolone resistance and 4-quinolone accumulation intermediate between wildtype and *marA* mutants (Cohen *et al*. 1989). Thus reduction in *OmpF* alone is insufficient to account for the phenotype of *marA* mutants. Changes in the efflux system between wildtype and mutant cells are also unlikely to explain these differences, because the kinetics of efflux of norfloxacin measured in everted inner membrane vesicles prepared from a *marA* mutant and its parent strain did not differ. The other factors are as yet unknown, but might relate to outer membrane protein changes in addition to *OmpF* reduction seen in *marA* mutants (Cohen *et al*. 1989).

Other Features of *nfxB*, *cfxB*, and *marA* Mutants

There appears to be an interaction between the *nfxB* and *marA* loci. Insertion of *Tn5* into the *marA* locus results in loss of the *MarA* resistance phenotype (George and Levy 1983b). Transduction of *marA:Tn5* into an *nfxB* mutant also results in loss of the *NfxB* resistance phenotype and may result in instability of the *nfxB* mutation (Hooper *et al*. 1989).

The *cfxB* mutation, in contrast to 4-quinolone resistance mutations in the *gyrA* gene, is genetically dominant in merodiploids (Hooper *et al*. 1989), suggesting the possibility that if such mutations were present on plasmids they might be more readily selectable. *marA* mutations were found to be co-dominant to wildtype alleles (George and Levy 1983b).

The frequency of selection of pleiotropic ("*Mar*-type") resistance also differs with the agent used for selection. The frequency with which mutants with norfloxacin resistance (at fourfold or more above the MIC) are selected using norfloxacin is substantially lower than that using either tetracycline or chloramphenicol at a range of similar relative selecting concentrations (Table 13.4) (Cohen *et al*. 1989). In addition, in *marA* mutants selection for "second-step" mutants with higher levels of resistance using norfloxacin may be accomplished at frequencies of 10^{-7}, in contrast to $<10^{-10}$ for the "first-step" selection with norfloxacin. These findings suggest the possibility that use of at least some non-quinolone antimicrobial agents may increase a background of low-level 4-quinolone resistance in some bacterial populations and that such low-level resistance may increase the likelihood of the development of higher, and more clinically important, levels of resistance.

Table 13.4. Selection of spontaneous norfloxacin resistance by three antimicrobial agents[a,b]

Selecting agent	Selecting concentration[c]	Frequency of resistance to norfloxacin[d]
Norfloxacin	2	$<1 \times 10^{-10}$
	3	$<1 \times 10^{-10}$
	5	$<1 \times 10^{-10}$
Tetracycline	2	$>4.1 \times 10^{-7}$
	3	1.1×10^{-7}
	5	4.9×10^{-8}
Chloramphenicol	2	1.3×10^{-7}
	3	3.6×10^{-8}
	5	$<1 \times 10^{-10}$

[a] Taken from Cohen *et al*. 1989 with permission.
[b] *E. coli* strain AG100.
[c] Multiple of MIC; MICs = 0.1 mg norfloxacin/l, 1.2 mg tetracycline/l, 6.9 mg chloramphenicol/l.
[d] Frequency of mutants with >4-fold increase in MIC of orfloxacin.

Occurrence of Pleiotropic 4-quinolone Resistance, Changes in Outer Membrane Proteins, and Changes in Accumulation in Species other than *E. coli*, including some Clinical Isolates

Pleiotropic patterns of resistance similar to those seen in *nfxB*, *cfxB*, and *marA* mutants in *E. coli* have also been seen in *Pseudomonas aeruginosa* (Rella and Haas 1982; Hirai *et al.* 1987; Mouton and Mulders 1987; Bayer *et al.* 1988; Robillard and Scarpa 1988; Chamberland *et al.* 1989; Legakis *et al.* 1989), the klebsiella-enterobacter group (Sanders *et al.* 1984; Gutmann *et al.* 1985; Kumada and Neu 1985; Lucain *et al.* 1989), *Serratia marcescens* (Traub and Kleber 1977; Gutmann *et al.* 1985; Traub 1985), *Salmonella paratyphi* A (Dang *et al.* 1988), and a variety of Gram-negative species (Piddock *et al.* 1989). In all of these reports the selection of resistance was carried out in the laboratory, and in many of these reports, initially susceptible clinical strains were used for selection. In a number of these studies, alterations (usually reductions) in outer membrane proteins (Sanders *et al.* 1984) and, in addition, reduced accumulation of 4-quinolones (Hirai *et al.* 1987; Dang *et al.* 1988; Chamberland *et al.* 1989; Legakis *et al.* 1989; Lucain *et al.* 1989; Piddock *et al.* 1989) or chloramphenicol (Gutmann *et al.* 1985) were also seen. In some studies with *P. aeruginosa*, an increase in a 54 kilodalton outer membrane protein was seen (Hirai *et al.* 1987), and in one study treatment with energy inhibitors produced an increase in 4-quinolone accumulation (Chamberland *et al.* 1989).

In a few instances, similar resistance seems to have developed in bacteria causing human infections. Strains of *P. aeruginosa* isolated from patients treated with 4-quinolones have developed pleiotropic resistance and reductions in some outer membrane proteins (Piddock *et al.* 1987; Daikos *et al.* 1988) and have exhibited reduced accumulation of enoxacin (Piddock *et al.* 1987). A norfloxacin-resistant strain of *E. coli* with reduced *OmpF* and norfloxacin accumulation has been isolated from a patient treated with norfloxacin (Aoyama *et al.* 1987). And finally, a strain of *S. marcescens* isolated from a patient treated only with ß-lactams and aminoglycosides developed resistance to ciprofloxacin and changes in outer membrane proteins (Sanders and Watanakunakorn 1986). Although resistance mechanisms likely to involve altered 4-quinolone permeation can occur in clinical settings, the relative frequencies of this mechanism of resistance and of that due to alterations in DNA gyrase, awaits the testing of a broader range of clinical isolates.

Conclusions

Our understanding of the mechanisms of 4-quinolone resistance is currently incomplete. Although the domains of the gyrase A protein involved in resistance are being identified, how these changes are translated into reduced 4-quinolone action remains to be defined. Even less information is available on the role of changes in the gyrase B subunit in 4-quinolone resistance.

Resistance associated with mutations other than those in the gyrase genes is even more complex and appears to involve regulatory mutations which affect expression of porins and possibly other outer membrane proteins as well as other factors as yet undefined. Reduced 4-quinolone accumulation in these mutants requires energy and appears to involve an energy-dependent efflux system located at least in part at the inner membrane, but the role of this efflux system in resistance has not yet been determined.

The medical importance of further studies to understand the mechanisms of 4-quinolone resistance is emphasized by the increasing recognition that development of resistance in some species, in particular *P. aeruginosa* (Desplaces *et al.* 1986; Giamarellou *et al.* 1986; Kresken and Wiedemann 1988) and *S. aureus* (Isaacs *et al.* 1988; Schaefler 1989; Shalit *et al.* 1989), may pose a limitation to clinical utility in some settings (Wolfson and Hooper 1989a,b).

References

Aoyama H, Fujimaki K, Sato K, Fujii T, Inoue M, Hirai K, Mitsuhashi S (1988) Clinical isolate of *Citrobacter freundii* highly resistant to new quinolones. Antimicrob Agents Chemother 32:922-924

Aoyama H, Sato K, Kato T, Hirai K, Mitsuhashi S (1987) Norfloxacin resistance in a clinical isolate of *Escherichia coli*. Antimicrob Agents Chemother 31:1640-1641

Bayer AS, Hirano L, Yih J (1988) Development of ß-lactam resistance and increased quinolone MICs during therapy of experimental *Pseudomonas aeruginosa* endocarditis. Antimicrob Agents Chemother 32:231-235

Bedard J, Wong S, Bryan LE (1987) Accumulation of enoxacin by *Escherichia coli* and *Bacillus subtilis*. Antimicrob Agents Chemother 31:1348-1354

Bourguignon GJ, Levitt M, Sternglanz R (1973) Studies on the mechanism of action of nalidixic acid. Antimicrob Agents Chemother 4:479-486

Chamberland S, Bayer AS, Schollaardt T, Wong SA, Bryan LE (1989) Characterization of mechanisms of quinolone resistance in *Pseudomonas aeruginosa* strains isolated *in vitro* and *in vivo* during experimental endocarditis. Antimicrob Agents Chemother 33:624-634

Cohen SP, Hooper DC, Wolfson JS, Souza KS, McMurry LM, Levy SB (1988a) Endogenous active efflux of norfloxacin in susceptible *Escherichia coli*. Antimicrob Agents Chemother 32:1187-1191

Cohen SP, McMurry LM, Levy SB (1988b) *marA* locus causes decreased expression of *OmpF* porin in multiple-antibiotic-resistant (*Mar*) mutants of *Escherichia coli*. J Bacteriol 170:5416-5422

Cohen SP, McMurry LM, Hooper DC, Wolfson JS, Levy SB (1989) Cross-resistance to fluoroquinolones in multiple-antibiotic-resistant (*Mar*) *Escherichia coli* selected by tetracycline and chloramphenicol: decreased drug accumulation associated with membrane changes in addition to *OmpF* reduction. Antimicrob Agents Chemother 33:1318-1325

Cullen ME, Wyke AW, Kuroda R, Fisher LM (1989) Cloning and characterization of a DNA gyrase A gene from *Escherichia coli* that confers clinical resistance to 4-quinolones. Antimicrob Agents Chemother 33:886-894

Daikos GL, Lolans VT, Jackson GG (1988) Alterations in outer membrane proteins of *Pseudomonas aeruginosa* associated with selective resistance to quinolones. Antimicrob Agents Chemother 32:785-787

Dang P, Gutmann L, Quentin C, Williamson R, Collatz E (1988) Some properties of *Serratia marcescens, Salmonella paratyphi* A, and *Enterobacter cloacae* with non-enzyme-dependent multiple resistance to ß-lactam antibiotics, aminoglycosides, and quinolones. Rev Infect Dis 10:899-904

Desplaces N, Gutmann L, Carlet J, Guibert J, Acar JF (1986) The new quinolones and their combination with other agents for therapy of severe infections. J Antimicrob Chemother 17(Suppl A):25-39

Fujimaki K, Fujii T, Aoyama H, Sato K-I, Inoue Y, Inoue M, Mitsushashi S (1989) Quinolone resistance in clinical isolates of *Serratia marcescens*. Antimicrob Agents Chemother 33:785-787

Gellert M (1981) DNA topoisomerases. Annu Rev Biochem 50:879-910

Gellert M, O'Dea MH, Itoh T, Tomizawa J-I (1976) Novobiocin and coumermycin inhibit DNA supercoiling catalyzed by DNA gyrase. Proc Natl Acad Sci USA 73:4474-4478

Gellert M, Mizuuchi K, O'Dea MH, Nash HA (1977) Nalidixic acid resistance: a second genetic character involved in DNA gyrase activity. Proc Natl Acad Sci USA 74:4772-4776

George AM, Levy SB (1983a) Amplifiable resistance to tetracycline, chloramphenicol, and other antibiotics in *Escherichia coli*: involvement of a non-plasmid-determined efflux of tetracycline. J Bacteriol 155:531-540

George AM, Levy SB (1983b) Gene in the major cotransduction gap of the *Escherichia coli* K-12 linkage map required for the expression of chromosomal resistance to tetracycline and other antibiotics. J Bacteriol 155:541-548

Giamarellou H, Galanakis N, Dendrinos C, Stefanou J, Daphnis E, Daikos GK (1986) Evaluation of ciprofloxacin in the treatment of *Pseudomonas aeruginosa* infections. Eur J Clin Microbiol 5:232-235

Gutmann L, Williamson R, Moreau N, Kitzis M-D, Collatz E, Acar JF, Goldstein FW (1985) Cross-resistance to nalidixic acid, trimethoprim, and chloramphenicol associated with alterations in outer membrane proteins of *Klebsiella, Enterobacter*, and *Serratia*. J Infect Dis 151:501-507

Hane MW, Wood TH (1969) *Escherichia coli* K-12 mutants resistant to nalidixic acid: genetic mapping and dominance studies. J Bacteriol 99:238-241

Helling RB, Adams BS (1970) Nalidixic acid-resistant auxotrophs of *Escherichia coli*. J Bacteriol 104:1027-1029

Helling RB, Kukora JS (1971) Nalidixic acid-resistant mutants of *Escherichia coli* deficient in isocitrate dehydrogenase. J Bacteriol 105:1224-1226

Hirai K, Aoyama H, Suzue S, Irikura T, Iyobe S, Mitsuhashi S (1986) Isolation and characterization of norfloxacin-resistant mutants of *Escherichia coli* K12. Antimicrob Agents Chemother 30:248-253

Hirai K, Suzue S, Irikura T, Iyobe S, Mitsuhashi S (1987) Mutations producing resistance to norfloxacin in *Pseudomonas aeruginosa*. Antimicrob Agents Chemother 31:582-586

Hooper DC, Wolfson JS, Souza KS, Tung C, McHugh GL, Swartz MN (1986) Genetic and biochemical characterization of norfloxacin resistance in *Escherichia coli*. Antimicrob Agents Chemother 29:639-644

Hooper DC, Wolfson JS, Ng EY, Swartz MN (1987) Mechanisms of action of and resistance to ciprofloxacin. Am J Med 82(Suppl 4A):12-20

Hooper DC, Wolfson JS, Souza KS, Ng EY, McHugh GL, Swartz MN (1989) Mechanisms of quinolone resistance in *Escherichia coli*: characterization of *nfxB* and *cfxB*, two mutant resistance loci decreasing norfloxacin accumulation. Antimicrob Agents Chemother 33:283-290

Horowitz DS, Wang JC (1987) Mapping the active site tyrosine of *Escherichia coli* DNA gyrase. J Biol Chem 262:5339-5344

Hrebenda J, Heleszko H, Brzostek K, Bielecki J (1985) Mutation affecting resistance of *Escherichia coli* K12 to nalidixic acid. J Gen Microbiol 131:2285-2292

Inoue Y, Sato K, Fujii T, Hirai K, Inoue M, Iyobe S, Mitsuhashi S (1987) Some properties of subunits of DNA gyrase from *Pseudomonas aeruginosa* PAO1 and its nalidixic-acid-resistant mutant. J Bacteriol 169:2322-2325

Isaacs RD, Kunke PJ, Cohen RJ, Smith JW (1988) Ciprofloxacin resistance in epidemic methicillin-resistant *Staphylococcus aureus*. Lancet ii:843

Kresken M, Wiedemann B (1988) Development of resistance to nalidixic acid and the fluoroquinolones after introduction of norfloxacin and ofloxacin. Antimicrob Agents Chemother 32:1285-1288

Kumada T, Neu HC (1985) *In vitro* activity of ofloxacin, a quinolone carboxylic acid compared to other quinolones and other antimicrobial agents. J Antimicrob Chemother 16:563-574

Kumar S (1976) Properties of adenyl cyclase and cyclic adenosine 3',5'-monophosphate receptor protein-deficient mutants of *Escherichia coli*. J Bacteriol 125:545-555

Kumar S (1980) Types of spontaneous nalidixic acid resistant mutants of *Escherichia coli*. Indian J Exp Biol 18:341-343

Lampe MF, Bott KF (1985) Genetic and physical organization of the cloned *gyrA* and *gyrB* genes of *Bacillus subtilis*. J Bacteriol 162:78-84

Legakis NJ, Tzouvelekis LS, Makris A, Kotsifaki H (1989) Outer membrane alterations in multiresistant mutants of *Pseudomonas aeruginosa* selected with ciprofloxacin. Antimicrob Agents Chemother 33:124-127

Lucain C, Regamey P, Bellido F, Pechère J-C (1989) Resistance emerging after pefloxacin therapy of experimental *Enterobacter cloacae* peritonitis. Antimicrob Agents Chemother 33:937-943

Misra R, Reeves PR (1987) Role of *micF* in the *tolC*-mediated regulation of *OmpF*, a major outer membrane protein of *Escherichia coli* K-12. J Bacteriol 169:4722-4730

Mizuno T, Chou M-Y, Inouye M (1984) A unique mechanism regulating gene expression: translational inhibition by a complementary RNA transcript (micRNA). Proc Natl Acad Sci USA 81:1966-1970

Mouton RP, Mulders SLTA (1987) Combined resistance to quinolones and ß-lactams after *in vitro* transfer on single drugs. Chemotherapy (Basel) 33:189-196

Nakamura S, Nakamura M, Kojima T, Yoshida H (1989) *gyrA* and *gyrB* mutations in quinolone-resistant strains of *Escherichia coli*. Antimicrob Agents Chemother 33:254-255

Nikaido H (1985) Role of permeability barriers in resistance to ß-lactam antibiotics. Pharmac Ther 27:197-231

Piddock LJV, Hall M, Griggs DJ, Wise R (1989) Selection and phenotypic characterization of the mechanism of resistance of enterobacteriaceae to quinolones. Rev Infect Dis 11(Suppl 5): S977-S978

Piddock LJV, Wijnands WJA, Wise R (1987) Quinolone/ureidopenicillin cross-resistance. Lancet ii:907

Rella M, Haas D (1982) Resistance of *Pseudomonas aeruginosa* PAO to nalidixic acid and low levels of ß-lactam antibiotics: mapping of chromosomal genes. Antimicrob Agents Chemother 22:242-249

Robillard NJ, Scarpa AL (1988) Genetic and physiological characterization of ciprofloxacin resistance in *Pseudomonas aeruginosa* PAO. Antimicrob Agents Chemother 32:535-539

Sanders CC, Watanakunakorn C (1986) Emergence of resistance to ß-lactams, aminoglycosides, and quinolones during combination therapy for infection due to *Serratia marcescens*. J Infect Dis 153:617-619

Sanders CC, Sanders WE Jr, Goering RV, Werner V (1984) Selection of multiple antibiotic resistance by quinolones, ß-lactams, and aminoglycosides with special reference to cross-resistance between unrelated drug classes. Antimicrob Agents Chemother 26:797-801

Schaefler S (1989) Methicillin-resistant strains of *Staphylococcus aureus* resistant to quinolones. J Clin Microbiol 27:335-336

Setlow JK, Cabrera-Juárez E, Albritton WL, Spikes D, Mutschler A (1985) Mutations affecting gyrase in *Haemophilus influenzae*. J Bacteriol 164:525-534

Shalit I, Berger SA, Gorea A, Frimerman H (1989) Widespread quinolone resistance among methicillin-resistant *Staphylococcus aureus* isolates in a general hospital. Antimicrob Agents Chemother 33:593-594

Shen LL, Kohlbrenner WE, Weigl D, Baranowski J (1989a) Mechanism of quinolone inhibition of DNA gyrase. Appearance of unique norfloxacin binding sites in enzyme-DNA complexes. J Biol Chem 264:2973-2978

Shen LL, Mitscher LA, Sharma PN, O'Donnell TJ, Chu DWT, Cooper CS, Rosen T, Pernet AG (1989b) Mechanism of inhibition of DNA gyrase by quinolone antibacterials: a cooperative drug-DNA binding model. Biochemistry 28:3886-3894

Smith JT (1984) Mutational resistance to 4-quinolone antibacterial agents. Eur J Clin Microbiol 3: 347-350

Sugino A, Bott KF (1980) *Bacillus subtilis* deoxyribonucleic acid gyrase. J Bacteriol 141:1331-1339

Sugino A, Peebles CL, Kreuzer KN, Cozzarelli NR (1977) Mechanism of action of nalidixic acid: purification of *Escherichia coli nalA* gene product and its relationship to DNA gyrase and a novel nicking-closing enzyme. Proc Natl Acad Sci USA 74:4767-4771

Sugino A, Higgins NP, Brown PO, Peebles CL, Cozzarelli NR (1978) Energy coupling in DNA gyrase and the mechanism of action of novobiocin. Proc Natl Acad Sci USA 75:4838-4842

Sugino A, Higgins NP, Cozzarelli NR (1980) DNA gyrase subunit stoichiometry and the covalent attachment of subunit A to DNA during DNA cleavage. Nucl Acids Res 8:3865-3874

Traub WH (1985) Incomplete cross-resistance of nalidixic and pipemidic acid-resistant variants of *Serratia marcescens* against ciprofloxacin, enoxacin, and norfloxacin. Chemotherapy (Basel) 31:34-39

Traub WH, Kleber I (1977) Selected and spontaneous variants of *Serratia marcescens* with combined resistance against chloramphenicol, nalidixic acid, and trimethoprim. Chemotherapy (Basel) 23:436-451

Ubukata K, Itoh-Yamashita N, Konno M (1989) Cloning and expression of the *norA* gene for fluoroquinolone resistance in *Staphylococcus aureus*. Antimicrob Agents Chemother 33: 1535-1539

Walton JR, Smith DH (1969) New hemolysin (γ) produced by *Escherichia coli*. J Bacteriol 93:304-305

Wolfson JS, Hooper DC (1989a) Bacterial resistance to quinolones: mechanisms and clinical importance. Rev Infect Dis 11(Suppl 5):S960-S968

Wolfson JS, Hooper DC (1989b) Fluoroquinolone antimicrobial agents. Clin Microbiol Rev 2: 378-424

Wolfson JS, Hooper DC, Ng EY, Souza KS, McHugh GL, Swartz MN (1987) Antagonism of wild-type and resistant *Escherichia coli* and its DNA gyrase by the tricyclic 4-quinolone analogs ofloxacin and S-25930 stereoisomers. Antimicrob Agents Chemother 31:1861-1863

Wolfson JS, Hooper DC (1989b) Fluoroquinolone antimicrobial agents. Clin Microbiol Rev 2: 378-424

Wolfson JS, Hooper DC, Ng EY, Souza KS, McHugh GL, Swartz MN (1987) Antagonism of wild-type and resistant *Escherichia coli* and its DNA gyrase by the tricyclic 4-quinolone analogs ofloxacin and S-25930 stereoisomers. Antimicrob Agents Chemother 31:1861-1863

Yamagishi JI, Yoshida H, Yamayoshi M, Nakamura S, Shimizu M. (1981) New nalidixic acid resistance mutations related to deoxyribonucleic acid gyrase activity. J Bacteriol 148:450-458

Yamagishi J, Yoshida H, Yamayoshi M, Nakamura S (1986) Nalidixic acid-resistant mutations of the *gyrB* gene of *Escherichia coli*. Mol Gen Genet 204:367-373

Yoshida H, Kojima T, Yamagishi J, Nakamura S (1988) Quinolone-resistant mutations of the *gyrA* gene of *Escherichia coli*. Mol Gen Genet 211:1-7

Discussion Summary

The demonstration that resistance to clinical levels of new generation 4-quinolones may involve mutations effecting an altered gyrase, in conjunction with a mutation which reduces the effective uptake of the drug, prompted questions about the stepwise development of resistance in an initially sensitive population of bacteria. It was considered important that reduced dosage regimens might confer an elevated risk of the development of such resistance, and there were queries as to whether the presence of one mutation increased the likelihood of selection of the double mutant. It was clearly not a completely simple exercise to measure the two mutation frequencies independently, but of more practical significance was the possibility that, since the membrane associated mutation conferred a degree of resistance to other unrelated agents as well as the 4-quinolones, other unrelated antibacterial agents might be able to act as selective agents for resistance to high levels of a 4-quinolone. It was reported that there did seem to be a connection between tetracycline resistance and resistance to 4-quinolones. In particular where tetracycline was administered before oxolinic acid where there seems a distinct possibility that the tetracycline pre-selects the membrane associated mutation and subsequent treatment of such a mutant population facilitates the selection of gyrase mutants which will then possess both mutations necessary for the expression of resistance to the newer 4-quinolones.

In this context a clear area of concern was the increasing veterinary use of some of the older generation 4-quinolones. Clearly there ought to be an associated risk of resistance in humans if we accept the premise that resistant organisms from animals will reach man through the normal food-chain. This was however also a significant question for an indirect effect since what might reach man through the food chain is not necessarily a ready-made 4-quinolone resistant mutant, but a cell which already possesses the membrane-associated mutation (selected initially by agents like tetracycline) now only requiring a single gyrase mutation to effect resistance to the most potent 4-quinolones. Evidently this was an important aspect since this indirect risk effect did not apply to the older generations of 4-quinolones which did not need the second mutation to effect clinical resistance - consequently there was no *increased* risk, a feature which clearly differentiates between the new and the old 4-quinolones.

The question of the stability of the resistant mutations in bacterial populations in the absence of direct selection was also raised in discussions of this presentation as well as in discussion of the presentation of Professor Smith on the frequency of mutation to resistance. Both Dr Hooper and Professor Smith were able to confirm that in many cases the mutations to resistance to the new 4-quinolones were quite stable and did not comply with the apocryphal instability of mutations to nalidixic acid resistance.

However the ability of 4-quinolone resistant mutants to compete with 4-quinolone-sensitive cells in mixed populations under non-selective conditions was considered to be a very worrying matter since mutants selected as resistant to the new generation 4-quinolones appeared perfectly competitive in simple mixed cultures. This phenomenon clearly differentiated them from *E. coli* mutants selected as resistant to ofloxacin, and other species selected as resistant to the older 4-quinolones. It was also suggested with reference to *Pseudomonas aeruginosa* that we are observing significant changes in the nature of resistant mutants over an extended time period. This appears to be so with regard to cystic fibrosis patients where in the early usage of ciprofloxacin it was common for ciprofloxacin-resistant infections to revert to ciprofloxacin-sensitive upon withdrawal of the drug, but now, a couple of years later, the ciprofloxacin-resistant organisms persist even after withdrawal of the drug. Clearly we are seeing the development of a problem which earlier studies of the 4-quinolones did not even hint at.

Chapter 14

In Vitro and *In Vivo* Mutation Frequencies to Resistance - do they Correlate in the Long Term?

J. T. Smith

Introduction

Very unusually among antibacterial agents the 4-quinolones are rarely if ever affected by plasmid-mediated resistance and, perhaps as a consequence of this, the frequency of clinical resistance to the 4-quinolones is much less than that seen with the other major groups of antibiotics and chemotherapeutic agents (Burman 1977). Despite nalidixic acid being discovered as long ago as 1962 it is only very recently that plasmid-mediated resistance to it has been claimed to exist (Panhotra *et al.* 1985; Munshi *et al.* 1987). This contrasts sharply with the situation seen with the penicillins, cephalosporins, aminoglycosides, tetracyclines, sulphonamides, macrolides, trimethoprim and chloramphenicol, which have been plagued by plasmid-mediated resistance much earlier after their introduction and, in some cases, even before their introduction. Worse still, resistance to these antibacterials shows every sign of becoming more frequent due no doubt to the transferable nature of the resistance genes. The 4-quinolones have so far escaped completely, or very lightly, from transferable drug resistance because even the plasmids claimed to confer nalidixic acid resistance do not confer resistance to modern 4-quinolones, although they seem to make their hosts more able to mutate to become ciprofloxacin-resistant (Munshi *et al.* 1987). However, whether the genes responsible for transferable resistance to nalidixic acid are plasmid-borne or located on the chromosome has been questioned (Crumplin 1987) and results published so far, claiming plasmid-mediated transferable 4-quinolone resistance, can all be explained by the plasmids containing mutator genes which increase the frequency of chromosomal (and hence non-transferable) mutations occurring conferring 4-quinolone resistance.

Another possible explanation is, it is known that the *nalB* impermeability mutation (Hane and Wood 1969) confers resistance to nalidixic acid but not to the modern 4-quinolone antibacterials (Smith 1984). When *E. coli* KL16 possesses this mutation it is able to mutate to resist 5X the MIC of norfloxacin at a frequency of 10^{-8} or of ciprofloxacin at a frequency of 10^{-9} (JT Smith unpublished observations) whilst *E. coli* KL16 without the *nalB* mutation does not exhibit any such resistant mutants in 10^{12} bacteria (Smith 1986). It has been observed that the *nalB* mutation, which confers only a very low level of resistance to nalidixic acid, can also confer an increase in the spontaneous mutation frequency to high-level nalidixic acid resistance ($\geq$10X MIC) from approximately 10^{-8} for the *wt* parent strain to approximately 10^{-5} for the *nalB* mutant strain (GC Crumplin personal communication). Such indications of a possible mutator effect associated with the *nalB* chromosomal mutant gene tend to support the possible existence of plasmid-borne mutator genes which lead to increased frequencies of chromosomal mutation to 4-quinolone resistance.

Apart from plasmids, the only mechanism that remains for bacteria to become resistant to the 4-quinolones is by chromosomal mutation. This in itself is unusual clinically as with other drugs chromosomal mutations, which occur readily in the laboratory, are generally not responsible for clinical antibiotic resistance. Because mutations so rarely cause clinical resistance to antibacterials the generally accepted concept that such mutants lack pathogenicity has arisen (Smith 1983).

Many studies on mutational 4-quinolone resistance have been made (reviewed by Smith 1986). The common theme in these publications is that whilst *E. coli* and other Gram negative bacteria are able to mutate to resist the older 4-quinolones at readily detectable frequencies *in vitro*, mutants either cannot be detected or occur at much lower frequencies with respect to their ability to resist norfloxacin, ofloxacin or ciprofloxacin. However, perhaps because the therapy of Gram positive infections has lagged behind that of Gram negative infections, fewer studies have been reported on the emergence of resistance to 4-quinolones *in vitro* with Gram positive pathogens. This study compares mutational resistance in eight species of Gram negative bacteria with that occurring in five species of Gram positive bacteria using ten different 4-quinolone antibacterials.

Materials and Methods

Media: The same single batch of Oxoid No 2 nutrient broth powder was used to prepare nutrient broth as a growth medium and as a diluent. The nutrient broth powder was also used to make nutrient agar plates which were prepared by adding Lab M agar at a final concentration of 1.5% to nutrient broth. A single batch of Lab M agar was also used throughout this study. These precautions were taken to ensure that the divalent metal ion content of all media used (i.e. liquid and solid) was standardized. Quadruple strength nutrient broth was prepared in 5ml quantities and sterilized in 1 oz bottles. Then sterile drug solution and sterile water were added to give a final total volume of 10ml. Drugs were dissolved in the minimal quantity of N/1 NaOH and immediately diluted with sterile water or, where appropriate, dissolved in sterile water. After sterilising and dissolving 1.5g of agar in 50 ml water by autoclaving, 10 ml of hot molten agar/water was added to each sterile bottle containing 10 ml of double-strength, nutrient broth and drugs. The contents were mixed and used to pour a single nutrient agar plate which was allowed to set then overdried. The concentration range used was a geometrically based progression increasing successively by 25%-50% at each incremental step. The ratios used were 1, 1.5, 2, 3, 4, 5, 7.5.

Bacterial strains: Staphylococcus aureus E3T, *Staphylococcus saprophyticus* MVK, *Staphylococcus warneri* C14LN19, *Streptococcus faecalis* NCTC775, *Escherichia coli* KL16, *Serratia marcescens* 120, *Salmonella typhimurium* LT2, *Shigella sonnei* ZBX4, *Providencia stuartii* EN215, *Klebsiella oxytoca* FSM, *Enterobacter aerogenes* 418 and *Pseudomonas aeruginosa* C17LN22 were subcultivated on drug-free nutrient agar. *Streptococcus pneumoniae* C3LN4 was subcultivated onto drug-free nutrient agar containing 7% laked horse blood. These subcultures were used as inocula to prepare liquid cultures in nutrient broth or in the case of *Strep. pneumoniae*, in nutrient broth + 7% laked horse blood, which were used to determine the sensitivity of these bacteria to 4-quinolones and to estimate the frequency at which mutational resistance to the 4-quinolones arose. When the sensitivity of the bacteria to 4-quinolones was tested the bacteria were grown overnight at 37°C and dilutions in nutrient broth, containing 20-200 colony forming units (CFU) inoculated onto nutrient agar containing various concentrations of the ten 4-quinolones studied. For *Strep. pneumoniae* only, 7% laked horse blood was added to the nutrient agar. After the inocula had been absorbed, the plates were inverted and incubated for up to 2 days at 37°C. The MIC was defined as the lowest drug concentration which completely inhibited colony formation. Mutation rates were only estimated for 4-quinolones that exhibited MIC values of less than 10μg/ml.

Mutation rates: The bacteria were grown at 37°C overnight in 200ml quantities of nutrient broth in 2 litre flasks, then shaken at 120 cycles per minute for 5 h. For *Strep. pneumoniae* 7% laked horse blood was added to the nutrient broth. The cultures were centrifuged down and the bacteria suspended at 20 times their original concentration in nutrient broth. Each of three to six 2.5 ml aliquots of this concentrate was added to 2.5 ml of molten nutrient agar which was then overlaid onto the surface of a nutrient agar plate (+7% laked horse blood for *Strep. pneumoniae*) containing a concentration of each 4-quinolone that resulted ultimately at 5, 10 or 20 times its MIC against sensitive parental strains. The number of aliquots tested was chosen so that at least 10^{12} bacteria were studied on any one occasion. In order to detect mutants occurring at relatively high frequencies, serial decimal dilutions of the concentrated cultures were also added to overlays which were similarly poured onto nutrient agar plates containing 4-quinolones and also onto nutrient agar lacking any antibacterial agent, in order to determine the viable counts of the concentrated cultures. When *Strep. pneumoniae* was studied 7% laked horse blood was added to all nutrient agars. The plates were incubated for up to 3 days at 37°C and colonies counted. When mutants were detected at least 20 colonies, if available, were subjected to one subcultivation on the same concentration of that 4-quinolone being tested before being scored. This procedure was necessary because not all colonies were able to survive a second subculture on the 4-quinolone concentration on which they were originally isolated.

Results

It can be seen (Table 14.1) that no mutants could be detected in the 10^{12} *E. coli* tested against ofloxacin, norfloxacin or ciprofloxacin even at only 5X their MIC values, while with all of the older drugs mutational resistance occurred to 10 or even 20 times their MIC values. When these results were first described (Smith 1986) a possible explanation for such differences between older and modern 4-quinolones was that norfloxacin, ofloxacin and ciprofloxacin possess two modes of action while the other 4-quinolones only possess one mode of action against sensitive bacteria (Smith 1984;

Ratcliffe and Smith 1985). This view comes from the generally accepted hypothesis that mutation frequencies are cumulative so that if the frequency against one mechanism of action is 10^{-8} then the frequency against two mechanisms would be 10^{-16}, which is beyond detection by the methods used here (10^{-12}).

Table 14.1. *Escherichia coli*

4Q	MIC (μg/ml)	Mutational 5X MIC	frequencies 10X MIC	at 20x MIC
NAL	3	10^{-8}	10^{-9}	10^{-8}
CIN	3	10^{-8}	10^{-8}	10^{-9}
PIP	0.75	10^{-9}	10^{-9}	-
FLU	0.4	10^{-9}	10^{-9}	-
ROS	0.2	10^{-8}	10^{-9}	10^{-9}
OXO	0.2	10^{-8}	10^{-9}	10^{-10}
NOR	0.04	-	-	-
OFL	0.03	-	-	-
CIP	0.004	-	-	-

4Q = 4-quinolone antibacterial ROS = rosoxacin
PIR = pirromidic acid   OXO = oxolinic acid
NAL = nalidixic acid   NOR = norfloxacin
CIN = cinoxacin   OFL = ofloxacin
PIP - = pipemidic acid   CIP = ciprofloxacin
FLU = flumequine
-       = no mutant detected in 10^{12} bacteria
(Legend refers to Tables 14.1 - 14.13)

Table 14.2. *Serratia marcescens*

4Q	MIC (μg/ml)	Mutational 5X MIC	frequencies 10X MIC	at 20x MIC
NAL	1.5	10^{-9}	10^{-9}	10^{-9}
PIP	0.75	10^{-9}	10^{-8}	-
FLU	0.75	10^{-9}	10^{-9}	10^{-10}
ROS	0.4	10^{-9}	10^{-10}	-
OXO	0.15	10^{-9}	10^{-8}	-
OFL	0.1	10^{-9}	-	-
NOR	0.075	-	-	-
CIP	0.03	-	-	-

The results with *S. marcescens* (Table 14.2) are similar to those seen with *E. coli* with the notable exception of ofloxacin where mutants were detected which resisted five but not 10 or 20 times its MIC. With this species both ofloxacin and ciprofloxacin have been shown to possess two modes of action (CS Lewin 1986, personal communication) so the hypothesis linking mutational frequencies to number of modes of action is questionable.

Tables 14.3 and 14.4 show the results with the two enteric pathogens, *Salmonella* and *Shigella*. Despite both genera being of similar sensitivity to 4-quinolones, mutational resistance was commoner in *Shigella* including that to norfloxacin and to ofloxacin while with *Salmonella* , the only modern 4-quinolone to exhibit mutational resistance was paradoxically ciprofloxacin.

Table 14.3. *Salmonella typhimurium*

4Q	MIC (µg/ml)	Mutational 5X MIC	frequencies 10X MIC	at 20x MIC
CIN	7.5	-	-	-
PIP	1.5	10^{-9}	-	-
FLU	1.5	-	-	-
ROS	0.4	10^{-9}	10^{-9}	10^{-10}
OXO	0.2	10^{-7}	10^{-8}	-
NOR	0.075	-	-	-
OFL	0.04	-	-	-
CIP	0.0075	10^{-8}	-	-

Table 14.4. *Shigella sonnei*

4Q	MIC (µg/ml)	Mutational 5X MIC	frequencies 10X MIC	at 20x MIC
NAL	1.5	10^{-8}	10^{-9}	10^{-9}
CIN	1.5	10^{-9}	10^{-9}	10^{-9}
PIP	0.75	10^{-9}	-	-
FLU	0.3	10^{-9}	10^{-10}	-
ROS	0.3	10^{-9}	10^{-9}	10^{-10}
OXO	0.1	10^{-8}	10^{-9}	-
NOR	0.03	10^{-9}	-	-
OFL	0.03	10^{-9}	-	-
CIP	0.0075	-	-	-

Table 14.5. *Providencia stuartii*

4Q	MIC (µg/ml)	Mutational 5X MIC	frequencies 10X MIC	at 20x MIC
CIN	10	10^{-7}	10^{-8}	10^{-9}
PIP	4	10^{-8}	10^{-9}	-
NAL	3	10^{-8}	10^{-8}	10^{-9}
FLU	1	10^{-10}	10^{-10}	-
ROS	0.75	10^{-10}	10^{-9}	10^{-9}
NOR	0.75	10^{-9}	-	-
OFL	0.4	-	-	-
CIP	0.2	10^{-8}	-	-
OXO	0.15	10^{-9}	10^{-9}	10^{-10}

The results with *Providencia* and *Klebsiella* (Tables 14.5 and 14.6) which, despite being very different in their sensitivities to 4-quinolones, show that the *Klebsiella* mutated at detectable frequencies to resist all nine drugs tested. The less sensitive *Providencia* did not exhibit mutational resistance to ofloxacin in the 10^{12} bacteria tested.

Somewhat along similar lines *Pseudomonas aeruginosa* and *Enterobacter aerogenes*, despite differing in their overall sensitivities to 4-quinolones, mutated to resist all 4-quinolones tested even at 10X the MIC of all drugs and in numerous instances mutants resisting 20X the MIC of several drugs occurred (Tables 14.7 and 14.8). Indeed, the more sensitive *Enterobacter* was better equipped to mutate to resist higher levels of norfloxacin and of ofloxacin than *Pseudomonas*. Even harder to understand than this is the finding (CS Lewin 1986, personal communication) that ofloxacin and ciprofloxacin exhibit two mechanisms of action against *E. aerogenes* and hence the hypothesis mentioned earlier with *E. coli* relating numbers of mechanisms of action with the inverse of mutation frequency is again rendered doubtful.

With the eight Gram negative genera tested there was a common theme that sensitivity to 4-quinolones bore no relationship to the ability of the bacteria to mutate to resist such drugs. This also applied to the five species of Gram positive bacteria tested (Tables 14.9 - 14.13). The most sensitive Gram positive organism tested was *S. aureus* (Table 14.13) and yet it was the only Gram positive bacterium to exhibit detectable mutational resistance to ofloxacin and ciprofloxacin.

Table 14.6. *Klebsiella oxytoca*

4Q	MIC (µg/ml)	Mutational 5X MIC	frequencies 10X MIC	at 20x MIC
CIN	7.5	10^{-8}	10^{-9}	-
NAL	3	10^{-5}	10^{-8}	10^{-8}
PIP	1	10^{-9}	10^{-10}	-
FLU	0.5	10^{-7}	10^{-10}	-
ROS	0.3	10^{-9}	10^{-10}	10^{-10}
OXO	0.15	10^{-5}	10^{-8}	10^{-9}
NOR	0.04	10^{-7}	-	-
OFL	0.04	10^{-7}	-	-
CIP	0.0075	10^{-6}	10^{-8}	-

Table 14.7. *Pseudomonas aeruginosa*

4Q	MIC (µg/ml)	Mutational 5X MIC	frequencies 10X MIC	at 20x MIC
PIP	5	10^{-7}	10^{-10}	-
ROS	2	10^{-6}	10^{-8}	10^{-9}
OXO	2	10^{-7}	10^{-8}	10^{-9}
OFL	0.5	10^{-7}	10^{-9}	-
NOR	0.3	10^{-6}	10^{-10}	-
CIP	0.04	10^{-7}	10^{-9}	-

Table 14.8. *Enterobacter aerogenes*

4Q	MIC (µg/ml)	Mutational 5X MIC	frequencies 10X MIC	at 20x MIC
CIN	3	10^{-7}	10^{-9}	-
PIR	1.5	10^{-6}	10^{-7}	10^{-7}
NAL	0.75	10^{-7}	10^{-7}	10^{-7}
PIP	0.75	10^{-7}	10^{-9}	-
FLU	0.075	10^{-7}	10^{-7}	10^{-7}
ROS	0.05	10^{-7}	10^{-7}	10^{-7}
OXO	0.05	10^{-7}	10^{-7}	10^{-7}
NOR	0.03	10^{-7}	10^{-7}	10^{-8}
OFL	0.015	10^{-7}	10^{-7}	10^{-7}
CIP	0.005	10^{-7}	10^{-7}	-

Table 14.9. *Streptococcus faecalis*

4Q	MIC(μg/ml)	Mutational 5X MIC	frequencies 10X MIC	at 20x MIC
NOR	4	-	-	-
OFL	1.5	-	-	-
CIP	1.5	-	-	-

Table 14.10. *Streptococcus pneumoniae*

4Q	MIC (μg/ml)	Mutational 5X MIC	frequencies 10X MIC	at 20x MIC
NOR	3	-	-	-
OFL	1	-	-	-
CIP	0.75	-	-	-

Table 14.11. *Staphylococcus saprophyticus*

4Q	MIC (μg/ml)	Mutational 5X MIC	frequencies 10X MIC	at 20x MIC
FLU	7.5	10^{-4}(33)	10^{-6}(20)	10^{-8}(63)
OXO	4	10^{-5}(43)	10^{-8}(0)	10^{-8}(0)
NOR	3	-	-	-
ROS	2	10^{-7}(20)	10^{-9}(0)	10^{-9}(0)
OFL	0.5	-	-	-
CIP	0.5	-	-	-

( ) = % loss of Novobiocin resistance

Table 14.12. *Staphylococcus warneri*

4Q	MIC(μg/ml)	Mutational 5X MIC	frequencies 10X MIC	at 20x MIC
PIR	5	10^{-6}	-	-
FLU	4	10^{-8}	10^{-11}	-
OXO	2	-	-	-
ROS	1	-	-	-
NOR	0.75	10^{-7}	10^{-8}	-
OFL	0.2	-	-	-
CIP	0.2	-	-	-

Table 14.13. *Staphylococcus aureus*

4Q	MIC (μg/ml)	Mutational 5X MIC	frequencies 10X MIC	at 20x MIC
PIR	4	10^{-5}(0)	10^{-7}(0)	-
FLU	2	-	-	-
OXO	1	10^{-6}(0)	-	-
NOR	0.75	10^{-8}(19)	10^{-9}(69)	-
ROS	0.4	10^{-8}(33)	-	-
OFL	0.1	10^{-9}(71)	-	-
CIP	0.1	10^{-10}(0)	-	-

( ) = % loss of coagulase

With *Strep. faecalis*, *Strep. pneumoniae* and *S. saprophyticus* no mutant was detected to ofloxacin, norfloxacin or ciprofloxacin and, with the last species mutants resisting, flumequine, oxolinic acid and rosoxacin were detected up to 20X their MIC values. Perhaps significantly several of these mutants of *S. saprophyticus* isolated had simultaneously lost novobiocin resistance while gaining 4-quinolone resistance. novobiocin resistance is a pathogenicity marker in this species as it is strongly associated with urease production (Dr M V Kelemen 1980, personal communication).

The remaining two Gram positive species (Tables 14.12 and 14.13) finally disprove the hypothesis that numbers of mechanisms of action are inversely related to mutation frequency because Lewin and Smith (1988) showed that for either *Staphylococcus* ofloxacin exhibits two mechanisms of action whilst ciprofloxacin only exhibits one mechanism of action against either species. Despite these results *S. warneri* did not mutate to resist either drug at detectable frequencies, while *S. aureus* mutated to resist both these modern 4-quinolones at detectable frequencies.

In addition it can be seen in Table 14.13 that a significant proportion of the *S. aureus* mutants isolated on norfloxacin, rosoxacin and particularly on ofloxacin had lost their ability to produce coagulase, which is the definitive pathogenicity marker for this species, but no ciprofloxacin-resistant mutant tested had lost the ability to produce coagulase. As coagulase is required by *S. aureus* for pathogenicity, these results would tend to favour ofloxacin over ciprofloxacin. Could this difference between ciprofloxacin and ofloxacin be explained by the findings that ciprofloxacin only exhibits a single mechanism (A) against this species while ofloxacin exhibits two mechanisms (A and B) against *S. aureus* (Lewin and Smith 1988)?

Finally the Gram positive bacteria taken as a whole contrast with the Gram negative bacteria in that mutational resistance, especially to the three modern drugs, occurred less readily in Gram positive organisms.

Discussion

The common theme in the many studies on mutational 4-quinolone resistance is that while *E.coli* and other Gram negative bacteria are able to mutate to resist the older 4-quinolones at readily detectable frequencies *in vitro*, mutants either cannot be detected or occur at much lower frequencies with respect to their ability to resist norfloxacin, ofloxacin or ciprofloxacin. However, despite these laboratory findings, some instances of rapid resistance development within patients can occur with *Pseudomonas aeruginosa* (Scully *et al.* 1986) *Citrobacter freundii*, *Pseudomonas maltophila* (Cheng *et al.* 1987) and *Campylobacter pylori* (Glupczynski *et al.* 1987). Another ominous common finding of these studies is that mutational resistance to any 4-quinolone also confers at least some resistance to all other 4-quinolones. There is one notable exception to this rule, the *nalC* mutation (Inoue *et al.* 1978), later known as the *nal-31* mutation (Yamagishi *et al.* 1986), which affects the B subunit of DNA gyrase. Smith (1984) showed that while these mutants were resistant to 4-quinolones lacking a C7 piperazine substituent, they were hypersensitive to 4-quinolones possessing such a substituent. This dichotomy may result from the mutation increasing the negative charge on the B subunits of DNA gyrase (Yamagishi *et al.* 1986) which would lead to an increase in the B subunit's attraction for the positively charged piperazine group, rendering the 4-quinolones possessing such a substituent more, rather than less, effective. Interestingly, the drugs that become hyperactive against the *nal-31* mutation are also those which are antagonized by acid pH values, while those drugs that are resisted by the *nal-31* mutation are also those which are more active at acid pH (Smith and Ratcliffe 1986).

However mutants containing altered B subunits of DNA gyrase seem to occur much less frequently than mutants with altered A subunits or with an altered uptake mechanism that results in impermeability. The resistance of the latter mutants has been shown to involve a reduction in the production of outer membrane proteins, particularly the *ompF* porin, which constitute pores through which 4-quinolones are transported (Hirai *et al.* 1986a; 1986b). Such uptake-deficient mutants have been variously identified in *Escherichia, Serratia, Klebsiella, Salmonella, Enterobacter, Providencia,* and *Pseudomonas* resistant to nalidixic acid (Rella and Haas 1982), enoxacin (Piddock *et al.* 1987), norfloxacin (Hooper *et al.* 1986; Hirai *et al.* 1986b) and ciprofloxacin (Sanders *et al.* 1984). Clinically, this mechanism of resistance is of particular concern because the *ompF* porin change also confers resistance to other antibacterials such as the tetracyclines, chloramphenicol, ß-lactams and the aminoglycosides and consequently cross-resistance between the 4-quinolones and these antibacterials may occur (Sanders *et al.* 1984; Hooper *et al.* 1986; Piddock *et al.* 1987). Unfortunately, unlike the *nal-31* mutation, resistance due to the reduced uptake of 4-quinolones appears to be quite common and such mutants have been identified not only *in vitro* but also *in vivo* (Sanders and Watanakunakorn 1986; Piddock *et al.* 1987). Indeed, Aoyama *et al* (1987) have described a clinical *E. coli* containing this resistance mechanism together with the production of a 4-quinolone-resistant gyrase that allows the organism to resist norfloxacin due to the combination of both resistance mechanisms, either of which alone was insufficient to cause therapeutic failure. Apart from *Pseudomonas aeruginosa* mutations resulting in an altered porin do not appear to confer resistance to clinical levels of the 4-quinolones unless accompanied by a concurrent mutation in the *gyrA* gene which confers a resistant A subunit of DNA gyrase.

It should be noted that so far such impermeability mutants have only been identified in Gram negative bacteria. Gram positive bacteria lack an outer membrane so it would seem that they would be unable to gain this type of impermeability resistance mechanism. Since Gram positive bacteria are generally less susceptible to the 4-quinolones than Gram negative bacteria, the lack of an outer membrane with its associated porins may be advantageous to bacteria with respect to susceptibility to these antibacterials.

Perhaps not surprisingly the results presented here show that *S. aureus* is able to mutate to 4-quinolone resistance at higher frequencies than most species of Gram negative bacteria, excepting *Klebsiella, Pseudomonas* or *Enterobacter* because similar findings with *S. aureus* and *Pseudomonas* have already been reported *in vivo* (Humphreys and Mulvihill 1985) and *in vitro* (Nakamura *et al.* 1983; Reeves *et al.* 1984a; 1984b; Limb *et al.* 1987). While the latter was also found to be the case by Cullmann *et al* (1985) and by Chin and Neu (1984), they could not detect any mutants that resisted modern 4-quinolones in *Strep. faecalis in vitro*, which also agrees with the results presented here. Perhaps these agreements with *Strep. faecalis* means it should not be surprising that this study failed to detect any 4-quinolone-resistant mutants in *Strep. pneumoniae.*

These *in vitro* studies do not answer what additional effects such mutations have had on bacterial pathogenicity in 11 out of the 13 species investigated and hence present the worst possible events that can possibly arise during 4-quinolone therapy. The results presented here are that, while mutational resistance to all 4-quinolones tested (except flumequine) occurred at detectable frequencies in *S. aureus*, a significant proportion of the mutants had lost the ability to produce coagulase, particularly in the case of ofloxacin. This loss of coagulase expression occurring simultaneously with the development of mutational 4-quinolone resistance is a clear indication that pathogenicity losses can occur. These could arise because changes in supercoiling caused by a mutationally altered DNA gyrase could turn off the expression of coagulase genes (Wang 1974; 1985). Another example of such a modification of the expression of a pathogenicity factor associated with resistance to the 4-quinolones is described later in this volume by Crumplin *et al*. However, the phenomenon causing the loss of pathogenicity from some of the *S. saprophyticus* mutants isolated in this study is more likely to be explicable in terms of plasmid curing events during the selection of the resistant mutants (Crumplin and Smith 1981; Weisser and Weidemann 1986; Briand *et al*. 1986). This may be a consequence of plasmid replicating systems being more susceptible to the drugs than the chromosomal replication system (Uhlin and Nordstrom 1985). Alternatively, the mechanism may be similar to that of the *gyrB* inhibitor novobiocin, which selects plasmid-free cells (Novick 1969) and this may occur because the sensitivity of host bacteria to nalidixic acid can be increased when they harbour R-plasmids (Crumplin and Smith 1981). It is as yet unclear whether this ability of the 4-quinolones to cure plasmids will have any clinical relevance (Weisser and Weidemann 1987).

An alternative mechanism which might be responsible for the loss of the novobiocin-resistance from *S. saprophyticus* may derive from the observation that *E. coli* carrying mutant *gyrA* alleles are relatively poor plasmid hosts (GC Crumplin, personal communication). Such strains harbouring a modified DNA gyrase may display relative deficiencies in either plasmid replication and/or segregation leading to the appearance of plasmid-free cells.

The overall conclusion concerning this study of 13 bacterial species for their ability to mutate to resist 4-quinolone antibacterials is that bacterial sensitivity to these drugs cannot predict the frequency with which mutants resisting them will develop. Similarly the multiplicity of mechanisms of action of 4-quinolones cannot be used to predict the development of mutational resistance to such drugs. On the other hand, these laboratory-based mutation frequency experiments, perhaps surprisingly, agree better with recent clinical results showing that *Pseudomonas* and *S. aureus* are exhibiting more frequent 4-quinolone resistance (Kresken and Wiedemann 1988; Blumberg *et al*. 1989). *In vitro Pseudomonas* and *S. aureus* together with *Klebsiella* and *Enterobacter* were better able to mutate to resist ofloxacin and ciprofloxacin than the other nine species tested. Only time will tell if the latter two genera will also pose clinical problems for 4-quinolone therapy.

References

Aoyama H, Sato K, Kato T, Hirai K and Mitsuhashi S (1987) Norfloxacin resistance in a clinical isolate of *Escherichia coli*. Antimicrob Agents Chemother 31 1640-1641

Blumberg HM, Rimland D, Wachsmuth IK (1989) Rapid development of ciprofloxacin resistance in methicillin-sensitive *S. aureus* (MSSA) and methicillin-resistant *S. aureus* (MRSA). 29th ICAAC Abstr 7 pp 102

Briand YM, Uccelli V, Laporte JR, Plessiat P (1986) Elimination of plasmids from *Enterobacteriaceae* by 4-quinolone derivatives. J Antimicrob Chemother 18: 667-674

Burman LG (1977) Apparent absence of transferable resistance to nalidixic acid in pathogenic Gram negative bacteria. J Antimicrob Chemother 3:509-516

Cheng AF, Li MKW, Ling TKW, French GL (1987) Emergence of ofloxacin-resistant *Citrobacter freundii* and *Pseudomonas maltophila* after ofloxacin therapy. J Antimicrob Chemother 20: 283-292

Chin NX, Neu HC (1984) Ciprofloxacin, a quinolone carboxylic acid compound active against aerobic and anaerobic bacteria. Antimicrob Agents Chemother 25:319-326

Crumplin GC (1987) Plasmid mediated resistance to nalidixic acid and new 4-quinolones? Lancet 2: 854-855

Crumplin GC, Smith JT (1981) The effect of R-factor plasmids on host-cell responses to nalidixic acid:I Increased susceptibility of nalidixic acid-sensitive hosts. J Antimicrob Chemother 7:379-388

Cullman W, Steiglitz M, Baars B, Opferkuch W (1985) Comparative evaluation of recently developed quinolone compounds - with a note on the frequency of resistant mutants. Chemotherapy 31:19-28

Glupczynski Y, Labbe M, Delmee M, Burette A, Avesini V, Bruck C (1987) Treatment failure of ofloxacin in *Campylobacter pylori* infection. Lancet 1:1096

Hane M, Wood T (1969) *Escherichia coli* K12 mutants resistant to nalidixic acid: Genetic mapping and dominance studies. J Bacteriol 99:238-241

Hirai K, Aoyama H, Irikura T, Iyobe S, Mitsuhashi S (1986a). Differences in susceptibilities to quinolones of outer membrane mutants of *Salmonella typhimurium* and *Escherichia coli*. Antimicrob Agents Chemother 29:535-538

Hirai K, Aoyama H, Suzue S, Irikura T, Iyobe S, Mitsuhashi S (1986b) Isolation and characterization of norfloxacin-resistant mutants of *Escherichia coli* K12. Antimicrob Agents Chemother 30:248-253

Hooper DC, Wolfson JS, Souza KS, Tung C, McHugh GI, Swartz MN (1986) Genetic and biochemical characterization of norfloxacin resistance in *Escherichia coli*. Antimicrob Agents Chemother 32: 639-644

Humphreys E, Mulvihill E (1985) Ciprofloxacin-resistant *Staphylococcus aureus*. Lancet 2:383

Inoue S, Ohue T, Yamagishi J, Nakamura S, Shimizu M (1978) Mode of incomplete cross-resistance among pipemidic, piromidic and nalidixic acids. Antimicrob Agents Chemother 14:240-245

Kresken M and Weidemann B (1988). Development of resistance to nalidixic acid and the fluoroquinolones after the introduction of norfloxacin and ofloxacin. Antimicrob Agents Chemother 32:1285-1288

Lewin CS, Smith JT (1988) Bactericidal mechanisms of ofloxacin. J Antimicrob Chemother 22 suppl C:1-8

Limb DI, Dabbs DJW, Spencer RC (1987) *In vitro* selection of bacteria resistant to the 4-quinolone agents. J Antimicrob Chemother 19:65-71

Munshi MH, Sack DA, Haider K, Ahmed ZV, Rahaman MM, Morshed MG (1987) Plasmid-mediated resistance to nalidixic acid in *Shigella dysenteriae* type 1. Lancet 2: 419-421

Nakamura S, Minami A, Katae H, Inoue S, Yamagishi J, Takase Y, Shimizu M (1983) *In vitro* antibacterial properties of AT-2266, a new pyridonecarboxylic acid. Antimicrob Agents Chemother 23: 641-648

Novick RP (1969) Extrachromosomal inheritance in bacteria. Bacteriol Rev 33:210-263

Panhotra BR, Desai B, Sharma PL (1985) Nalidixic acid-resistant *Shigella dysenteriae* 1. Lancet 2: 763

Piddock LJV, Wijnands WJA, Wise R (1987) Quinolone-ureidopenicillin cross-resistance. Lancet 2:907

Ratcliffe NT, Smith JT (1985) Norfloxacin has a novel bactericidal mechanism unrelated to that of other 4-quinolones. J Pharm Pharmacol 37:pp 92

Reeves DS, Bywater MJ, Holt HA (1984a) The activity of enoxacin against clinical bacteria isolates in comparison with that of five other agents, and other factors affecting that activity. J Antimicrob Chemother 14:(Suppl. C) 7-17

Reeves DS, Bywater MJ, Holt HA, White LO (1984b). *In vitro* studies with ciprofloxacin, a new 4-quinolone compound. J Antimicrob Chemother 13:333-346

Rella M, Haas D (1982) Resistance of *Pseudomonas aeruginosa* PAO to nalidixic acid and low levels of ß-lactam antibiotics: Mapping of chromosomal genes. Antimicrob Agents Chemother 22: 242-249

Sanders CC, Watanakunakorn C (1986) Emergence of resistance to ß-lactams, aminoglycosides and quinolones during combination therapy for infections due to *Serratia marcescens*. J Infect Dis 153: 617-619

Sanders CC, Sanders WE, Goering RB, Werner V (1984) Selection of multiple antibiotic resistance by quinolones, ß-lactams and aminoglycosides with special reference to cross-resistance between unrelated drug classes. Antimicrob Agents Chemother 26:797-801

Scully BE, Parry MF, Neu HC, Mandell W (1986) Oral ciprofloxacin therapy of infections due to *Pseudomonas aeruginosa*. Lancet 2: 819-822

Smith JT (1983) Mechanisms of drug resistance. In:DI Edwards,DR Hiscock (eds) Chemotherapeutic
    Strategy, Macmillan Press, London, pp 79-83
Smith JT (1984) Awakening the slumbering potential of the 4-quinolone antibacterials. Pharm J
    233: 299-305
Smith JT (1986) Frequency and expression of mutational resistance to the 4-quinolone antibacterials.
    Scand J Infect Dis 49:115-123
Smith JT, Ratcliffe NT (1986) Einfluss von pH-wert und magnesium auf die antibakterielle aktivitat von
    chinolonpraparatem. Infection 14:(Suppl 1) 31-35
Uhlin BE, Nordstrom K (1985) Preferential inhibition of plasmid replication by altered DNA gyrase
    activity in *Escherichia coli*. J Bacteriol 162:855-857
Wang JC (1974) Interactions between DNAs and enzymes: The effect of superhelical turns. J Molec
    Biol 87:797-816
Wang JC (1985) DNA topoisomerases. Ann Rev Biochem 54:665-697
Weisser J, Weidemenn B (1986) Elimination of plasmids by ofloxacin and enoxacin at near inhibitory
    concentrations. J Antimicrob Chemother 18:575-583
Weisser J, Weidemenn B (1987) Inhibition of R-plasmid transfer in *Escherichia coli* by 4-quinolones.
    Antimicrob Agents Chemother 31:531-534
Yamagishi J, Yoshida H, Yamayoshi M, Nakamura S (1986) Nalidixic acid-resistant mutations of the
    gyrB gene of *Escherichia coli*. Mol Gen Genet 204:367-373

Discussion Summary

The worrying question of the stability of mutants isolated as resistant to the newer
4-quinolones was discussed extensively with reference to this presentation and other
presentations. The clear differences between mutants isolated as resistant to new and old
4-quinolones could not be explained except by suggesting that some gyrase mutations
alter DNA superhelicity, whilst others appear to have no effect - consequently those
with altered superhelicity are likely to have altered gene expression control and be either
unstable or non-competitive, whilst those with no apparent change in superhelicity are
stable and grow near normally. If we suggest that the new agents tend to select mutants
with no obvious change in DNA structural control then this might explain the
differences.

The use of different levels of selecting drug (relative to the MIC of the parent) had
allowed Prof Smith to examine the question of whether resistance to low levels of a
4-quinolone predisposed the mutant to mutate to resistance to higher levels of the same
agent. The fact that the low-level resistant mutants for most species examined showed
greatly elevated frequencies of mutation to resistance to high levels of the drugs was
worrying in itself, but the fact that these elevated frequencies were much higher that the
basic mutation frequency for simple mutation to nalidixic acid resistance compounded
the concern.

The worries over the incidence and stability of resistance to the 4-quinolones was
however balanced by comments upon the observations of changes in potential
pathogenicity markers - particularly changes in coagulase production in *S. aureus* . It
was reported that these laboratory observations were effectively mirrored by clinical
studies in the US which indicated that resistance to ciprofloxacin and norfloxacin arose
at high frequencies whilst resistance to ofloxacin did not. Since Prof Smith had shown
that ciprofloxacin-resistant mutants remained coagulase positive (pathogenic) whilst a
high percentage of the ofloxacin-resistant mutants were coagulase-negative (non-
pathogenic) the clinical results are not surprising because only those mutants which are
still pathogenic are observable in the clinical situation - resistant mutants which are
non-pathogenic will behave as though they are still sensitive (i.e. the patient gets

better). Such changes in pathogenicity associated with the development of resistance was considered to be a still much under-studied area, and might well represent an important counterbalance to the development of resistance. Furthermore, in the *in vitro* study of resistances, changes in pathogenicity markers have to be taken into consideration before we can correlate *in vitro* data with *in vivo* situations with any validity.

Chapter 15

Measurement of the Frequency of Mutation to Resistance: The Importance of Testing under Physiological Conditions

M. Odell and G. C. Crumplin

It is a natural consequence of the extensive clinical use of any antibacterial agent that resistant organisms will be selected and that in the long term such selection of resistant organisms represents a significant constraint upon the clinically useful lifespan of an antibacterial agent. In fact it has even been said that "drug resistance has accompanied the development of chemotherapy like a faithful shadow and the history of chemotherapy is also a history of drug resistance" (Schnitzer and Grunberg 1957). For the majority of available antibacterial agents the origins of the resistance determinants may stem from either the spontaneous mutation of chromosomal markers or from the selection and disemination of plasmid-borne genes. The 4-quinolone antibacterial agents are unusual in that despite over 25 years of clinical use, there is still no substantiated incidence of a plasmid-borne marker conferring resistance to clinically obtainable levels of these agents. Hence, any consideration of the nature and development of resistance to this group of agents requires only examination of strains carrying chromosomal mutations. In the study of resistance to 4-quinolones, whether it be resistance to the archetypal molecule nalidixic acid or resistance to the new generations of 4-quinolones like ciprofloxacin, ofloxacin and norfloxacin, we therefore only have to consider the likely frequency of the selection of spontaneous mutants to resistance and the likely phenotypic properties of such resistant mutants.

In this context the new generations of 4-quinolones have to be compared with a structural analogue which has attained a long-standing reputation for the ease of the development of resistance *in vivo* (Ronald *et al.* 1966). This comparable agent which we must use as the benchmark is of course nalidixic acid. In order to be considered as a serious candidate for full development as a clinically useful antibacterial agent any new 4-quinolone must be shown to carry a significantly lower risk of the development of resistance than does nalidixic acid. The important question we must address is one of practicality, that is how best to carry out this form of risk assessment *in vitro* in a way which will provide data which relate properly to the situation which will exist in the *in vivo* situation? To this end we have in the past tailored the methodology to fit the convenience of the investigator by estimating spontaneous mutation frequencies under conditions of optimal growth of the bacteria in question - e.g. when using *Escherichia coli* all experiments are generally carried out with growth in rich and complete medium under aerobic conditions at 37°C. Subconciously we are of course making the assumption that for comparative purposes the infected human body can be considered as environmentally homogeneous.

Table 15.1 Spontaneous mutation frequencies to 4-quinolone resistance in *E.coli* KL16 (*wt*) with culture and selection carried out aerobically at 37°C

Selective agent	Spontaneous mut. freq.	Selective agent	Spontaneous mut. freq.
Nalidixic acid	5.0×10^{-8}	Oxolinic acid	1.0×10^{-9}
Cinoxacin	5.5×10^{-9}	Piromidic acid	3.7×10^{-8}
Pipemidic acid	8.9×10^{-9}	Acrosoxacin	1.55×10^{-8}
Flumequine	7.77×10^{-9}	Norfloxacin	1.1×10^{-11}
Enoxacin	1.1×10^{-11}	Amifloxacin	2.2×10^{-11}
Ofloxacin	1.1×10^{-11}	Ciprofloxacin	1.1×10^{-11}
S-25930	$<1.0 \times 10^{-12}$	S-25932	$<1.0 \times 10^{-12}$

Under such environmentally convenient conditions with *E.coli*, estimates of spontaneous mutation frequency to resistance to levels of a 4-quinolone $\geq$10X the MIC of the parent organism routinely yield figures not dissimilar to those shown in Table 15.1 for mutation in *E.coli* strain KL16. Here it can be seen that whilst the older generations of 4-quinolones yield mutation frequencies in the range 10^{-8} to 10^{-9}, resistance to the new 4-quinolones like ciprofloxacin, ofloxacin and norfloxacin occurs at much lower frequencies. On the basis of such evidence it would seem quite reasonable to predict that resistance to the new 4-quinolones is much less likely to occur (even under clinical conditions) than is the case with the older 4-quinolones. However, we should remember that such experiments in *E. coli* are not simply comparisons of mutation frequencies at a single genetic locus. As is shown in Figure 15.1 a large number of mutations in a wide variety of genes have been identified as contributing to the determination of the susceptibility of this species to the challenge of 4-quinolone antibacterial agents, hence it is quite likely that the mutation frequencies measured may reflect frequencies of both single mutational events and mutations at two or more loci. In statistical terms this makes an objective comparison difficult to make since we have no information available to indicate the proportions of single, double or even triple mutations in our samples. Despite this limitation upon the absolute validity of simple laboratory estimations, clinical experience gained with the new 4-quinolones does in general support the indication that resistance to the new generations of 4-quinolones is likely to arise much less often than does resistance to nalidixic acid.

However, close examination of the clinical literature on the effectiveness of 4-quinolones like ofloxacin and ciprofloxacin does yield indications that *in vivo* the situation might not be as straightforward as we would like to believe it to be. For example the case of ciprofloxacin when used in the treatment of *Pseudomonas* infections in patients with cystic fibrosis. In this case the *in vitro* estimates of the frequency of mutation to resistance in clinical isolates is in the order of 1×10^{-9} to

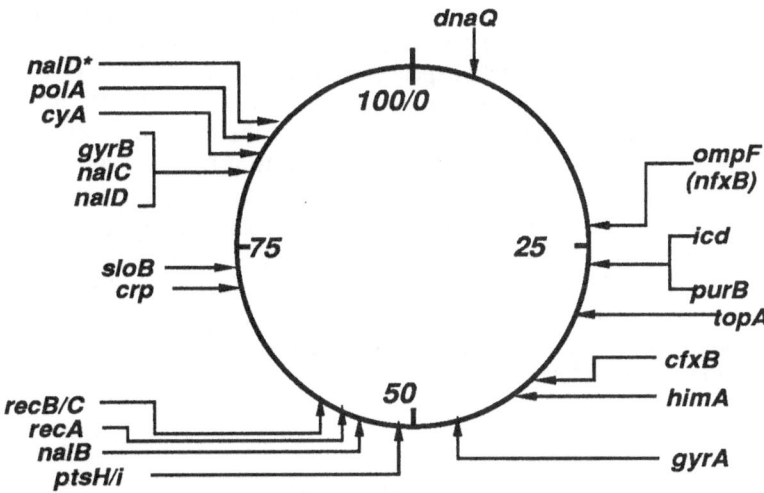

Fig. 15.1. Genetic map of *Escherichia coli* K12 chromosome showing location of mutations which affect the susceptibility of *E. coli* to the toxic effects of 4-quinolones.

1×10^{-10} for both *Ps. aeruginosa* and *Ps. cepacea* isolated from cystic fibrosis patients and for *Ps. aeruginosa* isolated from urinary tract infections in catheterized patients. In contrast to these *in vitro* estimates the clinical reports for the same species of organisms show resistant mutants arising in up to 37% of cases in cystic fibrosis (Roberts *et al.* 1985) but for *Pseudomonas aeruginosa* infections in other body sites the incidence of resistance is generally below 10%. Clearly the *in vitro* indications of a consistently low (and acceptable) frequency of mutation do not correlate with the clinical observations of markedly different incidences of resistance in different infection sites. Whilst it is obvious that the environment presented to *Pseudomonas* in the lungs of cystic fibrosis patients is rather special and differs significantly from environments available at other body sites, it seems equally obvious that the environmental differences are having a significant influence upon the selection of mutants resistant to ciprofloxacin. Inferential evidence of this nature has prompted us to consider the possibility that the environment under which we select mutants resistant to the toxic effects of the 4-quinolones may influence the apparent frequency of mutation to resistance.

Clearly the body of *Homo sapiens* does not represent a homogeneous environment for colonisation by bacteria, furthermore the local symptoms consequent upon acute infection may also differ from those prevailing prior to bacterial colonisation. Even at the most simplistic level there are significant variations in local conditions like temperature, oxygenation, pH, ionic strength, levels of nutrients and levels of metal ions - all of which may be influential upon the absolute bactericidal activity of the 4-quinolones when assessed *in vitro* (Ratcliffe and Smith 1983; Smith and Ratcliffe 1986). For example the lumen of the alimentary tract is largely anaerobic whilst other body sites are well oxygenated and *E. coli* in the gut lumen must exist under anaerobic conditions whilst the same organism in other body sites may have dissolved oxygen readily available. Consequently it would seem reasonable to suggest that we should consider carrying out *in vitro* estimations of spontaneous mutation frequency under conditions which reflect those likely to prevail at the site of an infection. From our understanding of basic biology we would expect that in the absence of environmental mutagenic factors either the spontaneous mutation frequency (SMF) *per se* approximates to a constant (K_1) or that at the very least the relative spontaneous mutation frequencies within a single gene to resistance to different 4-quinolones (i.e. ratio of SMFcip:SMFnal) approximates to a constant (K_2). The use of the comparison of spontaneous mutation frequencies derived from single standard condition studies is dependent upon the presumption of the existence of K_1 and/or K_2 as the basis of validity. In the light of such an assumption of constancy and the clinical indication of possible differences in mutation frequency in different body locations, we are at present undertaking an examination of the apparent validity of these constants with specific reference to the 4-quinolones.

The Effects of Temperature Variation

It has previously been shown that *E. coli* mutants resistant to certain 4-quinolones can be isolated much more readily in selections carried out at 30°C than in selections carried out at either 37°C or 25°C (Smith 1986). We have carried out isolations of mutants resistance to nalidixic acid, norfloxacin, pefloxacin, ofloxacin and ciprofloxacin in *Escherichia coli* CSH50, *Klebsiella aerogenes*, *Klebsiella pneumoniae*, *Proteus mirabilis* and *Serratia marcescens* with selection carried out at 32°C and 37°C. In all cases the starting culture from which selections were carried out was grown at 32°C with selection of mutants resistant to ≥10X the MIC. Attempts at isolating mutants from 10^{12} cfu of each species showed that for *E. coli* and *P. mirabilis* norfloxacin-resistant mutants could only be isolated at 32°C, pefloxacin-resistant mutants of *P. mirabilis* could only be isolated at 37°C, whilst in *S. marcescens* mutants resistant to nalidixic acid or pefloxacin could only be isolated at 32°C. Apart from these absolute distinctions, as can be seen in Table 15.2, it was in general easier to isolate 4-quinolone-resistant mutants at 32°C than it was if the selection was carried out at 37°C. Virtually identical results were obtained when the experiment was carried out with starter cultures grown at 37°C, thus indicating that the apparent constraint upon mutation frequency was the temperature of selection rather than the temperature at which the culture was grown. In such a selection system the bacteria are grown to a high cell density in the absence of any drug challenge before washing and plating out upon selective solid media. Since the challenge with the 4-quinolones is a potent bactericidal challenge, it is reasonable to presume that the mutational events which give rise to the selected resistant cells must have occured as spontaneous events during growth in the drug-free medium. Hence similar numbers of mutants available for selection will have been plated out onto the selective media for incubation at each temperature. This simple statistical consideration serves to emphasize that temperature can exert a marked influence upon the efficiency of 4-quinolones as selective agents. It is possible that this might reflect differences in the antibacterial potency of 4-quinolones at different temperatures both in terms of the MIC and the bactericidal potency. This possibility was investigated by measuring the MICs and monitoring bactericidal (as distict from bacteriostatic) activity at different temperatures. In no case was there any significant difference in either the MIC or the bactericidal potency which would account for the clear differences in the efficiency of selection of resistant mutants at the different temperatures.

Table 15.2 Spontaneous mutation frequencies to 4-quinolone resistance at 32°C and 37°C for five different species (figures derived from 10^{12} cfu. plated)

Bacterial species	Nalidixic acid		Norfloxacin		Ciprofloxacin		Pefloxacin		Ofloxacin	
	32°C	37°C	32°C	37°C	32°C	37°C	32°C	37°C	32°C	37°C
E. coli	10^{-8}	10^{-9}	10^{-10}	-	10^{-10}	10^{-11}	10^{-9}	10^{-10}	10^{-10}	10^{-11}
P. mirabilis	10^{-7}	10^{-9}	10^{-10}	-	10^{-10}	10^{-11}	-	10^{-9}	10^{-8}	10^{-10}
K. aerogenes	10^{-8}	10^{-9}	10^{-7}	10^{-9}	10^{-8}	10^{-9}	10^{-8}	10^{-10}	10^{-8}	10^{-10}
K. pneumoniae	10^{-7}	10^{-9}	10^{-9}	10^{-10}	-	-	10^{-9}	10^{-10}	10^{-10}	10^{-11}
S. marcescens	10^{-7}	-	10^{-10}	10^{-11}	10^{-10}	10^{-11}	10^{-9}	-	10^{-10}	10^{-11}

This preliminary examination of five different species of organism for mutation to resistance to five different 4-quinolones suggest that such variations are not confined to laboratory strains of E. coli and may in fact represent a fairly generalized phenomenon. This suggestion is further supported by measurements of more precise spontaneous mutation frequencies for mutation to ciprofloxacin-resistance in clinical isolates of Ps. aeruginosa and Ps. cepacea from cystic fibrosis patients showed that with selection at 37°C mutation frequencies were 1.5 x 10^{-10} and 6.0 x 10^{-10} respectively, and at 32°C they were 7.0 x 10^{-8} and 5.5 x 10^{-8} respectively. It might however be suggested that such apparent variation of mutation frequency with temperature of selection might be either a generalized phenomenon for all antibacterial agents or uniquely confined to the five 4-quinolones examined. In order to test these possibilities we carried out an extensive screen with S. marcescens using 25 different 4-quinolones and rifampicin and streptomycin as selective agents used at 25°C, 32°C, 37°C and 42°C. Figure 15.2 shows the results obtained with eight of the 4-quinolones and with coumermycin, rifampicin and streptomycin. As can be clearly seen, with the 4-quinolones there is a clear indication that selection at 32°C clearly favours the isolation of 4-quinolone-resistant mutants whilst there is no indication of an optimal temperature for the selection of mutants resistant to coumermycin, streptomycin or rifampicin. Essentially identical results were obtained for all of the 4-quinolones examined (see list appended to Fig. 15.2), thus indicating that the influence of temperature of selection was applicable to all 4-quinolones but not to the two unrelated antibacterial agents (rifampicin and streptomycin) or to coumermycin which is an inhibitor of the ATPase function of DNA gyrase. It seems evident from these studies that with specific reference to the 4-quinolones the apparent mutation frequency to resistance is not a constant and is related to the temperature at which the selection is carried out.

It is not easy to offer any rational explanation for these effects of the selection temperature particularly when we take into consideration the detailed phenotypes of some of the mutants. A particularly perplexing case relates to the selection of norfloxacin-resistant mutants of E. coli KL16. Whilst we were only able to isolate such mutants at 32°C, the mutants were equally resistant to the drug when subsequently tested at 37°C even though they could not be isolated at this temperature. At the present time we cannot offer any explanation of our inability to carry out the initial isolation at 37°C - clearly there is a need for extensive further study of this phenomenon.

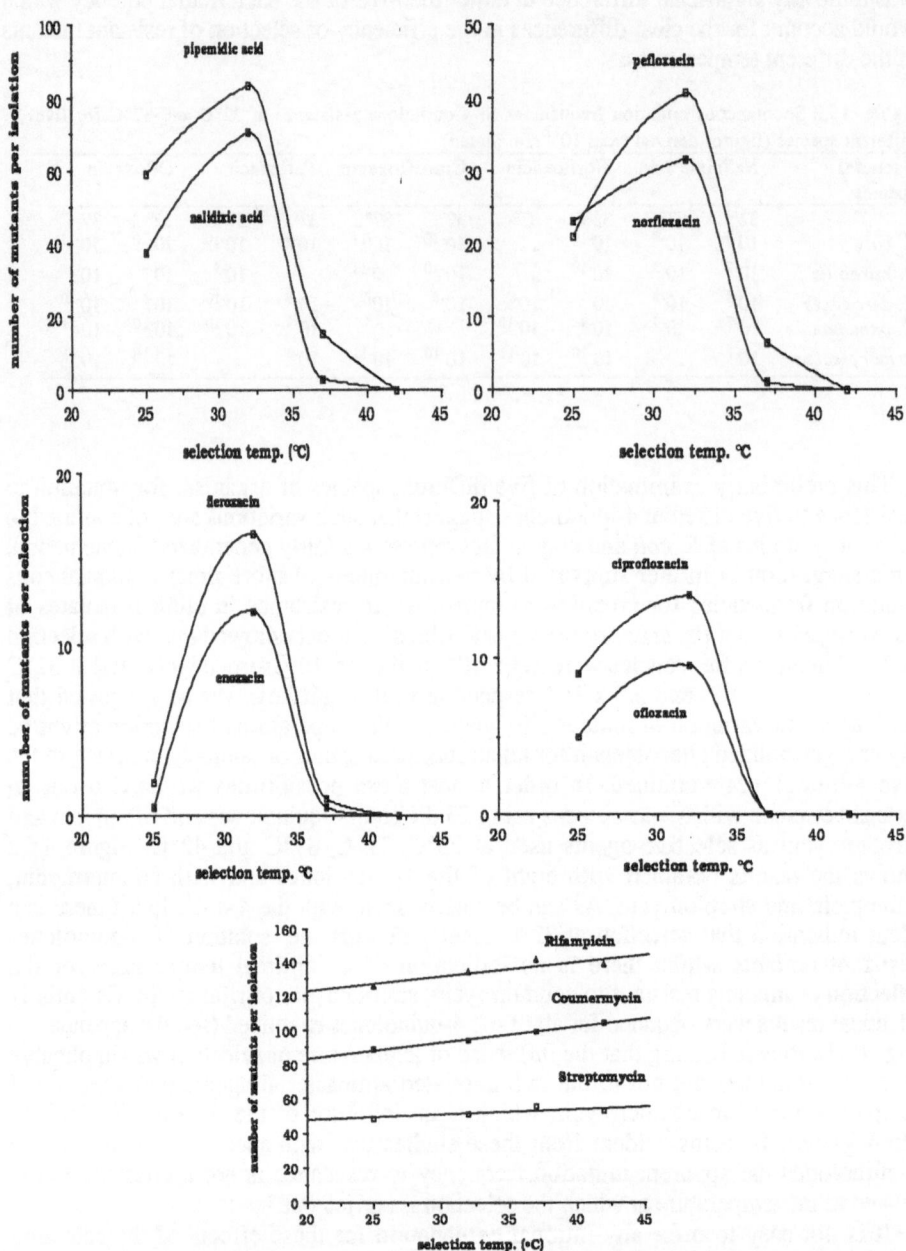

Fig. 15.2. The effects of selection temperatures upon the yield of mutants of *Serratia marcescens* resistant to the antibacterial effects of 4-quinolones and other unrelated agents.[4-quinolones used in the *Serratia* study and yielding results showing the same pattern of isolation yield as shown for eight 4-quinolones:-nalidixic acid, oxolinic acid, cinoxacin, piromidic acid, pipemidic acid, amifloxacin, acrosoxacin, S-25932, ciprofloxacin, norfloxacin, pefloxacin, enoxacin, CI-934, ofloxacin, fleroxacin, temafloxacin, difloxacin, hydroxy-nalidixic acid, CP-67015, flumequine (recaemic mixture and individual stereoisomers), and S-25930 (recaemic mixture and individual stereoisomers)].

The Effects of Aerobic and Anaerobic Environments

Examination of the possible influence of the availability of oxygen upon the frequency of mutation to resistance has been carried out in *E. coli* strain KL16 with particular reference to nalidixic acid and ofloxacin (Crumplin and Odell 1987). In this instance mutation frequencies were measured in cultures grown at 32°C under aerobic and anaerobic conditions with the selections carried out under the same conditions as the

Table 15.3. Spontaneous mutation frequencies of *Escherichia coli* strain KL16 under aerobic and anaerobic conditions

| Selection conditions | Nalidixic acid 50μg/ml | Drugs used to select resistant mutants | | | |
		Ofloxacin 2μg/ml	Streptomycin 100μg/ml	Rifampicin 100μg/ml	Ampicillin 40μg/ml
Aerobic	5×10^{-8}	$<3.4 \times 10^{-12}$	1.7×10^{-9}	4.85×10^{-8}	3.80×10^{-8}
Anaerobic	3×10^{-7}	1.2×10^{-7}	2.4×10^{-9}	2.50×10^{-8}	1.02×10^{-6}

growth of the starter culture. As can be seen in Table 15.3 the frequency of mutation to ofloxacin was four-five orders of magnitude higher under anaerobic conditions whilst there was not the same variation for the selection of mutants resistant to nalidixic acid. Comparative studies of the frequency of mutation to resistance to other unrelated agents (streptomycin, rifampicin and ampicillin) showed that for non-4-quinolones there was no significant variation (see Table 15.3). The results obtained with reference to nalidixic acid and ofloxacin show two interesting features which are of some significance:- Firstly, that mutation to ofloxacin-resistance is much higher under anaerobic conditions - suggesting that spontaneous mutation frequency is not a constant; and secondly that the relative spontaneous mutation frequencies for resistance to nalidixic acid and ofloxacin are not constant since under aerobic conditions

$$SMF^{nal} : SMF^{ofx} = 1.47 \times 10^4 : 1$$

and under anaerobic conditions

$$SMF^{nal} : SMF^{ofx} = 2.5 : 1.$$

These results clearly suggest that both of the approximate constants defined earlier are far from being constants, with reference to mutation to resistance to the 4-quinolones, but seem to be valid when considering mutation to resistance to unrelated antibacterial agents.

In order to establish whether in the case of the influence of availability of oxygen the variation is influenced primarily by the selection conditions or the growth conditions (since this is where the mutations isolated actually occur) we monitored the frequency of selection of mutants resistant to ofloxacin at 32°C in *E.coli* strain KL16 during the metabolic transition of aerobically grown cells to growth under anaerobic conditions. Figure 15.3 shows that with optical-density monitoring of the culture this strain adapted very slowly to the imposition of stringent anaerobic conditions with a resumption of growth being established after about 40 h (parallel viable count monitoring confirmed that growth and division recommenced at this time). Samples were removed from the culture at the timed indicated (I, II, and III) and estimates made of the incidence of ofloxacin-resistant organisms. As can be seen from the frequency figures shown on this figure, significant numbers of resistant mutants were only

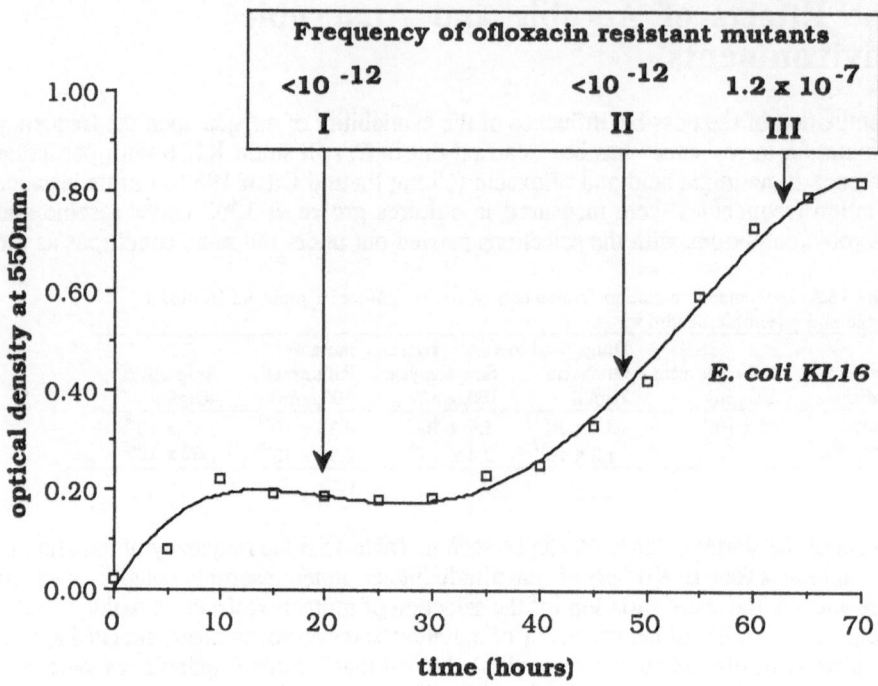

Fig. 15.3. The frequency of isolation of ofloxacin-resistant mutants during the adaptation period when aerobically grown *E. coli* are switched to a stringent anaerobic environment.

detected in the sample taken at time III when anaerobic growth was well established. The frequencies of ofloxacin-resistant mutants shown in this figure are derived from selections carried out under anaerobic conditions at 32°C. However, selections carried out under aerobic conditions at 32°C showed no significant differences. These results suggest that with respect to the influence of oxygenation upon the selection of ofloxacin-resistant mutants, the conditions prevailing during the growth of the cells is the limiting factor rather than the selection conditions as evident with respect to temperature variation.

In the light of the above observations of the variation in frequency of mutation to resistance to the 4-quinolones under conditions of varying temperature and oxygenation it is evident that the possible influences of other environmental variables should be investigated.

The Effect of Environment upon Resistant Phenotype

The study of resistant mutants isolated *in vitro* as models of potential mutants likely to arise in the clinical situation is based upon the presumption that similar mutants are likely to arise and be selected under similar drug-level selections irrespective of whether the selections are carried out *in vitro* or *in vivo*. Recent studies of 4-quinolone-resistant mutants clinical isolates have confirmed that these are *gyrA* mutants as are the overwhelming majority of *in vitro* isolates. However, no close comparison of mutants selected from a single strain *in vitro* and *in vivo* has been made so we still cannot be totally certain that mutants selected *in vitro* are valid models of clinical resistance. We have therefore attempted to begin to test the validity of such modelling by examining the phenotypes of 4-quinolone-resistant mutants isolated under different environmental conditions by examining the cross-resistance phenotypes of mutants of different species of Gram negative organisms isolated at different temperatures from the same initial culture.

Table 15.4. The effect of temperature of isolation on cross-resistance between ciprofloxacin and nalidixic acid (relative susceptibilities estimated by zone-radius around disc)

(a) *Proteus mirabilis*					
wt zone radius	Nalidixic acid = 13.0 mm		Ciprofloxacin = 12.00		
Isolated at 32°C	NAL-zone	CIP-zone	Isolated at 37°C	NAL-zone	CIP-zone
NALR	5.2	6.4	NALR	1.0	14.4
CIPR	3.8	4.2	CIPR	5.0	6.6
(b) *Klebsiella pneumoniae*					
wt zone radius	Nalidixic acid = 6.4 mm		Ciprofloxacin = 12.6		
Isolated at 32°C	NAL-zone	CIP-zone	Isolated at 37°C	NAL-zone	CIP-zone
NALR	4.4	9.6	NALR	4.2	14.8
(c) *Serratia marcescens*					
wt zone radius	Nalidixic acid = 10.0 mm		Ciprofloxacin = 15.4		
Isolated at 32°C	NAL-zone	CIP-zone	Isolated at 37°C	NAL-zone	CIP-zone
CIPR	5.2	8.0	CIPR	4.8	5.8

discs = Nalidixic acid 30µg   Ciprofloxacin 5µg

Cross-resistances between nalidixic acid and ciprofloxacin were estimated by the measurement of zones of inhibition around drug discs for mutants isolated at either 32°C or 37°C. Table 15.4 shows that for *P. mirabilis* nalidixic acid-resistant mutants isolated at 32°C were clearly resistant to both ciprofloxacin and nalidixic acid, whilst those isolated at 37°C were not only resistant to higher levels of nalidixic acid, but were also hypersensitive to ciprofloxacin. In the case of mutants originally isolated as ciprofloxacin-resistant, those isolated at 32°C were consistently resistant to higher levels of both drugs than were mutants isolated at 37°C. In *K. pneumoniae* mutants isolated as nalidixic acid-resistant at 32°C displayed a low level of resistance to ciprofloxacin whilst those isolated at 37°C were to some extent hypersensitive to ciprofloxacin. In *S. marcescens* mutants isolated on ciprofloxacin at 37°C were resistant to higher levels of both ciprofloxacin and nalidixic acid than were mutants isolated on ciprofloxacin at 32°C. These data are representative of all of the mutants isolated in this study and in all cases represent at least 10 mutants. However, such differences were not apparent when other 4-quinolones were used as the initial selective agent so whilst there are some differences in the nature of the mutants isolated under different temperature conditions there is no clear pattern as yet and, on the basis of this preliminary study, we can only suggest that *in vitro* selected mutants might not be truly representative of mutants derived under the different environmental conditions operating *in vivo*. Once again this suggests that further studies must be carried before we can be sure that *in vitro* derived mutants are, or are not, valid models of resistant mutants which are selected under clinical conditions.

Conclusion

The process of the selection of resistant mutants represents the summation of three independent variables; namely the frequency of the occurence of mutational events associated with the relevant gene(s); the frequency of such mutants which are "viable"; and the proportion of the viable mutants which are capable of surviving the level of challenge under the selection conditions. In the context of the selection of mutants resistant to a clinically used antibacterial agent we should of course be particularly interested in the selection of resistant mutants *in vivo* rather than under the artificial conditions imposed for convenience in the routine laboratory. From the preliminary data presented here it seems likely that the selection of mutants resistant to the toxic effects of 4-quinolones *in vivo* involves environmental factors which we have not previously considered. Consequently the mutation frequency *per se* and the relative mutation frequency (relative to mutation to resistance to other drugs) are far from simple expressions which can be derived from crude *in vitro* studies if we wish to relate them to potential frequencies in the clinical situation. The above results also include evidence to suggest that the selective "power" of any individual 4-quinolone is also subject to environmentally influenced variation.

Whilst our instincts would suggest that the estimation of the likelihood of mutation to resistance to a 4-quinolone in a particular infection *in vivo* should be carried out with the appropriate species of pathogen under the conditions likely to prevail at the site of infection - it has not been our practice to follow our instincts. Just because such environmental influences have not impinged upon our consideration of non-4-quinolone antibacterial agents we have assumed that they should have no effect upon such examination of 4-quinolones. Clearly this is not the case with environmental changes *in vitro* and I would therefore suggest that the site of infection environment may play as important a role in the selection of resistant mutants as it has already been shown to do with regard to the antibacterial activity of these agents in urine as compared to routine *in vitro* conditions (Smith and Ratcliffe 1986).

References

Crumplin GC and Odell M (1987) Development of resistance to ofloxacin. Drugs (suppl 1) 34:1-8

Ratcliffe NT and Smith JT (1983) Effects of magnesium on the activity of 4- quinolone antibacterial agents. J Pharm Pharmacol 35 pp 61

Roberts CM, Batten J and Hodson ME (1985) Ciprofloxacin-resistant *Pseudomonas*. Lancet i:1442

Ronald AR, Turck M and Petersdorf RG (1966) A critical evaluation of nalidixic acid in urinary tract infections. New England J Med 297:1081-1089

Schnitzer RJ and Grunberg E (1957) Drug Resistance of Micro-organisms. Pub Academic Press NY pp1

Smith JT (1986) Frequency and expression of mutational resistance to the 4-quinolone antibacterials Scand J Infect Dis (suppl) 49:115-123

Smith JT and Ratcliffe NT (1986) Einfloss von pH-wert und magnesium auf die antibakterielle activitat von chinolonproparoten. Infection14, suppl 1:31-35

Discussion Summary

The demonstration of significant variations in the apparent mutation rates to resistance to the 4-quinolones associated with variation of the growth and/or selection environment was considered to be a significant observation. It seemed to represent a potential explanation for a number of clinical experiences where resistance seemed to be a significant problem in certain patient populations and not in others. However, no truly quantitative data seemed to have been correlated from clinical studies. It was considered to be an area which was well worth investigating because of the scale of the variations observed in these preliminary *in vitro* studies. Whilst many workers would not have been at all surprised by a small, but statistically significant, variation, no-one could have anticipated variation by several orders of magnitude.

The discussions were principally confined to re-affirmations of the methodologies used in the studies until all (including the authors) were quite satisfied that there was no obvious artefactual situation. At this point it was suggested that we should view studies of the development of resistance carried out under the usual laboratory conditions with a degree of caution as we could not be certain whether such studies, when related to the clinical situation, represented under-estimations or over-estimations of the likelihood of the development of resistance. Clearly the 4-quinolones had once again revealed unexpected features which differentiated them from other antibacterial agents and, despite the volume of studies carried out over many years, we still could not predict the outcome of extensive clinical usage in all types of infection.

The possibility that the likelihood of resistance developing may vary between different body sites, and that we could not, with any certainty, extrapolate between different species, generated an air of insecurity. However, this could not dispell the general concensus amongst the clinicians attending that the new generation 4-quinolones clearly displayed a lower risk of clinical failure through resistance development than did nalidixic acid which was still perceived as presenting a significant risk of the development of resistance. In the light of these studies, and observations reported by Professor Smith, we clearly still had a lot to learn about both the development of resistance and the consequences (long-term and short-term) of resistance development.

Chapter 16

Plasmid-Borne Resistance to 4-Quinolones a Real or Apparent Absence?

P. Courvalin, C. Poyart-Salmeron and E. Derlot

Introduction

The first quinolone, nalidixic acid, was introduced into clinical practice in the early 1960s. The 6-fluorinated quinolones, which are significantly superior, in particular in intrinsic activity and antibacterial spectrum, have been increasingly used since the 1970s. It is the purpose of this review to analyse why, in the early 1990s, plasmid-mediated resistance to quinolones has not yet been proven.

Apparent Absence of Plasmid-Mediated Resistance to Quinolones

It is a general rule that, following introduction into medical practice of a new antibiotic, sooner or later plasmid-mediated resistance to this compound emerges. Bacterial resistance of this type is likely to preceed the launching of the "new" molecule but the use (often indiscriminate) of the antibiotic favours dissemination and hence detection of resistance. Also, and quite obviously, availability of the drug is a prerequisite for screening and finding, by the medical microbiologist, of the "new" resistance phenotype. As usual, there are a few exceptions to the rule (Table 16.1) which include quinolones.

Table 16.1. Antibiotics with no known plasmid-mediated resistance

Nitrofurans	
Novobiocin	
Polypeptides	bacitracin
	colistin
	polymyxin B
Quinolones	
Rifampicin	

It is noteworthy that novobiocin (coumermycin), which also inhibits DNA gyrase, belongs to this group. There has been a single report of plasmid-mediated resistance to nalidixic acid in bacteria (Munshi *et al.* 1987). A *Shigella dysenteriae* type I isolate, responsible for an epidemic in southern Bangladesh, was found to harbour a 30-kilobase plasmid conferring resistance to nalidixic acid only. However, upon careful reinspection of the data, it appeared that nalidixic resistant mutants of the recipient rather than authentic transconjugants were obtained. This observation, possibly due to the fact that the plasmid acts as a mutator factor specific for nalidixic acid resistance (ZU Ahmed 1989, personal communication) confirms that chromosomal mutation is the mechanism responsible, so far, for bacterial resistance to quinolones.

Is there a (Bacterial) Need for Plasmid-Mediated Resistance to Quinolones?

As already mentioned, bacteria may be resistant to quinolones, as well as to other groups of antibiotics (Table 16.2), following chromosomal mutations.

Table 16.2. Antibiotics to which bacteria are resistant by chromosomal mutations

Aminoglycosides	
ß-lactams	
Chloramphenicol	
Erythromycin	
Fosfomycin	
Fusidic acid	
Novobiocin	
Polypeptides	bacitracin
	colistin
	polymyxin B
Quinolones	
Rifampicin	

Certain mutations (e.g. *gyrA*) lead to cross-resistance to all commercially available quinolones, and to quinolones only. The degree of cross-resistance depends upon the intrinsic activity of the quinolone considered: the higher the activity, the lower the level of resistance. This observation implies that "genetic" and "clinical" resistance do not always correlate. Mutations of this class, which are distinct, although clustered (Yoshida *et al.* 1988), are stable and not deleterious to the host. Considering the heritable character of chromosomal mutations, one can easily imagine that such mutants do not require transferable resistance to survive and disseminate.

Other mutations (e.g. *nfxB* and *norC*) confer resistance to quinolones and to structurally unrelated antibiotics such as ß-lactams, chloramphenicol, tetracycline, and trimethoprim (Hirai *et al.* 1986; Hooper *et al.* 1986). These permeability mutants are resistant to low levels of quinolones but can be selected for by the other groups of antibiotics. Even more strikingly, *marA* mutants resistant to fluoroquinolones can be isolated at a 1000-fold-higher frequency on tetracycline and chloramphenicol than on norfloxacin (Cohen *et al.* 1989).The *marA* mutants could, in turn, lead to second step mutants highly resistant to fluoroquinolones.

Taking these findings into account and considering the multiplicity of mutations leading to quinolone resistance (Table 16.3) it seems that there is no need for acquisition of extrachromosomal DNA for bacteria to cope with the increasing selective pressure exerted by quinolones.

Table 16.3. Quinolone resistance mutations in *E. coli*[a]

Selecting agent	Mutation
Nalidixic acid	*gyrA*
	gyrB
	nalB
	nalD
	crp
	cya
	icd
	purB
	ctr
Norfloxacin	*nfxB*
	norB[b]
	norC
Ciprofloxacin	*cfxB*[b]
Tetracycline or chloramphenicol	*marA*[b]

[a] Adapted from D. Hooper
[b] Possibly allelic mutations

Along this line, it is of interest to note that no plasmid-mediated resistance has developed towards antibiotics (quinolones, rifampicin, fusidic acid) exhibiting high levels and high frequencies of mutation. Although this does not apply to fosfomycin.

Nevertheless, acquisition of a plasmid conferring high level resistance to quinolones (and most likely to many other groups of antibiotics by different mechanisms) would give rise to quinolone resistant cells at a much higher frequency than chromosomal mutations, in particular when two independent events are required. It would also provide the recipient cell with a very efficient way to escape the bactericidal activity of a number of combinations of antibiotics.

Potential Mechanisms of Plasmid-Mediated Resistance to Quinolones

The predictable mechanisms of plasmid-borne resistance to quinolones are listed in Table 16.4.

Table 16.4. Conceivable mechanisms of plasmid-mediated resistance to quinolones

Type	Biochemical mechanism	Origin
By pass	Both subunits of DNA gyrase altered	Gram-positive bacteria
	Insensitive DNA gyrase	Naturally-resistant bacteria[a]
Drug detoxification	N-oxidation	Organism possessing a biotransformation pathway
Impermeability	Porin reduction	
		Naturally resistant bacteria
	Decreased import	
		Chromosomal mutation
	Increased efflux	

[a] e.g. *Acinetobacter baumannii*, *Clostridium difficile*, *Enterococcus* spp. *Pseudomonas cepacia*, *P. maltophilia*.

By-Pass

The by-pass mechanism consists in the presence, in the host bacterium, of an additional target insensitive to the antibiotic. A prerequisite, for this mechanism to occur, is that the plasmid-borne allele is dominant over its chromosomal counterpart. The results of transcomplementation tests (Table 16.5) indicate that plasmid-mediated resistance to quinolones cannot be achieved by providing either one of the subunits of the DNA gyrase altered.

Table 16.5. Dominance relationships among *gyr* mutations

Genotype	Phenotype
$gyrA^S$-$gyrA^R$	$gyrA^S$
$gyrB^S$-$gyrB^R$	$gyrB^S$
$gyrA^S$+$gyrB^S$-$gyrA^R$+$gyrB^R$	$gyrA^R$+$gyrB^R$a

[a] Speculated from Ukaba *et al.* 1989

Recessivity of the resistant mutant allele *versus* the wild-type (susceptible) chromosomal gene may also explain lack of plasmid-mediated resistance to rifampicin by altered RNA polymerase. However, on the basis of preliminary data (Ubukata *et al.* 1989), and despite the fact that the mechanism(s) of resistance to quinolones in naturally resistant bacterial genera or species is not known, one can envisage that a merodiploid encoding a susceptible and a resistant (in both subunits) DNA gyrase might be resistant to quinolones. A similar mechanism accounts, in the vast majority of the clinical isolates, for resistance to sulphonamide and trimethoprim, which, like quinolones, are synthetic compounds.

Drug Inactivation

As just mentioned, quinolones are entirely synthetic molecules and their presence in nature is therefore unlikely. Thus, it is difficult to conceive that bacteria have developed resistance by enzymatic modification to an antibiotic to which they have never been exposed.

There is increasing circumstancial evidence that antibiotic resistance plasmid genes in human pathogens originate in antibiotic producing microorganisms. Conjugal transfer of genetic material from prokaryotes to eukaryotes has recently been demonstrated (Heinemann and Sprague 1989). However, transfer in the opposite direction has not yet been reported. The apparent polarity of the DNA transfer does not favour the aquisition by bacteria of antibiotic resistance determinants from the producing organism *Homo sapiens*.

Cellular Impermeability

Decreased drug penetration can be achieved by several mechanisms (Table 16.4). It is unlikely that altered porins will be plasmid-borne. When chromosomally encoded, this mechanism confers only low level resistance to quinolones. If plasmid-mediated, the level of resistance should be even lower since the additional gene would have to compete with the chromosomal wild-type allele. The presence of the incoming gene on a multicopy plasmid would tend to increase, by favourable competition, the resistance level. However, *in vitro* cloning constructions of this sort indicate that the presence in a cell of large amounts of a cellular membrane structural component is often toxic for the host. Gene substitution, by homologous recombination, tend generally to be the final outcome of such unstable merodiploids. Finally, modified pre-existing cellular structures are rarely infectious.

Antibiotic resistance by impermeability is often associated with reduced growth yield and rate of the microorganism and, in competition studies with the parent in the absence of selective pressure, the resistant isolate is often rapidly outgrown (Courvalin *et al*.1985). It is not very tempting for a plasmid to pick up a gene conferring bacterial resistance to quinolones in these conditions. The new host would rapidly be counter-selected in the absence of antibiotic, a situation that, even nowadays, a bacterium can still experience.

Quinolones Oppose Plasmids

Certain resistance plasmids increase the quinolone susceptibility of the host (Crumplin and Smith 1981). Quinolones tend to eliminate plasmids (Table 16.6); they inhibit their replication (Uhlin and Nordstrom 1985) and their transfer.

Table 16.6. Plasmid curing agents

Intercalating dyes	Acridine orange
	Acriflavine
	Ethidium bromide
	Quinacrine
Inhibitors of DNA gyrase	Quinolones
	Novobiocin
	Coumermycin
Inhibitors of transcriptase	Rifampicin
Surfactants	Sodium dodecyl sulfate
Physical agents	Ultra-violet light (254 nm)
	Temperature (42°C)

In addition, certain *gyrB* mutations decrease the ability of bacteria to act as donors, recipients or to stably maintain extrachromosomal DNA (Friedman *et al*. 1984; Wolfson *et al*. 1982).

Prevention of Plasmid-Mediated Resistance to Quinolones: a Shakespearian Situation

Apart from the classical approaches to minimize emergence and subsequent spread of bacterial resistance to antibacterials (Wolfson and Hooper 1989), a question specific to this group of antibacterials remains: to use or not to use? Clearly intensive utilisation of these very potent antibacterials leads to a substantial increase in resistance rate (Wolff *et al*. 1985). However, considering the above mentioned activities of quinolones on plasmid physiology one may conceive situations where the nett result of usage of these drugs would lead to a decrease in incidence of resistance in the bacterial population. Unfortunately, efficiency of plasmid curing and inhibition of transfer is never 100%, in particular *in vivo* (e.g. animal models, Chabbert 1970). Because of high frequencies of intergeneric plasmid transfer, the increase in resistance rate is most often extremely rapid after cessation of therapy; the *in vivo* efficacy of the most potent plasmid curing agent is therefore most transient.

Antibiotic combinations may not constitute the best approach to the prevention of plasmid mediated-resistance to quinolones. As already pointed out, resistance plasmids often confer resistance to a large variety of structurally unrelated antibiotics. Combination therapy against strains harbouring such replicons turns out to actually be, most of the time, monotherapy.

Conclusion: une Aussi Longue Absence

We have seen, in this highly speculative review, that various factors can account for the apparent absence of plasmid-mediated resistance to quinolones in bacteria, thirty years after introduction of the first molecules in clinical settings. However, because of the many advantages of this family of drugs, there is a tremendous selective pressure towards resistance. A similar evolution has recently been observed with glycopeptides and 5-nitroimidazoles which, in both cases led, after a lag phase of several decades, to plasmid-mediated transferable resistance (Leclercq *et al.* 1988; Breuil *et al.* 1989).

References

Breuil J, Dublanchet A, Truffaut N, Sebald M (1989) Transferable 5-nitroimidazole resistance in the *Bacteroides fragilis* group. Plasmid 21:151-154

Chabbert YA (1970) Effet de l'acriflavine *in vivo* sur les entérobactéries hébergeant des résistances transférables. 6th International Congress of Chemotherapy, Tokyo, 1969. Progess in Antimicrobial and Anticancer Chemotherapy, Univ of Tokyo Press Edit, 733

Cohen SP, McMurry LM, Hooper DC, Wolfson JS, Levy SB (1989) Cross-resistance to fluoroquinolones in multiple-antibiotic-resistant *(Mar)* *Escherichia coli* selected by tetracycline or chloramphenicol: decreased drug accumulation associated with membrane changes in addition to *OmpF* reduction. Antimicrob Agents Chemother 33:1318-1325

Courvalin P, Derlot E, Chabbert YA (1985) Mécanismes et conditions d'émergence *in vitro* des résistances bactériennes aux quinolones. In: Pocidalo JJ, Vachon F, Regnier B (eds) Les Nouvelles Quinolones. Arnette, Paris, France, pp 73-82

Crumplin GC, Smith JT (1981) The effect of R-factor plasmids on host-cell responses to nalidixic acid I. Increased susceptibility of nalidixic acid-sensitive hosts. J Antimicrob Chemother 7:379-388

Friedman DI, Plantefaber LC, Olson EJ, Carver D, O'Dea MH, Gellert M (1984) Mutations in the DNA *gyrB* gene that are temperature sensitive for Lambda site-specific recombination, Mu growth, and plasmid maintenance. J Bacteriol 157:490-497

Heinemann JA, Sprague GF Jr (1989) Bacterial conjugative plasmids mobilize DNA transfer between bacteria and yeast. Nature 340:205-209

Hirai K, Aoyama H, Suzue S, Irikura T, Iyobe S, Mitsuhashi S (1986) Isolation and characterization of norfloxacin-resistant mutants of *Escherichia coli* K-12. Antimicrob Agents Chemother 30:248-253

Hooper DC, Wolfson JS, Souza KS, Tung C, McHugh GL, Swartz MN (1986) Genetic and biochemical characterization of norfloxacin resistance in *Escherichia coli*. Antimicrob Agents Chemother 29:639-644

Leclercq R, Derlot E, Duval J, Courvalin P (1988) Plasmid-mediated resistance to vancomycin and teicoplanin in *Enterococcus faecium*. N Engl J Med 319:157-161

Munshi MH, Sack DA, Haider K, Ahmed ZU, Rahaman MM, Morshed MG (1987) Plasmid-mediated resistance to nalidixic acid in *Shigella dysenteriae* type 1. Lancet 2:419-421

Ubukata K, Itoh-Yamashita N, Konno M (1989) Cloning and expression of the *norA* gene for fluoroquinolone resistance in *Staphylococcus aureus*. Antimicrob Agents Chemother 9:1535-1539

Uhlin BE, Nordström K (1985) Preferential inhibition of plasmid replication *in vivo* by altered DNA gyrase activity in *Escherichia coli*. J Bacteriol 162:855-857

Wolff M, Buré A, Pathé JP, Pangon B, Regnier B, Rouger-Barbier D, Vachon F (1985) Evolution des résistances bactériennes à la péfloxacine dans le service de réanimation de l'hôpital Claude Bernard. In: Pocidalo JJ, Vachon F, Regnier B (eds) Les Nouvelles Quinolones. Arnette, Paris, France, pp 213-226
Wolfson JS, Hooer DC, Swartz MN, McHugh GL (1982) Antagonism of the B subunit of DNA Gyrase eliminates plasmids pBR322 and PMG110 from *Escherichia coli*. J. Bacteriol 152:338-344
Yoshida H, Kojima T, Yamagishi J, Nakamura S (1988) Quinolone-resistant mutations in the *gyrA* gene of *Escherichia coli*. Mol Gen Genet 211:1-7

Discussion Summary

The range of possible ways in which a plasmid-borne gene (or group of genes) could effect a resistance to the toxic effects of the 4-quinolones had been clearly presented by Dr Courvalin, and there was some discussion as to whether all of the theoretical options were practicable with reference to this unusual group of agents. However, of greater concern was the fact that in the last two to three years there had again been published reports of R factor plasmids in *Shigella dysenteriae* which were transmissible and conferred resistance to nalidixic acid. Three aspects of these reports concerned many members of the audience:-

1   Was it possible that such resistances could also confer resistance to the newer generations of 4-quinolones which were superceding nalidixic acid? - It was however made clear by various workers who had had the opportunity to examine these plasmids that they did not confer such cross-resistance to clinical levels of the newer agents. Furthermore it seemed unlikely because it had already taken nature more than 25 years to find a way around nalidixic acid toxicity and the newer agents were not simply more potent agents, but also introduced additional mechanisms of action into the inhibitory equation.

2   Were these really plasmid-borne resistances to the 4-quinolones? In the light of the few earlier reports in the literature from Lebek, Jonson, and Babcock, none of which had been found to be anything more than the plasmid mobilisation of chromosomal genes, there was extreme scepticism. However, it was accepted that the most recent reports from Kashmir and Bangladesh did represent something particularly interesting and scientifically exciting. The few people who had attempted to repeat the initial observations had been unable to show clearly that there was a plasmid borne marker which effected 4-quinolone resistance in recipient cells. The most likely explanation of the observations was the suggestion that the plasmids did not carry a 4-quinolone resistance gene *per se*, but seemed to carry a gene (or group of genes) which acted as a mutator which markedly predisposed the host cells to mutate to resistance to nalidixic acid. Whether or not this was gene-specific or a generalized phenomenon was as yet unknown; and anyway this was still only a hypothesis. Also unknown was the mechanism of resistance involved in the original isolates - clearly this was an area ripe for further detailed study.

3   Was there something peculiar about *S. dysenteriae*? This was thought to be unlikely, and it was generally thought that we had become aware of this phenomenon primarily because the disease was endemic in this part of the world, and local resistance to other routinely used agents meant nalidixic acid was being used extensively because this drug is a lot cheaper than the new 4-quinolones and can't be afforded by such third-world countries. Furthermore S. dysenteriae is very sensitive to nalidixic acid.

It was universally accepted that the whole area of plasmid-borne resistance to the 4-quinolones was both an open question and a scientific can of worms. Also, whilst we could still say that there was no proven incidence of plasmid-borne resistance to the 4-quinolones to date, we could not rule out the possibility of such a phenomenon arising. However, it was also considered that because of the mechanistic problems involved in generating such a resistance, the demands upon bacteria were likely to be such that even if it were to occur, there is a reasonable chance that, in the absence of stringent and persistant selection with the drug, such a plasmid is unlikely to represent more than a localized and short-term problem.

Anaerobiosis and Staphylococcal Resistance to Certain 4-Quinolones

D. J. C. Knowles, G. T. Grimes and L. S. Holmes

Introduction

Many of the fluorinated 4-quinolones are very potent antibacterial agents acting on the same target, DNA Gyrase, as nalidixic acid1. Unlike nalidixic acid and the older compounds, their level of activity is such as to include a broad range of both Gram negative and Gram positive bacteria.

Of the frequently encountered pathogens, staphylococci are among the least sensitive to these agents, albeit with MIC_{90}s of 0.5 to 4.0µg/ml. Furthermore, it is relatively easy to select staphylococcal mutants *in vitro* with decreased sensitivity to 4-quinolones and the same is known to occur *in vivo* (Kaetz *et al.* 1987). Indeed treatment failures associated with small decreases in staphylococcal susceptibility have been reported (Humphreys and Mulvihill 1985).

Although much is known about resistance mechanisms in Gram negative organisms especially *E. coli* (Wolfson and Hooper 1989), little has been published on resistance in Gram positive bacteria. In this poster we report on the existence of a number of isolates of *S. aureus* which show decreased susceptibility to only certain 4-quinolones under aerobic, but not anaerobic conditions.

Methods: Bacteria *E. coli* KL16, *gyrA nalA* and *nalC* mutants were kindly provided by JT Smith, London School of Pharmacy. *S. aureus* JAP4 is a methicillin-resistant isolate from Japan and CK5a is a veterinary isolate. 84/10600, 84/10601, 85/1102, 85/1608, 85/1741, 85/1949, 85/1951 and 85/1955 were obtained from J Naidoo, PHLS Colindale and are representative of clinical staphyloccal isolates showing reduced sensitivity to ciprofloxacin. These isolates all come from different sources within the UK with the exception of 84/10600 and 84/10601 which were sequential isolates from a patient undergoing ciprofloxacin therapy. Cultures were maintained on nutrient agar slopes at 4°C and were cultured for 18 h at 37°C in 5ml Isosensitest broth (Oxoid).

Sensitivity Tests: Conventional agar and microdilution MICs were carried out in Isosensitest medium (Oxoid) in duplicate - one set being incubated aerobically, the other anaerobically. All quinolones were dissolved in 0.1M NaOH and diluted in distilled water. Bacterial inocula were prepared, usually in a series of 10-fold dilutions, and inoculated using a multipoint inoculator (Denley Instruments). In preliminary experiments, one set of cultures and dishes were preincubated in an anaerobic cabinet (Don Whitley MKIII) for 4 h at 37°C before dilution and inoculation. Subsequently, identical results were obtained when the dishes were placed in the anaerobic cabinet within 10 min of inoculation. MICs were read after 18 h at 37°C, unless otherwise stated.

Selection of norfloxacin-resistant variants: A gradient of norfloxacin concentrations across IST plates was prepared by dispensing solutions at 1mg/ml and 10mg/ml using the variable cam of a Model DU spiral plater (Don Whitley). After 4 h, the plates were flooded with a washed suspension of *S. aureus* Oxford or JAP4 containing 5x109 cfu/ml. The plates were then drained and incubated aerobically at 37°C for 18 to 42 h. Five colonies growing within the inhibition zone were resuspended in small volumes of IST broth and flooded onto a further set of gradient plates. This cycle was repeated three times and after one transfer onto antibiotic-free medium, the colonies were tested for sensitivity to norfloxacin both aerobically and anaerobically.

Results

Two laboratory strains of *S. aureus* (CK5a and JAP4) were found to be more resistant to norfloxacin aerobically (MIC = 64μM, 20μg/ml) than anaerobically. This was in contrast to fully sensitive strains such as Oxford, which showed no differential susceptibility, and to *E. coli*, which tended to be less sensitive anaerobically.

Table 1. Effect of anaerobiosis on norfloxacin MICs

Organism	Aerobic MIC	Anaerobic MIC	Aerobic:Anaerobic MIC
S. aureus Oxford	2	2	1:1
S. aureus CK5a	64	4	16:1
S. aureus JAP4	64	8	8:1
E.coli KL16	0.5	1	0.5:1
E.coli gyrA	4	4	1:1
E.coli nalC	0.06	0.13	0.5:1

MIC (μM) on IST agar with 10^5 cfu/spot

This differential susceptibility was also seen in broth microdilution tests and was not associated with pH changes.

Table 2. Effect of anaerobiosis on norfloxacin MICs

Organism	aerobic MIC		Anaerobic MIC	
	pH 6.0	pH 7.3	pH 8.0	pH 7.3
S. aureus Oxford	4	1	1	1
S. aureus CK5a	16	16	16	2
S. aureus JAP4	8	16	16	1
E.coli KL16	16	4	1	4

MIC (μM) in IST broth with 10^5 cfu/well

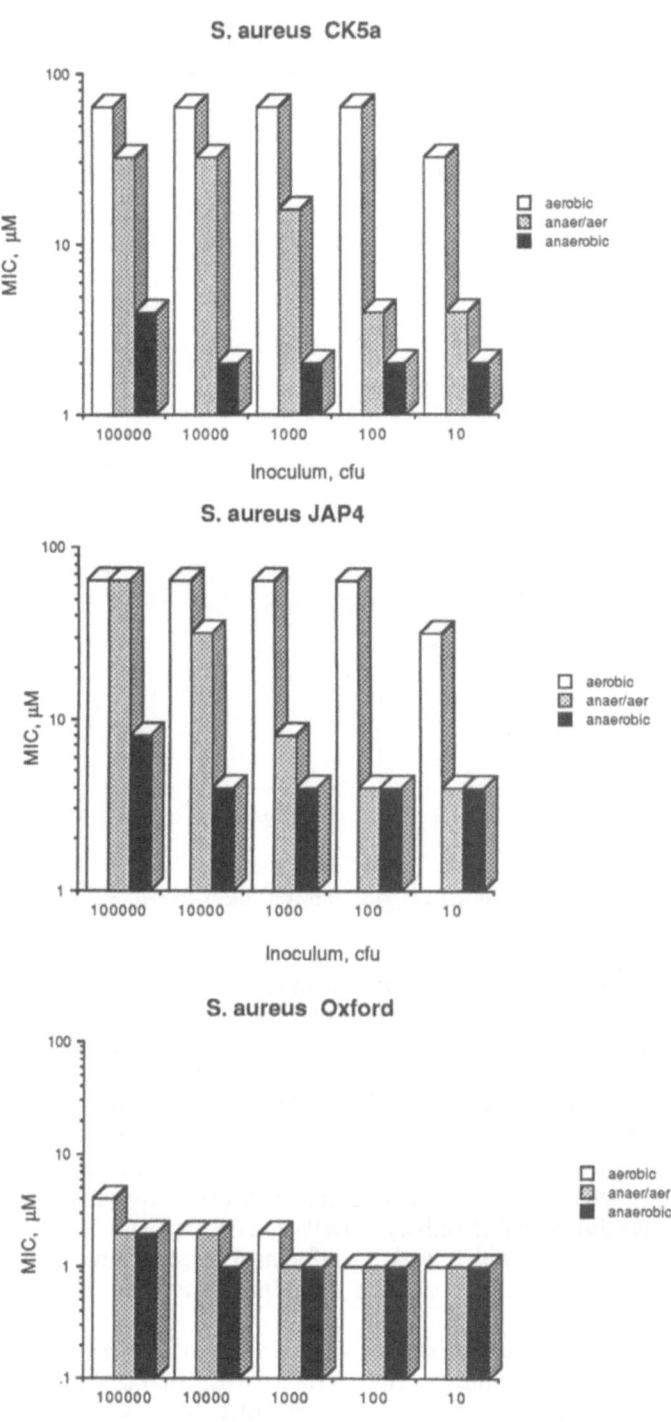

Fig. 1. Aerobic and anaerobic susceptibility of *S. aureus* strains to norfloxacin.

Varying the inoculum from 10^5 cfu to 10 cfu showed that the populations of CK5a and JAP4 bacteria were homogeneous in this differential susceptibility to norfloxacin. In this experiment, agar MICs were recorded after:

42 h aerobic incubation,
18 h anaerobic then 24 h aerobic incubation.
42 h anaerobic incubation

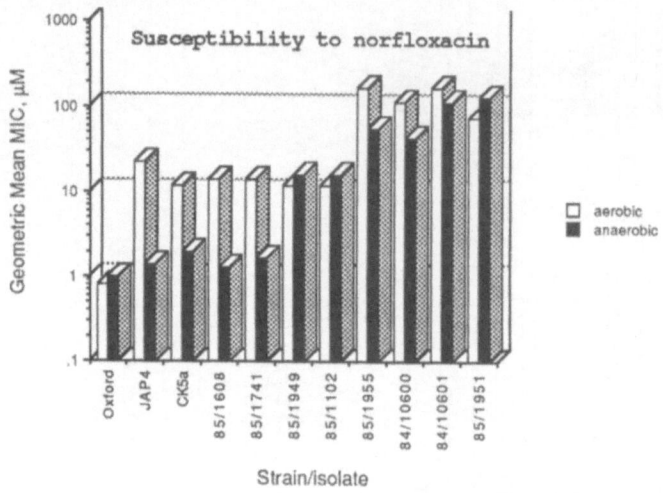

Fig. 2 Aerobic and anaerobic susceptibility of ciprofloxacin resistant strains to norfloxacin.

There was an inoculum effect in the second treatment suggesting that *circa* 99% of the bacteria that had been inhibited anaerobically by concentrations of norfloxacin below the aerobic MIC were unable to recover when subsequently incubated aerobically.

To determine whether the resistance phenotype of CK5a and JAP4 was to be found among clinical isolates showing reduced susceptibility to 4-quinolones, representative ciprofloxacin resistant staphylococcal strains were obtained from PHLS, Colindale.

Four showed intermediate resistance to norfloxacin and, of these, two (85/1608 and 85/1741) shared the same differential sensitivity as JAP4 and CK5a. Four other isolates showed a higher level of resistance with little or no differential susceptibility.

(Agar MIC Results are pooled from treatments using inocula ranging from 10^5 to 10 cfu/spot)

Attempts to select more highly norfloxacin resistant variants of JAP4 resulted in a reduction of the differential susceptibility. Thus aerobic MICs increased only 2 to 4-fold to 64 or 128μM (20 to 40 μg/ml) whereas anaerobic MICs increased 8 to 16-fold.

Selection of resistant variants of the fully sensitive Oxford strain did not give rise to any differentially resistant organisms; aerobic and anaerobic MICs were both increased to *circa* 32μM (10μg/ml).

Table 3. Norfloxacin MICs of resistant variants of S. aureus.

Organism	Aerobic MIC	Anaerobic MIC	Aerobic:Anaerobic MIC
JAP4	32	2	16:1
JAP4/1	64	32	2:1
JAP4/2	64	32	2:1
JAP4/3	64	32	2:1
JAP4/4	128	16	8:1
JAP4/5	128	16	8:1
Oxford	2	2	1:1
Oxford/1	32	64	0.5:1
Oxford/1	32	32	1:1
Oxford/1	32	32	1:1
Oxford/1	32	32	1:1
Oxford/1	32	32	1:1

MIC (µM) on IST agar with 10^5 cfu/spot

Table 4. Extent of Cross-Resistance between different 4-quinolones

4-Quinolone	MIC (µM)	Relative MIC	
	Oxford aerobic	Oxford aerobic	Oxford anaerobic
Norfloxacin	1.5	1.0	0.7
Norfloxacin	1.2	1.0	1.0
Ciprofloxacin	0.8	1.0	0.7
Pipemidic acid	84.4	1.0	1.0
Pefloxacin	1.3	1.0	1.0
Ofloxacin	1.2	1.0	1.0
Flumequine	16.0	1.0	1.3
Oxolinic acid	16.0	1.0	2.0

4-Quinolone	Relative MIC			
	JAP4aerobic	JAP4 anaerobic	CK5 aaerobic	CK5 aanaerobic
Norfloxacin	21.1	1.3	16.0	1.0
Norfloxacin	21.1	2.0	12.1	1.5
Ciprofloxacin	9.2	0.9	16.0	1.3
Pipemidic acid	>12.1	1.3	8.0	2.0
Pefloxacin	1.7	1.5	2.0	1.5
Ofloxacin	2.3	1.0	2.0	1.7
Flumequine	0.9	1.5	1.1	1.5
Oxolinic acid	1.0	2.0	1.0	1.5

4-Quinolone	Relative MIC			
	85/1102 aerobic	85/1102 anaerobic	84/10601 aerobic	84/10601 anaerobic
Norfloxacin	13.9	12.1	222.8	128
Norfloxacin	12.1	13.9	193.8	168.7
Ciprofloxacin	6.1	6.1	84.4	73.5
Pipemidic acid	>12.1	>12.1	>12.1	>12.1
Pefloxacin	12.1	9.2	84.4	255.9
Ofloxacin	8.0	8.0	42.2	73.4
Flumequine	36.8	24.3	24.3	32.0
Oxolinic acid	10.6	9.2	64.0	48.5

Geometric mean MIC using inocula from 10 to 10^5 cfu/spot (expressed as relative to aerobic MIC for S. aureus Oxford).

Four staphylococcal strains were chosen to determine the degree of their cross-resistance to other 4-quinolones.

The results represent the geometric mean agar MIC data from treatments using inocula ranging from 10^5 to 10 cfu/spot and are expressed relative to the aerobic MICs of *S. aureus* Oxford.

Top block shows actual aerobic MIC data for Oxford and relative anaerobic MICs.

Middle block shows relative MIC data for JAP4 and CK5a - representative of differential susceptibility phenotype. Note that:

[1] Resistance was not expressed anaerobically to any quinolone tested
[2] Aerobic cross-resistance was incomplete; both strains were 8 to 21 times more resistant than Oxford to norfloxacin, ciprofloxacin and pipemidic acid but virtually as sensitive to pefloxacin, ofloxacin, flumequine and oxolinic acid.

Bottom block shows relative MIC data for 85/1102 (intermediate resistance to norfloxacin) and 84/10601 (high level resistance to norfloxacin). Note that:

[1] Equivalent resistance is expressed aerobically and anaerobically.
[2] Cross-resistance among the other quinolones tested is complete; 84/10601 being generally more resistant than 85/1102.

Conclusions

1 Clinical staphylococcal isolates showing at least two phenotypically distinct forms of quinolone resistance are described; one form ("differentially susceptible") was not expressed anaerobically and gave rise to only partial cross-resistance among the 4-quinolones. The other form was expressed both aerobically and anaerobically and resulted in complete cross-resistance among the 4-quinolones tested.
2 It is not known whether the differentially susceptible strains arose as a result of selection pressure in response to 4-quinolone usage or whether they represent a naturally occurring sub-population which have been highlighted by the use of 4-quinolones and consequent sensitivity testing.
3 Attempts to select for norfloxacin resistance *in vitro* generated variants which did not display this phenotype. When selection was carried out in JAP4 (differentially susceptible background), organisms were obtained whose increase in anaerobic resistance to norfloxacin was greater than the increase in their aerobic resistance - resulting in a partial reduction in the difference between aerobic and anaerobic MICs. As such, these organisms resembled some of the more highly norfloxacin resistant clinical isolates.
4 The extent of cross-resistance among the 4-quinolones in the differentially susceptible strains is intriguing. Those compounds, whose activity is affected appear to be those with an unsubstituted 7-piperazine moiety. The striking contrast between norfloxacin and pefloxacin highlights the importance of this secondary amino group in norfloxacin.

References

Hooper DC. and Wolfson JS (1989) Mode of action of the Quinolone Agents: Review of Recent Information. Rev Infect Dis 11 Suppl 5:S902-S911

Humphreys H and Mulvihill E (1985) Ciprofloxacin resistant Staphylococcus aureus. Lancet ii:383

Kaatz GW, Barriere SL, Schaberg DR and Fekety R (1987) The emergence of resistance to ciprofloxacin during treatment of experimental Staphylococcus aureus endocarditis. J Antimicrob Chemother 20:753-758

Wolfson JS and Hooper DC (1989) Bacterial Resistance to Quinolones: Mechanisms and Clinical Importance. Rev Infect Dis 11 Suppl 5:S960-S968

Gram Negative Mutants Resistant to 4-Quinolones: Are they Competent Pathogens?

G. C. Crumplin

Introduction

Nalidixic acid, the archetypal 4-quinolone antibacterial agent, was limited in its use by a reputation for the ease of development of resistance both in vitro and in vivo (Ronald *et al*. 1966). The ease with which even the most inexperienced investigators can isolate spontaneous mutants resistant to therapeutic levels of the drug in the laboratory merely serves to reinforce this reputation and colours our perception of possible problems to come with the new generations of 4-quinolones.

Despite this reputation, the examination of the incidence of nalidixic acid resistant organisms amongst urinary tract isolates over a long time span yields a rather unexpected picture. The data shown in Figure 1 a-d are drawn from the records of a single hospital laboratory over a 17 year period and represent the incidence of resistant isolates amongst some 40 000 isolates of the major Gram negative urinary pathogens. The most interesting feature is the fact that the incidence of nalidixic acid-resistance remains remarkably constant over the period whilst the incidence of resistance to the other agents shown progressively increases.

The obvious factors which might account for this difference between agents are the possibility that nalidixic acid was not used clinically, and the fact that there is no record of plasmid-mediated resistance to nalidixic acid whilst the other agents are not exempt from such transmissible resistance. However, in this instance nalidixic acid was in regular use and other similar surveys support these observations. However, since these data are derived from pre-treatment isolates, what we are looking at is not the incidence of resistance arising during treatment, but the incidence of the initiation of infection by resistant organisms. Consequently, since intuition would suggest that there is no reason why there should be a lower proportion of nalidixic acid-resistant organisms available to colonize the urinary tract, it might be suggested that the nalidixic acid-resistant organisms might be less likely to initiate an acute infection.

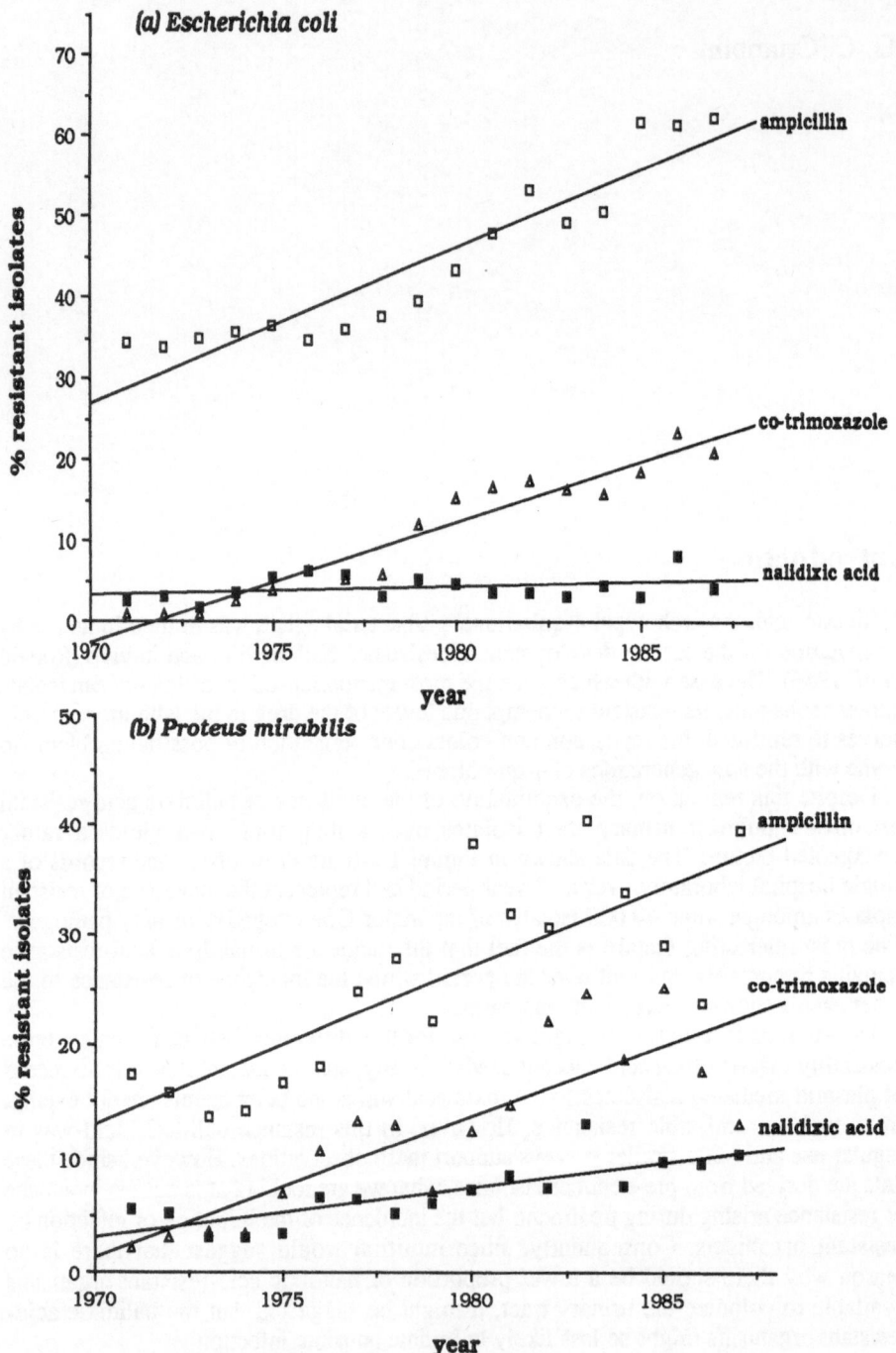

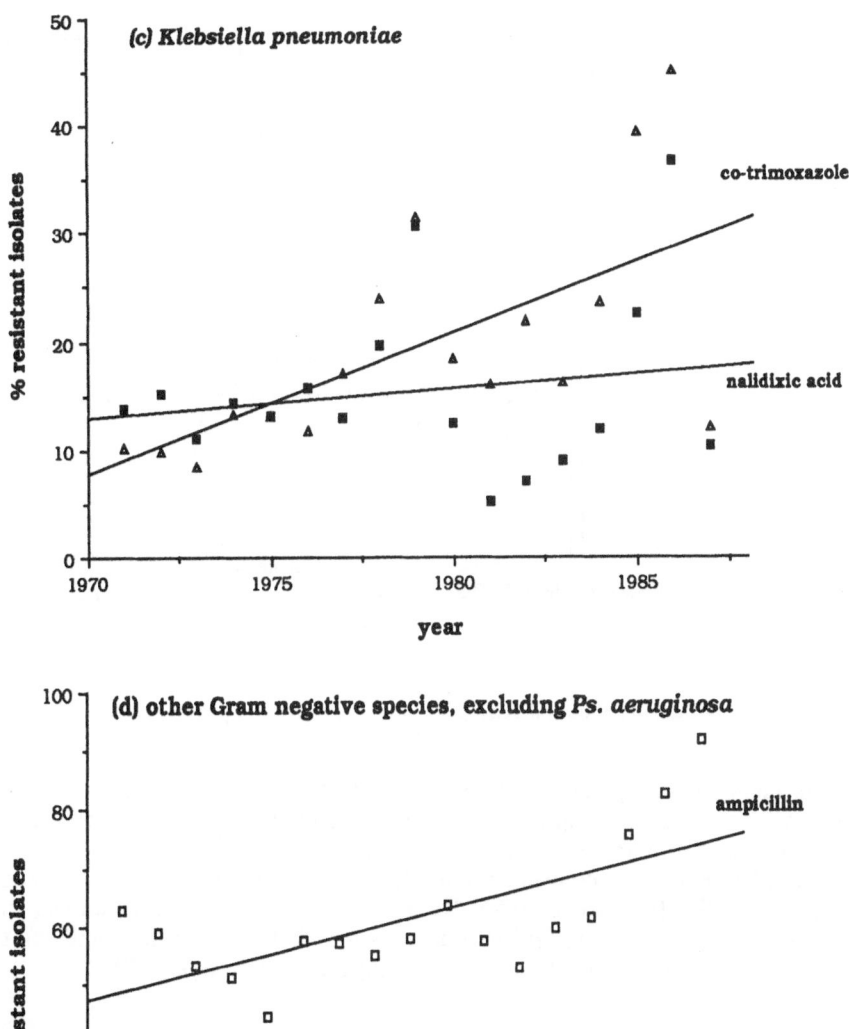

Fig. 1.a-d. Incidences of resistant clinical uropathogenic isolates from a single hospital laboratory.

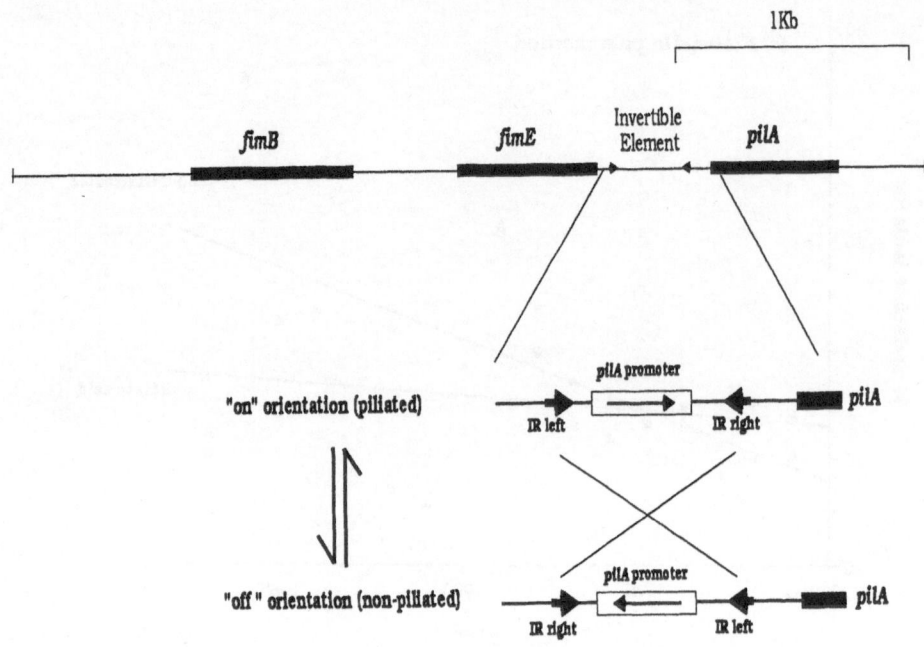

Fig. 2. Organization of the pilin structural gene (pilA) and its promoter in phase-variation

Can this suggested interpretation be tested, and if so, is a similar situation possible with the new generations of 4-quinolones which have the same basic mechanisms of action and mechanisms of resistance as nalidixic acid?

Using type 1 piliation as a convenient indicator of the ability of bacteria to colonize the urinary tract (Hultgren *et al.* 1985; Keith *et al.* 1986), we have examined the phenotypes of mutants isolated as resistant to a number of 4-quinolones with regard to the presence or absence of pili and the ability to agglutinate guinea pig RBCs.

Type 1 Piliation

Type 1 pili, or fimbrae, are a class of filamentous, proteinaceous structures present on the cell surface of many strains of *Escherichia coli* and other species of Gram negative bacteria. Piliation confers a distinct and specific adhesive function on the host bacteria. Type 1 piliation can be conveniently monitored *in vitro* by the mannose-sensitive haemagglutination of guinea pig RBCs. The synthesis of type 1 pili by *E. coli* is controlled by a cluster of genes located at 98 min on the chromosome (Brinton 1959; 1965), and the phenotype is metastable like the flagellar phase-variation of *Salmonella* spp (Brinton 1959). The phase-variation is determined by an invertible element which contains the promoter of the pilin structural gene *pilA* (see Fig. 2) (Abraham *et al.* 1985; Freitag *et al.* 1985; Klemm 1986). Whilst the inversion of the element is known to be dependent upon the presence of Integration Host Factor (*himA* and *himB* gene products) the actual invertase enzyme is as yet unknown (Blomfield 1988).

As a convenient test organism we used the laboratory strain of *E. coli* strain CSH50. Recent work has shown this to be a hyper-piliated mutant strain which, unlike *wt pil*⁺ *E.coli*, conveniently displays haemagglutination when grown on solid media.

The Selection of 4-Quinolone-Resistant Mutants

Spontaneous mutants of CSH50 were isolated from overnight cultures of *pil*⁺ cells grown in nutrient broth at 32°C by plating out onto Isosensitest Agar containing one of the following concentrations of a 4-quinolone:-

Nalidixic acid	50 µg/ml
Norfloxacin	2.5 µg/ml
Pefloxacin	2.5 µg/ml
Ofloxacin	2.0 µg/ml
Ciprofloxacin	1.0 µg/ml

Plates were incubated at 32°C and isolated colonies picked off for subculture and subsequent examination for haemagglutination and cross-resistance studies.

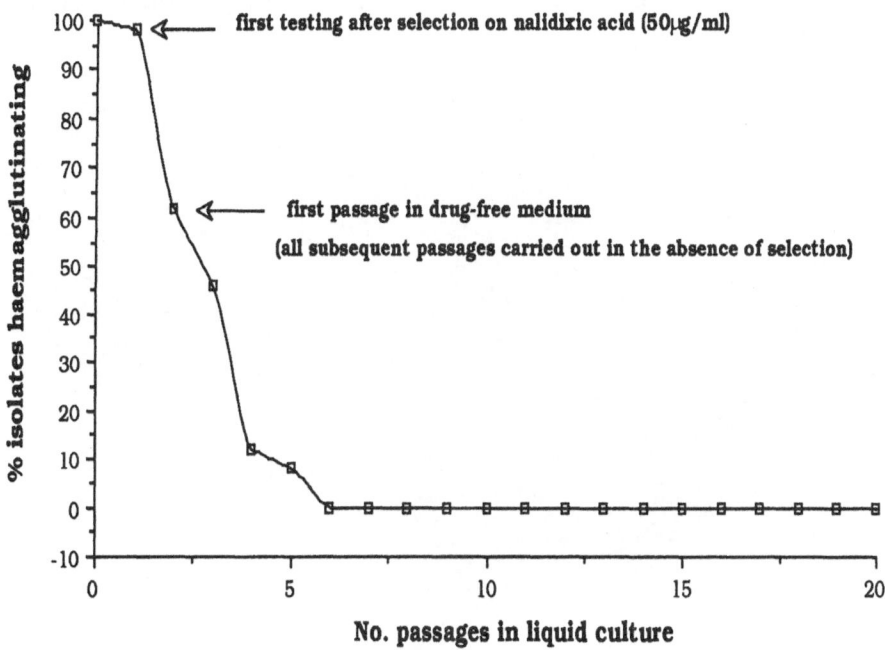

Fig. 3. *Escherichia coli* CSH50: % nalidixic acid-resistant mutants showing mannose-sensitive haemagglutination of guinea-pig RBCs.

Examination of Resistant Mutants for Haemagglutination

Mutants of CSH50 were tested for the mannose-sensitive haemagglutination of guinea-pig RBCs immediately upon isolation (before subculturing). Almost without exception, every isolate was haemagglutination-positive. However, subsequent subculturing of the mutants on agar either in the presence or the absence of the selecting drug showed a rapid increase in the frequency of colonies which failed to haemagglutinate. Such results were obtained with mutants selected upon all of the 4-quinolones used in this study. Fig. 3 shows a typical result obtained with a mutant isolated as being resistant to ≥50 µg/ml of nalidixic acid. Essentially similar results were obtained with mutants isolated as resistant to norfloxacin, pefloxacin, ofloxacin, and ciprofloxacin.

Examination of 4-Quinolone-Resistant Mutants by Electron-Microscopy

The failure of subcultured 4-quinolone-resistant cells to haemagglutinate might be explained by a failure to transcribe the adhesion structural gene which is independent of the pilin genes (Maurer and Orndorff 1985;1987), consequently isolates were examined by electron microscopy for the presence of pili.

Sample bacteria were washed in water before applying samples to a grid coated with Formvar and a carbon support film and allowed to air dry. Samples were then rotary shadowed with platinum at an angle of 3° before viewing under the electron microscope.

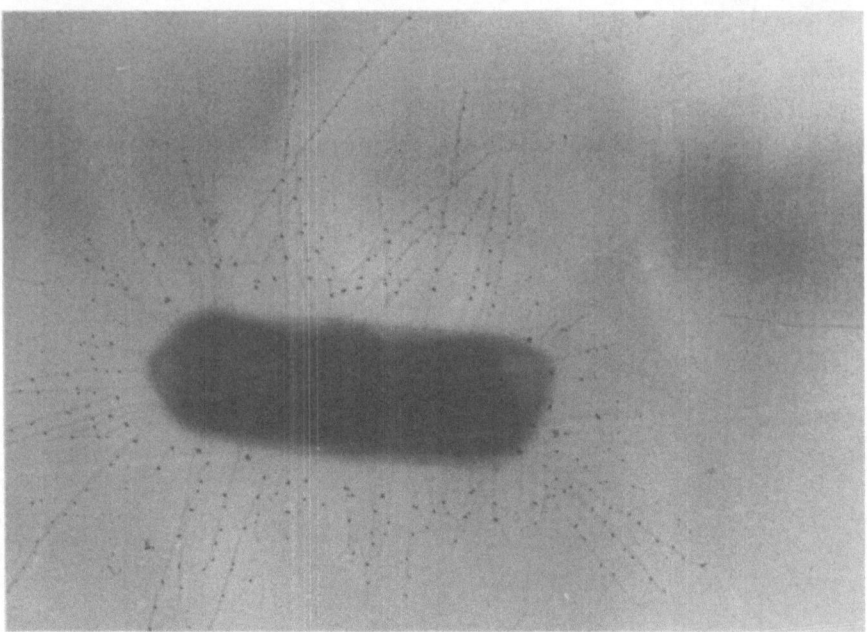

Fig. 4a. Typical piliated 4-quinolone-sensitive *E. coli* CSH50.

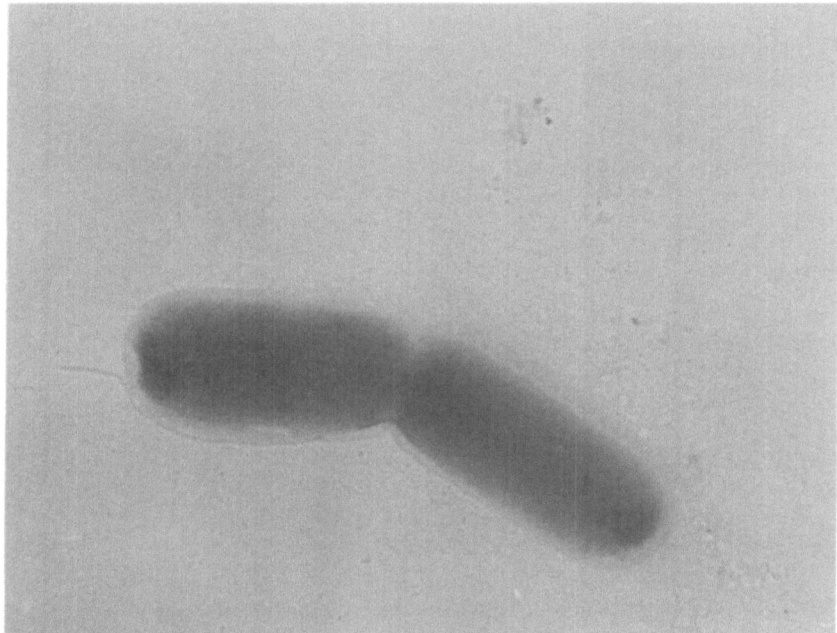

Fig 4b. Typical non-piliated 4-quinolone-sensitive *E.coli* CSH50.

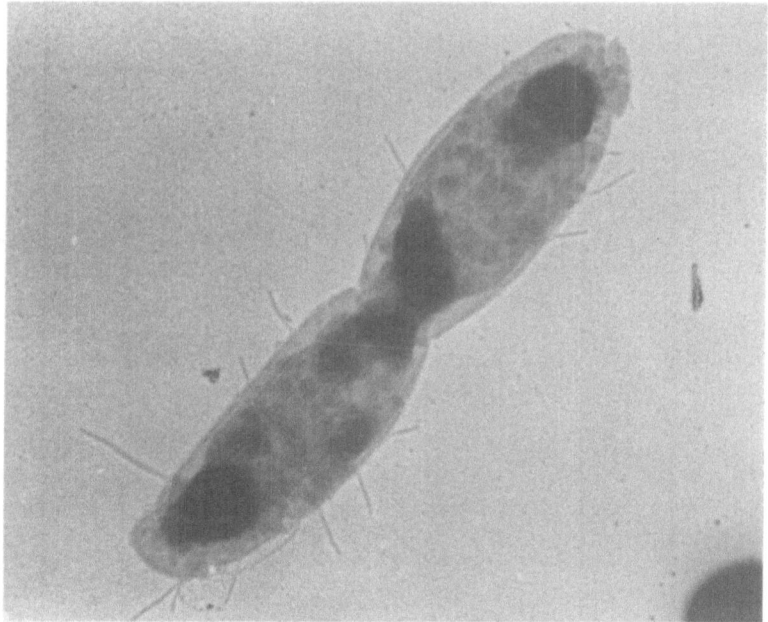

Fig 4c. Typical pefloxacin-resistant *E.coli* CSH50 immediately after isolation.

Fig 4d. Typical ofloxacin-resistant *E.coli* CSH50 immediately after isolation.

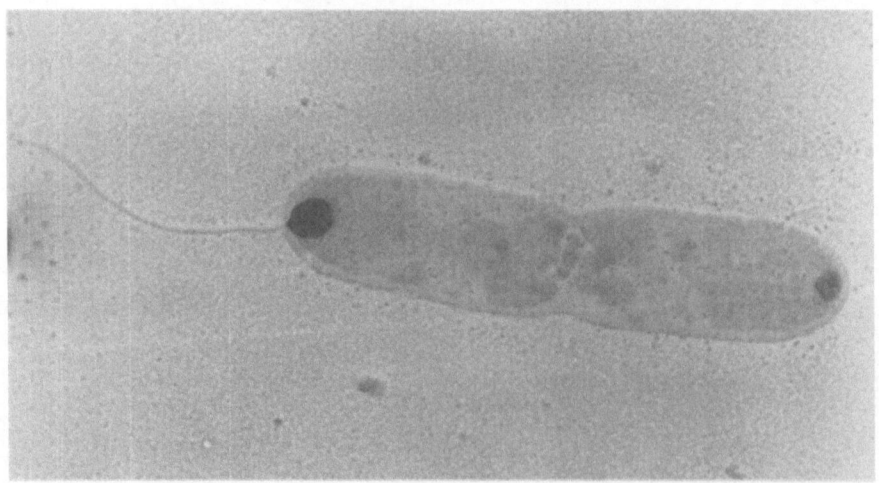

Fig 4e. Typical pefloxacin-resistant *E. coli* CSH50 after five drug-free subcultures.

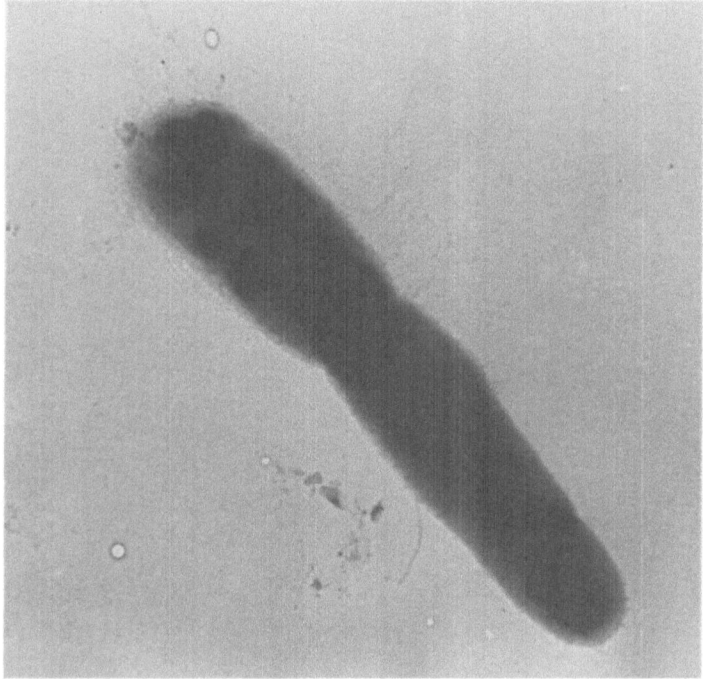

Fig 4f. Typical pefloxacin-resistant *E.coli* CSH50 after ten drug-free subcultures.

Figure 4 a-f shows typical cells from the parent CSH50 strain (piliated and non-piliated) as well as cells upon immediate isolation as pefloxacin-resistant and after repeated subcultures in the absence of the selecting 4-quinolone.

As can be seen in Figures 4c and 4d, the resistant mutant immediately after isolation shows what appear to be reduced numbers of pili, and pili which seem rather stunted. Figures 4e and 4f show typical cells sampled from five and ten drug-free subcultures respectively. In these cases the cells are lacking any obvious pili.

These data indicate that the loss of the haemagglutination phenotype is associated with the loss of surface pili.

Examination of the subcultures for the emergence of haemagglutination-positive colonies as evidence of phase-variation failed to find any indication that the 4-quinolone-resistant mutants were capable of phase-variation.

4-Quinolone-Resistant Mutants Derived from Non-Piliated Parent

When 4-quinolone-resistant mutants were isolated from a non-piliated parent culture, the resistant isolates were invariably haemagglutination-negative, and upon repeated subculturing remained haemagglutination-negative and showed no evidence of phase-variation. Electron-microscopic examination of cells from such cultures confirmed them as having no detectable pili.

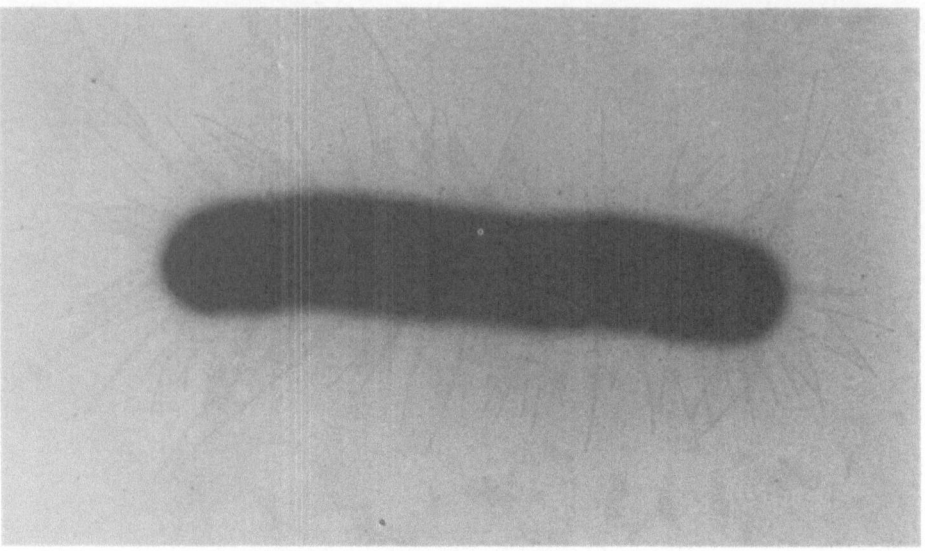

Fig. 5. A typical piliated cell from a pefloxacin-sensitive revertant derived from a non-piliated pefloxacin-resistant parent

Properties of 4-Quinolone-Sensitive Revertants

In order to try and explain the possible mechanism(s) of the apparent loss of phase-variable piliation and haemagglutination we attempted to isolate 4-quinolone-sensitive revertants by replica plating. Five revertant isolates were obtained from nalidixic acid-resistant strains, two from pefloxacin-resistant strains and one from ciprofloxacin-resistant strains.

Each of these isolates showed phase variation in liquid cultures with apparently normal haemagglutination of guinea pig RBCs. Examination of cells from haemagglutinating colonies on solid medium by electron microscopy showed the cells to be normally pilated (see Fig. 5).

Discussion

The results presented above can be interpreted as indicating that 4-quinolone-resistant mutants of *Escherichia coli* CSH50 are defective in the phase variation of piliation. They appear to display only a uni-directional phase-variation, being able to carry out the "on" → "off" switch but not the "off" → "on" switch. The appearance of apparently normal piliation and phase-variation in 4-quinolone-sensitive revertants suggests that the mutation(s) conferring the 4-quinolone-resistance (most likely to be DNA gyrase

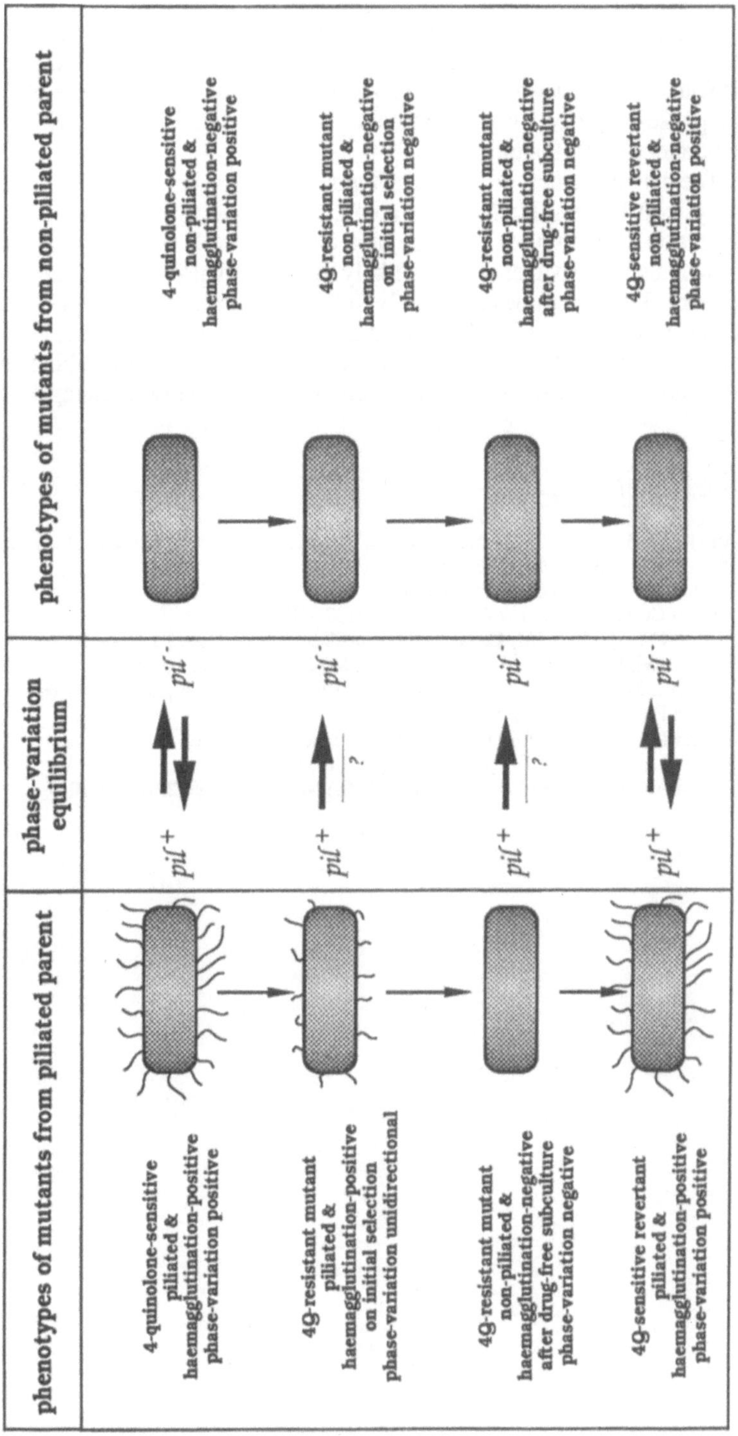

Fig. 6. Summary of piliation phenotypes of strains derived in this study.

phenotypes of mutants from piliated parent	phase-variation equilibrium	phenotypes of mutants from non-piliated parent
4-quinolone-sensitive piliated & haemagglutination-positive phase-variation positive	pil^+ ⇄ pil^-	4-quinolone-sensitive non-piliated & haemagglutination-negative phase-variation positive
4Q-resistant mutant piliated & haemagglutination-positive on initial selection phase-variation unidirectional	pil^+ ↑ ? pil^-	4Q-resistant mutant non-piliated & haemagglutination-negative on initial selection phase-variation negative
4Q-resistant mutant non-piliated & haemagglutination-negative after drug-free subculture phase-variation negative	pil^+ ↑ ? pil^-	4Q-resistant mutant non-piliated & haemagglutination-negative after drug-free subculture phase-variation negative
4Q-sensitive revertant piliated & haemagglutination-positive phase-variation positive	pil^+ ⇄ pil^-	4Q-sensitive revertant non-piliated & haemagglutination-negative phase-variation positive

mutations) interfere with either the inversion of the invertible element, or result in failure to transcribe and translate the *pilA* gene. This interpretation of the data is summarized in Figure 6.

A repeat study using the normally piliated clinical isolate *Escherichia coli* strain Mg1655 has been initiated. Thus far Nalidixic acid-resistant and norfloxacin-resistant isolates have been obtained and none of these show any evidence of haemagglutination or phase-variation. These results support the data obtained with the hyper-piliated laboratory CSH50 strain.

The data presented supports the hypothesis that 4-quinolone-resistant *Escherichia coli* may well be less likely to be able colonize the urinary tract and initiate an acute infection. Since type 1 piliation as a phase variable phenotype is also found in other Gram negative species it seems likely that this phenomenon is not restricted to a single species. Further studies are clearly indicated to examine other pathogenicity associated characteristics in 4-quinolone-resistant mutants in order to assess both the likelihood of such organisms initiating an infection, or sustaining an established infection if the resistance-conferring mutation(s) occur during the course of therapy.

References

Abraham JM, Freitag CS, Clements JR and Eisenstein BI (1985) Proc Natl Acad Sci USA 8:5724-5727
Blomfield IC (1988) PhD Thesis (Univ. York)
Brinton CC (1959) Nature 183:783-786
Brinton CC (1965) Trans NY Acad Sci 27:1003-1054
Freitag CS, Abraham JM, Clements JR and Eisenstein BI (1985) J Bacteriol 163:668-675
Hultgren SJ, Porter TN and Schaeffer AJ (1985) Infect Immun 50:370-377
Keith BR, Maurer L, Spears PA and Orndorff PE (1986) Infect Immun 53:693-696
Klemm P (1986) EMBO J 5:1389-1393
Maurer L and Orndorff PE (1985) FEMS Microbiol Let 30:59-66
Maurer L and Orndorff PE (1987) J Bacteriol 169:640-645
Ronald AR, Turck M and Petersdorf RG (1966) New Eng J Med 297:1081-1089

Quinolone-Resistant *E. Coli* Mutants Overproduce a 60 kDa Protein Highly Homologous to *GroEL*

P. Hallett, A. Mehlert and A. Maxwell

Introduction

DNA gyrase is a member of a group of enzymes known as topoisomerases which are responsible for the control and modification of the topological state of DNA *in vivo*. *E. coli* DNA gyrase is composed of two A subunits and two B subunits of molecular masses 97 kDa and 90 kDa, coded for by the *gyrA* and *gyrB* genes respectively. DNA gyrase has been established as the likely target of the quinolone group of synthetic antibacterial drugs, which include nalidixic acid (NAL), oxolinic acid (OXO), norfloxacin (NFX) and ciprofloxacin (CFX). The quinolones have been found to be particularly useful in the treatment of urinary tract and enteric infections. Due to their success, there has been an extensive search for derivatives with broader spectra of bactericidal activities and improved pharmacological properties. Although it was previously thought that quinolone drugs bind to the A subunit, mutants conferring quinolone resistance have been mapped to both the *gyrA* and *gyrB* genes (Yamagishi *et al.* 1986; Yoshida *et al.* 1988). Other evidence has suggested that the drugs may preferentially bind to the gyrase-DNA complex (Shen *et al. 1989*).

During an investigation into gyrase-quinolone interactions, we have generated mutants resistant to quinolones in *E. coli* strains that over-produce the DNA gyrase A subunit. Many of these mutants have been found to over-express a 60 kDa protein which is highly homologous in N-terminal amino-acid sequence to the *E. coli* heat-shock protein *GroEL*.

Results

The *gyrA* gene has been cloned onto a high copy number, ampicillin-resistant plasmid, pMK90 (Mizuuchi *et al* 1984), such that *gyrA* is under the control of the heat inducible

Table 1

Strain	Mutagen	MIC (µg/ml)			
		NAL	OXO	NFX	CFX
PHB8.5	bisulphite	6.2	0.86	0.76	0.24
PHB13.21	bisulphite	9.2	0.7	0.56	0.17
PHE39.34	EMS	12.8	0.98	0.6	0.23
PHH67.11	HA	15.2	1.0	0.71	0.24
PHG1	GMM	8.0	0.92	0.8	0.26
N4186		2.76	0.02	0.01	0.003
RW1053		2.5	0.03	0.02	0.005

EMS = Ethyl Methyl Sulphonate; HA = Hydroxylamine; GMM = Gap mis-repair mutagenesisNAL = nalidixic acid; OXO = oxolinic acid; NFX = norfloxacin; CFX = ciprofloxacin.

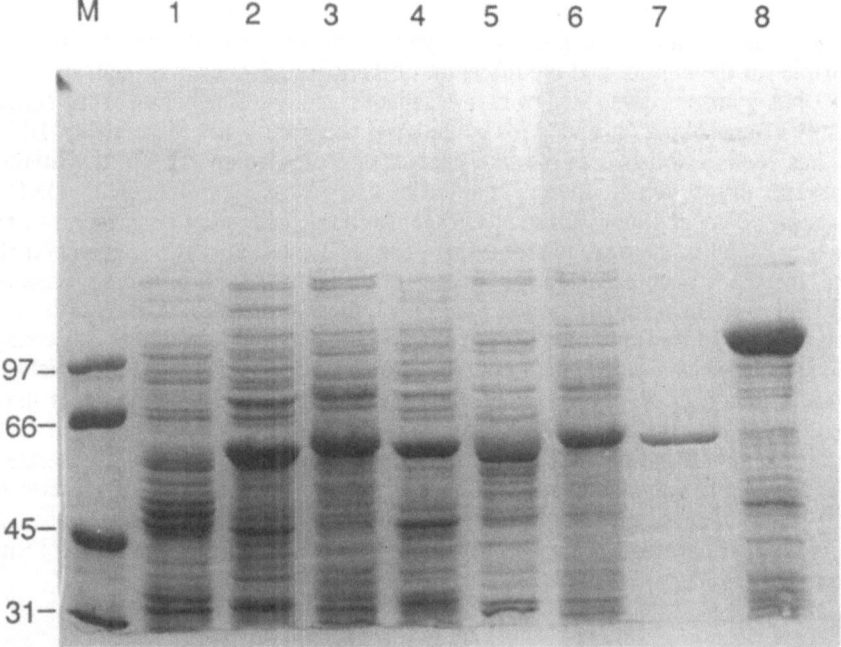

Fig. 1. Expression of a 60 kDa protein in quinolone-resistant strains.
M = molecular weight markers; 1 = RW1053; 2 = PGH1; 3 = PHB13.21; 4 = PHB8.5;
5 = PHH67.11; 6 = PHE39.34; 7 = 60 kDa protein purified from PHB13.21; 8 = N4186.

1 pL promoter. Exponential growing cultures of *E. coli* strain RW1053 containing this plasmid (= N4186) were subjected to different mutagens, including sodium bisulphite, hydroxylamine, and ethyl methyl sulphonate. In addition, gap misrepair mutagenisis was used to generate mutants at single sites in pMK90. Supercoiled plasmid was randomly nicked using DNase I and ethidium bromide, the nick enlarged with exonuclease III and then repaired using dNTP's and error-prone AMVRT. Following mutagenisis, ampicillin-resistant bacteria were screened for CFX-resistance. For many of the mutants generated the MIC's were determined for nalidixic acid, oxolinic acid, norfloxacin and ciprofloxacin. The results are listed in Table 1.

Mutants with high MIC's with respect to CFX were analysed for evidence of over-expression of a mutant *gyrA* protein by preparing cell extracts which were then analysed by SDS PAGE. Fig. 1 shows an example of such a gel. Many of these cell extracts showed little evidence of a protein band at 97 kDa corresponding to *gyrA* protein, but instead showed over-production of a protein of 60kDa. One particular mutant, PHB13.21, derived from sodium bisulphite mutagenisis, strongly expressed the 60kDa protein and was selected for further study.

The 60 kDa protein from PHB13.21 was partially purified using Polymin P and ammonium sulphate precipitation, then used in various assays. One possibility was that the 60kDa protein represented a truncated *gyrA* protein. Several experiments have discounted this. Purified 60 kDa protein did not support ATP-dependent supercoiling of relaxed pBR322 DNA in the presence of wild-type *gyrB* protein. Other evidence, including fingerprinting using Staphylococcal V8 protease, indicated there was no apparent relationship between the 60 kDa protein and *gyrA* protein.

N-terminal sequence determination of the 60 kDa protein revealed no homology to the *gyrA* protein, but did however show a high degree of homology to *E. coli groEL* gene product. Fig. 2 shows a comparison of the N-terminal sequences of 60 kDa, *groEL*, and *gyrA* protein.

Over the first 16 amino acids only five differences occurred between the 60 kDa and *groEL* proteins.

GroEL is a 57 kDa protein of unknown function, although genetic studies show it is an essential gene in *E. coli* (Fayet *et al.* 1989). It is a known heat-shock protein, and it has been speculated that its function in the cell may be concerned with the prevention of aggregation of proteins which have become partially unfolded as a consequence of heat-shock or other stresses (Pelham 1986).

The relationship between the 60 kDa and *groEL* proteins have been further examined. Under denaturing conditions it was found that both proteins have very similar molecular masses (Fig. 4), and under non-denaturing conditions the 60 kDa protein migrates slightly slower than the *groEL* protein (Fig. 3). Additionally, the isoelectric

gyrA   :   S D L A R E I T P V N I E E E L

60 kDa :   A A K D V V F G G E A R A R M V

groEL   :   A A K D V K F G N D A R V K M L

Fig. 2. Non-denaturing gradient gel of *groEL* and 60 kDa proteins
M = Jack bean urease.

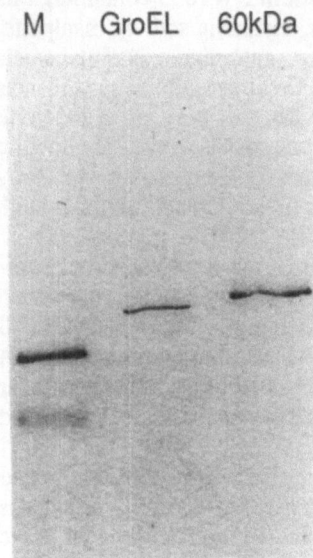

Fig. 3. Staphylococcal V8 protease digest of *groEL* and 60 kDa proteins. For each time point indicated, the sample on the left is *groEL* protein, the sample on the right is 60 kDa protein.

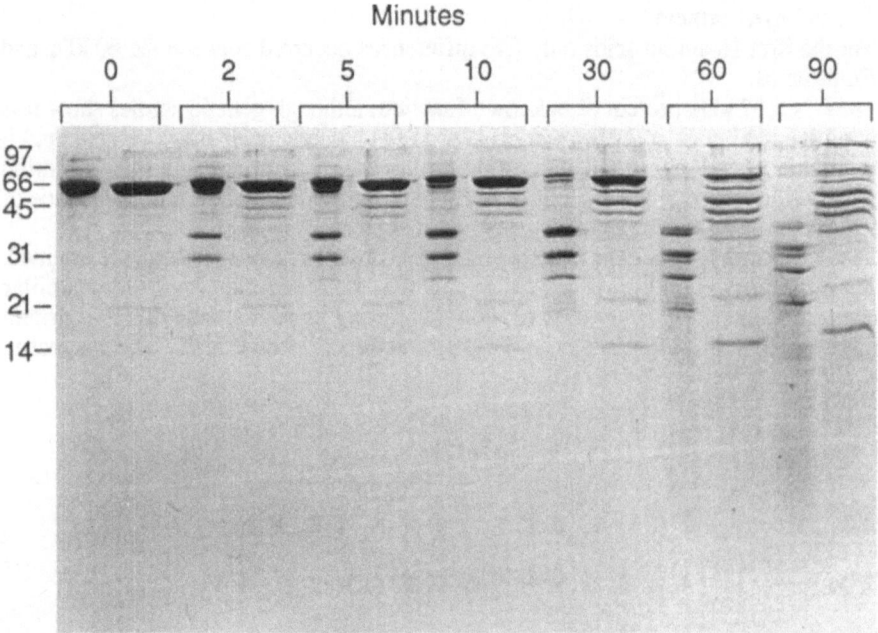

Fig. 4. Staphylococcal V8 protease digest of GroEL and 60 kDa proteins. For each time point indicated, the sample on the left is GroEL protein, the sample on the right is 60 kDa protein.

points of the 60 kDa and *groEL* proteins were found to be slightly different at 5.5 and 5.4 respectively (data not shown). Also Staphylococcal V8 protease digestions of *groEL* and 60 kDa proteins show there to be few common peptides despite the significant N-terminal sequence homology between both proteins.

It seems therefore that the 60 kDa protein is similar but not identical to *groEL* protein. It is not at present clear what the function of the 60 kDa protein is in the *E. coli* cell or its relationships with DNA gyrase and quinolone resistance. These are issues for further investigation.

Conclusions

We have shown that:

1   There is overproduction of a 60 kDa protein in certain *E. coli* strains resistant to quinolone drugs.
2   The 60 kDa protein is unrelated to the *gyrA* protein.
3   The 60 kDa protein has high homology to the *E. coli groEL* protein.

References

Fayet O, Ziegelhoffer T, Georgopoulos C (1989) The *groES* and *groEL* heat shock gene products of *E. coli* are essential for bacterial growth at all temperatures. J Bacteriol 171:1379-1385

Mizuuchi K, Mizuuchi M, O'Dea H, Gellert M (1984) Cloning and simplified purification of *Escherichia coli* DNA gyrase A and B proteins. J Biol Chem 259:199-201

Pelham H (1986) Speculations on the functions of the major heat shock and glucose related proteins. Cell 46:959-961

Shen LL, Mitscher LA, Sharma PN, O'Donnell TJ, Chu DTW, Cooper CS, Rosen T, Pernet AG (1989c) Mechanism of inhibition of DNA gyrase by quinolone antibacterials. A cooperative drug-DNA binding model. Biochemistry 28:2886-2894

Yamagishi J, Yoshida H, Yamayoshi M, Nakamura S (1986) Nalidixic acid-resistant mutations of the *gyrB* gene of *Escherichia coli*. Mol Gen Genet 204:367-373

Yoshida H, Kojima T, Yamagishi J, Nakamura S (1988) Quinolone-resistant mutations of the *gyrA* gene of *Escherichia coli*. Mol Gen Genet 211:1-7

Subject Index